Recent Developments in Carbonate Sedimentology in Central Europe

Edited by

German Müller · Gerald M. Friedman

With 168 Figures

Springer-Verlag New York Inc. 1968

German Müller, Dr. rer. nat., o. Professor für Mineralogie und Petrographie an der Universität Heidelberg

Gerald M. Friedman, Ph. D., Professor of Geology, Rensselaer Polytechnic Institute, Troy, N. Y./USA

 Library of Congress Catalog Card Number 69-17255. Printed in Germany.

Title No. 1525

Preface

In the field of sedimentary research, ever increasing emphasis has been put on the investigations of carbonates and carbonate rocks during the past 30 years.

It is thus quite natural that in Central Europe — where classical carbonate investigations have already been carried out 100 years ago — numerous scholars turned to the study of this sediment type.

On the occasion of a visiting professorship of G. M. Friedman at the Laboratorium für Sedimentforschung, Heidelberg University, a seminar on "Recent Developments of Carbonate Sedimentology in Central Europe" was held in July 1967. 90 persons involved in carbonate investigations participated, and 35 lectures were held.

The present volume contains 30 papers summarizing the different subjects of the seminar. Of course these contributions only represent a small part of the work actually performed in the field of carbonate research in Central Europe. We believe, however, that they give a general survey of the work and the working methods employed in the different sectors of carbonate investigations.

One of the purposes of the present symposium is to acquaint the English speaking countries with our recent results which usually — due to the language barrier — are not accessible to them.

We wish to thank all those who submitted their manuscripts for publication, and H. U. Schmincke, W. C. Park, and E. H. Schot who did a lot of the translation into English. Thanks are also due our draftsman, Fred Rückert, who re-arranged a great number of the drawings and designed the cover of the book.

Heidelberg and Troy, November 1968

German Müller
Gerald M. Friedman

Contents

A. Processes of Carbonate Formation and Diagenesis

B. Microtexture and Microporosity of Carbonate Rocks

C. Geochemistry of Carbonates and Carbonate Rocks

D. Regional Carbonate Petrology

I. Freshwater Carbonates and Carbonate Rocks

II. Marine Carbonate Rocks

E. Applied Carbonate Petrology

Precipitation of Oöids and Other Aragonite Fabrics in Warm Seas

R. G. C. Bathurst*

Abstract

The precipitation of $CaCO_3$ from tropical sea water yields aragonite in four fabrics: oöids, needles, cement in grapestone, and intragranular filling. Reasons for this varied precipitation are tentatively considered. The *stability* of aragonite in sea water seems to depend on adsorption of hydrated Mg^{2+} on calcite so inhibiting its growth and on the, probably, less negative free energy of formation of magnesian calcite compared with calcite. In oölitic coats the *tangential orientation* of optic axes may depend upon "slow" adsorption of fragments of aragonitic monolayer already present in the water. The role of *agitation in oöid growth* is uncertain. The *maximum size of oöids* at about 1 mm may reflect a balance between growth and abrasion. The *geographic distribution* of the four fabrics seems to depend mainly on the amount of movement of the bed load.

A. Introduction

The inorganic precipitation in tropical sea water of such varied forms of aragonite as oölitic coats, needle-muds, the intergranular cement of grapestone, and intragranular drusy filling in skeletal body chambers, poses numerous problems. These are not only of great interest but of peculiar difficulty for the geologist, because he can expect little help from the existing body of conventional knowledge in physical chemistry. In addition to the fundamental thermodynamic control there is a large element of kinetic regulation of the several growth processes, particularly in reactions involving adsorption on crystal faces. The more important matters which call for further enquiry can conveniently be considered under five headings, although it will be obvious that they are intimately related and overlapping. (1) Why, in sea water, is the inorganically precipitated mineral aragonite and not calcite? (2) Why, in marine oöids, do the optic axes of the aragonite crystals in the oölitic coats have a preferred tangential orientation? (3) What role does the agitation of oöids play in their growth? (4) What controls the maximum size of oöids? (5) Why does aragonite grow attached to an oöid in one place while a few kilometres away it is precipitated on separate needles, yet elsewhere as a grapestone cement and, more generally, as needles oriented normally to the surface within the intragranular pores of skeletal grains?

In the present state of knowledge these questions must remain to various degrees unanswered. Chemical and physical data relating to the environments in which the various fabrics grow are scarce and there is an urgent need for field and laboratory measurements. Nevertheless, it is worthwhile to draw attention to the problems which need to be solved, to summarize present knowledge, and to suggest a few working hypotheses which may encourage the future development of some useful researches.

* The Jane Herdman Laboratories of Geology, University of Liverpool, England.

B. The Stability of Aragonite in Sea Water

The shallow waters of some contemporary tropical or neartropical seas, such as the Great Bahama Bank (Cloud, 1962), are supersaturated for both aragonite and calcite so that, if the chemical environment were a simple one, calcite would be precipitated since it is the least soluble polymorph of $CaCO_3$. In fact, the inorganic precipitate is everywhere aragonite, and this anomalous situation has attracted the interest of research workers who have approached the problem from two main directions. Pobeguin (1954) and Cloud (1962) especially have emphasized the relation between the form of the precipitated polymorph and the level of supersaturation, whereas Lippmann (1960), Kitano (1962) and Berner (1966) have stressed the importance of the adsorption of hydrated Mg^{2+} cations on the crystal face. Pobeguin (1954) examined the simplest possible system to discover the basic controls of polymorphic precipitation. She precipitated $CaCO_3$ using the double decomposition of NH_4Cl and $Ca(CH_3COO)_2$ at 16° to 18°C and atmospheric pressure. By varying the forms of the vessels in which the reaction was carried out she arranged for the two solutions to mix by diffusion at different rates, and her mixing times were related to mineralogy as follows:

10 to 15 days:	calcite only,
6 to 8 days:	calcite, with a little vaterite and a little aragonite,
24 to 26 hours:	calcite, vaterite and much aragonite.

Calcite began to appear as soon as the solution reached supersaturation, but aragonite formed only in a solution that was supersaturated for calcite and near supersaturation for aragonite. From these experiments Pobeguin concluded (1954, p. 95), "De ces expériences, nous pouvons, semble-t-il, tirer la conclusion suivante: dans les conditions envisagées, le facteur essentiel déterminant la précipitation de l'une ou l'autre forme de calcaire, en particulier de calcite ou d'aragonite, est *la vitesse de diffusion des ions* CO_3 et Ca, c'est-à-dire, en quelque sorte, la quantité d'ions en présence simultanément, en un point donné et à un moment donné." It is plain, therefore, that the higher supersaturation favours the more soluble polymorph. Cloud (1962, p. 123) believed that the order of precipitation vaterite-aragonite-calcite as supersaturation decreases is an expression of Ostwald's rule that precipitation takes place in steps from the least stable to the most stable solid phase. Brooks et al. (1951), on the other hand, stated that this is known to be untrue for certain substances other than $CaCO_3$ and proposed, instead, that the phase which appears is that which differs least from the preceding phase (solution or solid) both in composition and structure.

However, simple rate of diffusion through the solution is not the only factor that can regulate the type of lattice construction. Brooks and his co-workers (1951) demonstrated that certain substances adsorbed on the crystal face could retard its growth. In one of their experiments they dripped N/10 solutions of $CaCl_2$ and Na_2CO_3 into distilled water, to which seed crystals of calcite could be added. Various quantities of $MgCl_2$ were added. As the concentration of $MgCl_2$ was increased the nucleation of calcite was progressively inhibited and, eventually, both discrete nucleation and overgrowth of calcite on seeds were suppressed. At the same time crystals of the metastable hexahydrate, $CaCO_3 \cdot 6\,H_2O$, appeared in increasing quantity and later the metastable monohydrate as well. Brooks et al. concluded that the growth of calcite was retarded owing to the adsorption of Mg^{2+} on the crystal face: this is in agreement with the well known tendency for crystal growth to be controlled not only by general adsorption but by selective adsorption at specific sites. Other authors have found that the Mg^{2+} ion is particularly influential (LIPPMANN, 1960; KITANO, 1962; USDOWSKI, 1963). LIPPMANN has suggested that it is the hydrated Mg^{2+} which is adsorbed on the growing calcite lattice so as to prevent further growth. This is in line with the earlier work of DAVIES and JONES (1955) on the growth rate of AgCl seed crystals from solution. This reaction seemed to them to be interface-controlled and they postulated an adsorbed monolayer of hydrated ions at the crystal surface. They suggested that growth took place as a result of simultaneous dehydration of Ag^+ and $Cl^-$and that it was the rate of this dehydration process that controlled the rate of growth.

LIPPMANN (1960) supposed that the reluctance of calcite to grow in aqueous solutions containing Mg^{2+} in addition to Ca^{2+} depends on the fact that the standard free energy of hydration of the Mg^{2+} is greater than that of the Ca^{2+}. Where the activity of Mg^{2+} in the solution is high enough, some Mg^{2+} must be incorporated in the calcite lattice (making a magnesian calcite) and so a mixture of Ca^{2+} and Mg^{2+} with attached H_2O dipoles will gather at the surface of a nucleus (as in the Stern diffuse double layer). As the Ca^{2+} ions are dehydrated they will join with $CO_3{}^{2-}$ ions to build a calcite lattice, but this process will be hindered by the presence of still hydrated Mg^{2+} ions, so much so that aragonite will be precipitated in preference to calcite. In considering this idea of LIPPMANN's it is useful to bear in mind also that the free energies of formation of high-magnesian calcites will probably be less than that of calcite because the high-magnesian calcites are more soluble and because the free energy of formation of magnesite ($MgCO_3$) is less than that of calcite. Therefore, two reactions concern us: the endothermic dehydration of cations and the exothermic formation of a crystal lattice.

It is possible to make a rough and ready analysis of the effect upon the precipitation of aragonite and high-magnesian calcite that is produced by the greater free energy of hydration of Mg^{2+} over Ca^{2+} and by the assumed smaller free energy of formation of high-magnesian calcite compared with aragonite, and to show that what thermodynamic knowledge is available predicts the precipitation of aragonite to the exclusion of calcite.

LIPPMANN has estimated the standard free energies of hydration as -501 kcal/mol for Mg^{2+} and -428 kcal/mol for Ca^{2+}. The U.S. National Bureau of Standards gives the standard enthalpies of hydration as -456 kcal/mol for Mg^{2+} and -377 kcal/mol for Ca^{2+}. The differences between the values for Mg^{2+} and Ca^{2+} in each case are closely similar, so Lippmann's free energy figures, with their smaller difference, will be used in the following argument since they give less support to his hypothesis.

For convenience it will be assumed that the magnesian calcite has 10% Mg^{2+} in its cations. If all the cations adsorbed on the crystal face were Mg^{2+} then the free energy of hydration would be increased above that for Ca^{2+} by $501 - 428 = 73$ kcal/mol, but, as Mg^{2+} is only 10% of the cations, the increase will be 7.3 kcal/mol. Thus the free energy of hydration of the combined adsorbed cations (Ca^{2+}_{90} Mg^{2+}_{10}), which will go to construct the crystal lattice of the magnesian calcite, will be $428 + 7.3 = 435.3$ kcal/mol compared with 428 kcal/mol for the Ca^{2+} taking part in the growth of pure calcite or aragonite, assuming that the aragonite lattice will accept no Mg^{2+} at all.

The standard free energies of formation of magnesian calcites are not known, but the value for magnesite ($MgCO_3$) is -246 kcal/mol, lower 23.8 by than the value for calcite which is —269.8 kcal/mol. We can guess crudely that the standard free energy of formation for our magnesian calcite will be smaller than that for calcite by a tenth of the difference between the two, thus $-269.8 + 2.38 = -267.4$ kcal/mol.

Regarding crystal growth as a two stage process involving endothermic dehydration ($+ \Delta G_h$) and exothermic lattice building ($- \Delta G_f$), we can now view the growth of aragonite and our magnesian calcite (with 10% Mg^{2+}) in terms of ΔG, the change in Gibb's free energy.

$$\begin{aligned} \Delta G \text{ aragonite} &= + \Delta G_h - \Delta G_f, \\ &= +428 - 269.6, \\ &= +158.4 \text{ kcal/mol.} \\ \Delta G \text{ magnesian calcite} &= + \Delta G_h - \Delta G_f, \\ &= +435.3 - 267.4, \\ &= +167.9 \text{ kcal/mol.} \end{aligned}$$

Thus the energy budget favours precipitation of aragonite by about 20 kcal/mol. This is a crude and inelegant method of estimation but it serves to illustrate the principle. The standard free energies of formation of pure calcite and pure aragonite are very close, differing only in the first decimal place, and comparatively slight changes in the environment, such as the introduction of Mg^{2+} into the solution, would be expected to upset the thermodynamic balance.

A demonstration of the scale of adsorption of both Mg^{2+} and Ca^{2+} ions on aragonite and calcite grains has been given by BERNER (1966). He soaked grains in solutions with ion ratios, either Mg^{2+}/Ca^{2+} or Ca^{2+}/Mg^{2+}, of about 50/1. He then measured the ratio Mg^{2+}/Ca^{2+} of the ions adsorbed on the grain surfaces and found that this was very close to the ratio in the solution. In the sediments of Florida Bay the ratio Mg^{2+}/Ca^{2+} has a mean value of about 3/1 but BERNER warned that the distribution of Mg^{2+} is likely to be uneven, perhaps with a preference for magnesian calcite and other favoured sites. FYFE and BISCHOFF (1965) found evidence for the blocking of active sites on a calcite face by Mg^{2+} ions, and TAFT (in CHILINGAR et al., 1967, p. 158) mentioned a critical concentration of Mg^{2+} in the solution which must be exceeded if the growth of calcite is to be inhibited.

Summarizing the reasons for the inorganic precipitation of aragonite instead of calcite in sea water, it seems that various factors can lead to the necessary high supersaturation of $CaCO_3$ in the water. One of these factors is a change in the rate of diffusion of ions to the crystal face, this rate being in turn influenced by, amongst other things, the adsorption of hydrated Mg^{2+} cations on calcite. BROOKS et al. (1951, p. 162) emphasized that, if nucleation of a stable phase is retarded by the adsorption of foreign ions, then supersaturation is promoted in the adjacent solution and the precipitation of metastable phases may in this way be encouraged. The possibility of intermediate steps in the growth of $CaCO_3$ crystals, involving short-lived metastable phases, has yet to be examined.

Since going to press, four highly relevant papers have appeared by STOCKMAN, GINSBURG and SHINN (1967), WEYL (1967), BISCHOFF and FYFE (1968) and BISCHOFF (1968). They are listed in the references.

C. The Tangential Orientation of Oölitic Aragonite

The aragonite fabric of the marine, inorganic, oölitic lamellae (as distinct from the lamellae with randomly oriented crystals resulting from algal boring and micritization: Newell et al., 1960; Bathurst, 1966) is peculiar in having a preferred tangential orientation of *c* axes (Illing, 1954, p. 38; Newell et al., 1960, p. 490; Bathurst, 1967, p. 98). This orientation is perpendicular to that found where aragonite has grown with *c* radially on the surfaces of grains in beach rock (Ginsburg, 1953), in the subsurface as cement (Cullis, 1904) or in the chambers of Foraminifera or in the tubes of *Halimeda* in the Bahamas today. The anomalous tangential growth needs to be explained.

It is significant that attempts to precipitate oöids in stagnant conditions in the laboratory have yielded only spherulites with *c* radial (Lalou, 1957; Usdowski, 1963). Dr. Monaghan has told me that his experiments with Dr. Lytle (Monaghan and Lytle, 1956) also gave spherulites of this type. On the other hand Bucher (1918) stressed the influence on the crystal fabric of foreign substances in the emulsion in which concretions grow, and Usdowski (1963) has shown how the crystal fabric depends on, amongst other things, the Mg^{2+}/Ca^{2+} ratio and the salinity of the solution. So far, however, there is no evidence showing the extent to which chemical factors determine (if they do at all) the tangential orientation of marine oölitic lamellae.

Various workers, among them Rusnak (1960) and Usdowski (1963), have suggested that the tangential fabric is the result of normal growth of a radial-fibrous fabric modified in conditions of severe abrasion. In the initial stages of drusy growth the crystals are variously oriented, but the radial fibres are more easily broken and thus the tangential fibres are selectively preserved.

Another factor which may be important here is the process whereby new crystals become attached to the oöid surface. New nuclei must be continually supplied to the system, otherwise growth would yield only a single large crystal. In other words, new crystals, if they are to have a new orientation, cannot grow simply by extending pre-existing crystals in the oöid surface. On the contrary, a new, allochthonous piece of crystal lattice must become adsorbed on the surface. A crystal can only nucleate in two ways, either by formation of an entirely new nucleus as a result of ionic collision and the chance building of a tiny piece of stable monolayer (comprising for $CaCO_3$ perhaps no more than 5 unit cells, as suggested by the work of Christiansen and Nielson, in Nielson, 1955) or by oriented overgrowth on an existing lattice of appropriate ionic construction (not necessarily the same mineral), onto which the new lattice can be fitted, if necessary, by suitable elastic adjustment (van der Merwe, 1949). In the natural environment, in sea water, there is plenty of suspended lattice debris of colloidal dimensions, largely the product of grain breakage and abrasion. Thus crystal overgrowth must be, for energy considerations, the normal process whereby nucleation takes place. Thus we must conceive of tiny pieces of lattice, perhaps no more than pieces of monolayer, becoming attached to the faces of the crystals in the oöid surface, but with orientations different from those of the host lattices. It may be at this early stage that the final, tangential, orientation of the fully-grown fabric is determined. It seems reasonable to assume that, because of the strong (010) and (110) cleavages of aragonite, the suspended pieces of lattice debris will be flattened in a plane containing the *c* axis. Further, it seems proper to assume that these pieces would tend to become attached with their shortest axes perpendicular to the oöid surface, since the area of attachment would thus be greatest. Consequently, the newly attached lattice fragments would, from the first, be oriented with their *c* axes tangentially arranged over the host surface of the oöid. Some initial irregularity of orientation would, however, be caused by the uneven topography of the surface, and this irregularity would be enhanced where many newly attached lattice fragments interfered with each other, preventing the ideal tangential orientation of the attached fragments. Thus where nucleation is rapid (high density) a radial-fibrous orientation is to be expected, caused by competition between nuclei, by the natural tendency for radially oriented crystals to obstruct the growth of those more tangentially oriented. Where nucleation is slow (low density), there should be ample opportunity for the newly attached lattice fragments to become oriented with *c* tangential. The oölitic cortex of a Recent Bahamian oöid may have been growing for 2,000 years. The coat may be about 100 μ thick and so, as the unit cell is about 7 Å across, this means that the rate of accretion is equivalent to a layer of about 70 unit cells every year. It is not yet possible to say to what degree this is a competitive situation for the crystals. The fraction of its life that an oöid spends actually growing (as distinct from lying buried in an oölite shoal) may be small. It is possible that the annual accretion of a layer of 70 unit cells thickness takes place in a few hours. Much would depend also on the ability of an untidy accumulation of monolayered fragments to become reorganised (recrystallized?) in the interval between one growth period and the next.

During the last few years a new factor has become apparent in our physico-chemical picture of carbonate sedimentation — the omnipresent mucilage (SHEARMAN and SKIPWITH, 1965; BATHURST, 1967b). Coating the carbonate grains in the Persian Gulf and Bahamas there is a transparent, gelatinous mucilage of organic origin: this appears either as a sheath, or as a filling in the many algal bores which ramify within the grain (SHEARMAN and SKIPWITH, 1965), or as a matrix which binds grains together to a depth of about 0.25 cm below the surface of the sediment (BATHURST, 1967b). The significance of this mucilage in the study of oöid growth is twofold: the sheath may isolate the oöid surface from the sea water, also it may act as an adhesive to which colloidal aragonitic debris can become attached. The first possibility suggests the existence of some degree of selective ion transport through the membrane, the second prompts one to look again at SORBY's (1879, p. 74) idea that oöids grow by the simple adherence of debris to the outer surface — as a snowball — or, indeed, as an oncolite.

The question of ion transfer awaits experimental tests. The snowball hypothesis does not, in fact, stand up to close scrutiny for a number of reasons. The oöid is pure aragonite whereas the local sediment, in the Bahamas, contains some high-magnesian calcite. There is no sign, in the oölitic coat, of a random accumulation of detritus, as in an oncolite: the mosaic fabric and the tangential optical orientation are uniform. BATHURST (1967b) has described the mucilage as "elastic" and so it is unlikely to be at the same time adhesive. The high polish of the surfaces of growing oöids is a result of abrasion (NEWELL et al., 1960, p. 490; DONAHUE, 1965): this is inconsistent with crude mechanical adhesion. (The balance between precipitation and abrasion is discussed in a later section). Finally, oöids with tangential aragonite grow in caves, in the dark, where algae which are a possible source of the mucilage could not live.

D. The Rôle of Agitation in Oöid Growth

The four essential conditions for oöid growth have been given by CAYEUX (1935, p. 261) and DONAHUE (1965). They may be summarized as (a) a supersaturated solution, (b) availability of detrital nuclei, (c) agitation of grains, and (d) a splash cup. Supersaturation is necessary for growth, a detrital nucleus provides the initial substrate for oölitic growth, agitation prevents the oöid from becoming cemented to the floor and ensures that all of the growth surface is presented to the supersaturated solution. The splash cup keeps the oöid from straying out of the growth-promoting environment: its marine equivalent is the system of bathymetric and hydraulic controls which keep the oöid on the active oölite shoal.

The amount of agitation which is required for oöid growth is not obvious. The nuclei of oöids in Laguna Madre, Texas, described by FREEMAN (1962), are unevenly covered by the oölitic cortex and remain partly bare. This suggests that the oöids rotate at a rate which is slow compared with the rate of growth. NEWELL et al. (1960, p. 490) have also recorded locallized, rounded protuberances on some Bahamian oöids. BATHURST (1967a) has discovered that oölitic coats about 3 μ thick have grown on carbonate grains in Bimini Lagoon, Bahamas, in quiet conditions where abrasion is so low that the oöids are not polished, as they are in the classic area around Browns Cay (NEWELL et al., 1960). If the main requirement is that agitation shall prevent the oöids from becoming cemented together and shall expose all of the oöid surfaces at some time or other to the supersaturated solution, then very little agitation is needed because the growth rate is so slow, only about 30 to 180 μ in 2,000 years. On the other hand, it may be necessary that the agitation, when it occurs, shall be violent so that the oöid can be suspended in the turbulent solution. In this environment a high level of supersaturation is maintained at the oöid surface because local depletion of ions is promptly made good by efficient mixing, and growth is not hampered by dependence on diffusion rates as in the stagnant subsurface environment.

E. The Upper Size Limit of Marine Oöids

Marine oöids, both Recent and ancient, rarely exceed 1 mm in diameter. CAROZZI (1957) has suggested that, when an oöid has reached a critical size it can no longer be moved by the water, and thus ceases to grow. Although there must be some truth in this, in that agitation of larger grains is more difficult, it is certainly clear that oöids of all sizes on the Browns Cay oölite shoals, Bahamas, move in the tidal currents. This I have seen myself. Moreover, virtual stagnation has not prevented the growth of thin (3 μ) oölitic coats on the grains in Bimini Lagoon (BATHURST, 1966), though it seems likely that it has resulted in a remarkable slowing of growth, yielding a 3 μ cortex instead of one about 100 μ thick. Another factor, it seems, may also play a part in restricting oöid growth. As an oöid grows its mass increases as the third power of the radius, but the resistance of the water to the motion of the oöid rises only as the square of the radius. Oöids of 1 mm diameter will obey closely the impact law (RUBEY, 1933). Therefore, as they grow, the force exerted by one oöid on another will increase. Thus the amount of mechanical damage, of abrasion, that oöids can inflict on one another must increase as they grow. The surface area affected by collision remains proportionally the same, the damage simply becoming more severe. It must be supposed that a time will come when the rate of loss by abrasion will equal the rate of accretion by oölitic growth.

Studies of oöid shape support this hypothesis. Where oölitic growth continues over a non-spherical nucleus, such as a rod or plate, the outer surface of the oöid becomes increasingly spherical. Although growth takes place all over the surface, net accumulation is greater on the flatter surfaces away from domes and edges. The greater susceptibility to abrasion of curved surfaces and edges over flat surfaces is well known and obvious. Donahue (1965, p. 252) notes that some cave pearls (oöids) are preferentially polished over their "high points", giving rise to "locallized zones of polish along raised areas."

F. Oöid, Needle, Grapestone, or Intragranular Cement?

In different parts of the Great Bahama Bank, and of the Trucial coast embayment in the Persian Gulf, there are deposits of oöids, aragonite needle-muds and grapestone. Why should the form of the precipitate vary geographically in this manner?

Before this problem can be discussed it is necessary to look briefly at the growth of needle-muds and grapestone.

The growth of aragonite needles must necessarily be based, as with oöids, on the overgrowth of pre-existing crystal surfaces. In sea water so rich in colloidal carbonate debris it is thermodynamically essential that this be so. The supply of solute is practically infinite, its removal being continually made good by tidal mixing of bank with deep waters. The limitation on crystal size would appear to be controlled by the maximum possible *rate* of supply of solution, by competition between the growing crystals, and finally by the rate of burial of crystals and their removal, temporarily or permanently, from the growth environment.

Grapestone consists of calcarenite grains cemented together by micritic aragonite (Illing, 1954). In the Berry Islands, Bahamas, these grains are densely bored by unicellular algae (Bathurst, 1966) and it may be that the intensive photosynthesis helps in the precipitation of $CaCO_3$. These grains, or aggregates of grains, are also immersed in the mucilage already referred to, which binds the top 0.25 cm of the Recent calcarenite into a sub-tidal mat which resists tidal scour (Bathurst, 1967b). Thus the grains are maintained in close contact for long periods in an environment depleted in CO_2, and it is not surprising that they become cemented together by a precipitate of aragonite.

Turning now to the variation in form of the precipitate — oöid, needle or cement — it would appear that growth of oöids and needles is promoted where the turbulence is sufficient to put the particles into suspension in a supersaturated solution. On the oölite shoals, in particular, the level of supersaturation must be further raised by the escape of CO_2 across the turbulent water-air interface (Bolin, 1960; Kanwisher, 1963) and by the addition of abrasional debris. Tentatively, therefore, it is possible to consider a series oöid-needle-grapestone which represents a gradient of decreasing grain agitation, though *not* of decreasing water turbulence. On the oölite shoal the loss of CO_2 is rapid, sand sized nuclei are common (shell fragments, faecal pellets), but clay sized crystals are scarce. Precipitation takes places, therefore, on calcarenite grains. In the quieter areas of the needle-muds, turbulence is lowest, though adequate to put the needles into suspension. Needles are plentiful and are readier sites for precipitation than the heavy calcarenite grains which move less freely in the quieter waters. In the grapestone region, though turbulence is known to be greater than in the needle-mud region, the grains are bound together by the subtidal mat in an environment that is probably losing CO_2 relatively rapidly. The surfaces of the grains are being continually destroyed through boring and the pores are full of mat, so that oriented, radial growth of aragonite cement is impossible.

Lastly, reference must be made to the ubiquitous precipitation, from tropical sea water, of drusy fillings of aragonite in the chambers of Foraminiferida, the tubes of *Halimeda*, the empty bores left by algae (BATHURST, 1966) and in other intragranular voids. All these microenvironments are nearly enclosed spaces, a few microns or tens of microns across, they are all at some time the sites of decomposition of organic tissue, and their contained water is near-stagnant. This group of intragranular environments, to which one should add the pores in faecal pellets, is one of considerable biochemical complexity and must await a sophisticated joint attack by the biologist and the petrologist together.

G. The Need for Experimental Studies

Many of the matters discussed in this paper are concerned with the physics and chemistry of surfaces, a field into which geologists have been regrettably hesitant to venture, though happily some fruitful and adventurous steps have been taken by some workers in recent years, particularly in the area of aragonite stability. The growth of oöids is surely, above all, a problem in surface chemistry and should be amenable to imaginative experimental study from this point of view. The various hypotheses advanced in the earlier pages of this paper are only tentative and, in the testing of them, there lies great scope for profitable future research.

Acknowledgements

I gratefully acknowledge the help of my wife in the preparation of the manuscript, the advice of Dr. G. SKIRROW on chemical matters, of Dr. T. P. SCOFFIN on carbonate sedimentology generally, and valuable discussions with my colleagues of the Heidelberg Kolloquium.

References

BATHURST, R. G. C.: Boring algae, micrite envelopes and lithification of molluscan biosparites. Geol. J. **5**, 15—32 (1966).
— Oölitic films on low energy carbonate sand grains, Bimini lagoon, Bahamas. mar. Geol. **5**, 89—109 (1967a).
— Sub-tidal gelatinous mat, sand stabilizer and food, Great Bahama Bank. J. Geol. **75**, 736—738 (1967b).
BERNER, R. A.: Diagenesis of carbonate sediments: interaction of magnesium in sea water with mineral grains. Science **153**, 188—191 (1966).
BISCHOFF, J. L.: Catalysis, inhibition, and the calcite-aragonite problem. II. The vaterite-aragonite transformation. Am. J. Sci. **266**, 80—90 (1968).
—, and W. S. FYFE: Catalysis, inhibition, and the calcite-aragonite problem. I. The aragonite-calcite transformation. Am. J. Sci. **266**, 67—79 (1968).
BOLIN, B.: On the exchange of carbon dioxide between the atmosphere and the sea. Tellus **12**, 274—281 (1960).
BROOKS, R., L. M. CLARK, and E. F. THURSTON: Calcium carbonate and its hydrates. Phil. Trans. Roy. Soc. London Ser. A. **243**, 145—167 (1951).
BUCHER, W. H.: On oölites and spherulites. J. Geol. **26**, 593—609 (1918).
CAROZZI, A.: Contribution à l'étude des propriétés géométriques des oolithes. Bull. Inst. natn. genev. **58**, 1—52 (1957).
CAYEUX, M.: Les Roches Sédimentaires: Roches Carbonatées, 463 p. Paris: Masson 1935.

Chilingar, G. V., H. J. Bissell, and R. W. Fairbridge: Carbonate Rocks, Physical and Chemical Aspects, 393 p. Amsterdam: Elsevier 1966.

Cloud, P. E., Jr.: Environment of calcium carbonate deposition west of Andros Island Bahamas. Prof. Pap. U.S. geol. Surv. **350**, 1—138 (1962).

Cullis, C. G.: The mineralogical changes observed in the cores of the Funafuti borings. In: T. G. Bonney (Editor): The Atoll of Funafuti, p. 392—420. London: The Royal Society 1904.

Davies, C. W., and A. L. Jones: The precipitation of silver chloride from aqueous solutions. Part 2. — Kinetics of growth of seed crystals. Trans. Faraday Soc. **51**, 812—817 (1955).

Donahue, J.: Laboratory growth of pisolite grains. J. Sediment. Petrol. **35**, 251—256 (1965).

Freeman, T.: Quiet water oölites from Laguna Madre, Texas. J. Sediment. Petrol. **32**, 475—483 (1962).

Fyfe, W. S., and J. L. Bischoff: The calcite-aragonite problem. In: Pray, L. C., and R C. Murray (Editors): Dolomitization and Limestone Diagenesis. A Symposium. Soc. Econ. Palaeont. Miner. spec. Publs. **13**, 89—111 (1965).

Ginsburg, R. N.: Beach rock in South Florida. J. Sediment. Petrol. **23**, 85—92 (1953).

Illing, L. V.: Bahaman calcareous sands. Bull. Am. Assoc. Petrol. Geologists **38**, 1—95 (1954).

Kanwisher, J.: On the exchange of gases between the atmosphere and the sea. Deep-Sea Research **10**, 195—207 (1963).

Kitano, Y.: The behavior of various inorganic ions in the separation of calcium carbonate from a bicarbonate solution. Bull. Chem. Soc. Japan **35**, 1973—1980 (1962).

Lalou, C.: Etude expérimentale de la production de carbonates par les bactéries des vases de la baie de Villefranche-sur-Mer. Ann. inst. océanogr. (Monaco) **33**, 202—267 (1957).

Lippmann, F.: Versuche zur Aufklärung der Bildungsbedingungen von Kalcit und Aragonit. Fortschr. Mineral. **38**, 156—160 (1960).

Monaghan, P. H., and M. A. Lytle: The origin of calcareous ooliths. J. Sediment. Petrol. **26**, 111—118 (1956).

Newell, N. D., E. G. Purdy, and J. Imbrie: Bahamian oölitic sand. J. Geol. **68**, 481—497 (1960).

Nielsen, J.: The kinetics of electrolyte precipitation. J. Colloid Sci. **10**, 576—586 (1955).

Pobeguin, T.: Contribution à l'étude des carbonates de calcium, précipitation du calcaire par les végétaux, comparaison avec le monde animal. Ann. sci. nat. Botan. et biol. végétale **15**, 29—109 (1954).

Rubey, W. W.: Settling velocities of gravel, sand, and silt particles. Am. J. Sci. **25**, 325—338 (1933).

Rusnak, G. E.: Some observations of recent oolites. J. Sediment. Petrol. **30**, 471—480 (1960).

Shearman, D. J., and P. A. d'E. Skipwith: Organic matter in Recent and ancient limestones and its rôle in their diagenesis. Nature (London) **208**, 1310—1311 (1965).

Sorby, H. C.: The structure and origin of limestones. Proc. geol. Soc. **25**, 56—95 (1879).

Stockman, K. W., R. N. Ginsburg, and E. A. Shinn: The production of lime mud by algae in South Florida. J. Sediment. Petrol. **37**, 633—648 (1967).

Usdowski, H. E.: Der Rogenstein des norddeutschen Unteren Buntsandsteins, ein Kalkoölith des marinen Faziesbereichs. Fortschr. Geol. Rheinld. Westf. **10**, 337—342 (1963).

van der Merwe, J. H.: Misfitting monolayers and oriented overgrowth. Discussions Faraday Soc. **5**, 201—214 (1949).

Weyl, P. K.: The solution behaviour of carbonate materials in sea water. Exploration Production Research Division, Shell Oil Company, Publication **428**, 178—228 (1967).

The Fabric of Carbonate Cement and Matrix and its Dependence on the Salinity of Water

GERALD M. FRIEDMAN*

With 8 Figures

Abstract

Lithification of carbonate sediments occurs under both marine and freshwater conditions. Mineralogically, texturally, and genetically four different types of cement and matrix are recognized. These are: 1. marine aragonite as fibrous cement or micrite, 2. marine high-magnesian calcite micrite, 3. freshwater low-magnesian calcite drusy mosaic or wedge-shaped crystals or as fibrous cement or rods, and 4. freshwater low-magnesian calcite micrite. Aragonite and high-magnesian calcite are the minerals which form cement or matrix in the marine environment, whereas low-magnesian calcite makes up cement or matrix in the freshwater environment. Aragonite forms fibrous cements under hypersaline or very shallow marine conditions, such as in deep-sea carbonate sediments of the Red Sea, or in beachrocks or reefs. It may also be deposited as micrite, as in grapestones. High-magnesian calcite forms a micrite cement in the deep-sea environment or in beachrocks or a fibrous cement in reefs. Low-magnesian calcite forms a drusy calcite mosaic under fresh water conditions or occurs in the shape of rods or wedge-shaped crystals in freshwater limestones, such as in tufa or cave pearls. It occurs as a cement in marine rocks in which it has been emplaced post-depositionally by fresh water. Low-magnesian calcite micrite is formed in fresh water, such as in lakes, where it makes up oncolites or massive limestones. Texturally fresh water low-magnesian calcite micrite is indistinguishable from aragonite micrite or from marine high-magnesian calcite micrite or its low-magnesian diagenetic equivalent.

A. Introduction

The most important aspect of carbonate diagenesis is lithification. During lithification, the interparticle porespace is progressivly occluded, but at the same time moldic porosity may be developed (FRIEDMAN, 1964). The conditions under which lithification takes place are far-reaching. Lithification commonly occurs concomitantly with freshwater dissolution of aragonite; the solution becomes progressively supersaturated with calcium bicarbonate and deposits calcite in the interparticle porespace when CO_2 or H_2O are expelled (FRIEDMAN, 1964). Freshwater lithification may mean that the diagenetic changes which lead to consolidation occur in a subaerial environment. Yet this does not always fit the stratigraphic model. Stratigraphically, the basins in which thick sections of carbonate sediments accumulate are continuously sinking and no evidence may be available which indicates even periodic pulses of uplift or retreat of sea-level. Hence from the stratigraphic point of view, exposure of the carbonate sediments in their basin of deposition to a subaerial environment may seem to many stratigraphers a remote possibility. However, this possibility is not as far-fetched as it may seem. A basin of deposition, like the present-day Bahama Platform, is extremely

* Rensselaer Polytechnic Institute, Troy, N.Y.

shallow with a depth of only 10 to 40 feet over a large part of the platform. Slight sea-level oscillations or extremely minor fluctuations of the basin may lead to exposure of the entire basin with resultant freshwater lithification of carbonate sediments. There is, however, another way of explaining freshwater lithification. This writer has pointed out (1964) that freshwater lithification is not necessarily synonymous with subaerial environment. Lithification can occur in the subsurface where fresh waters are more common than geologists seem to admit. Hence carbonate sediments can lithify in the subsurface with resultant mineralogical and textural characteristics comparable to those developed under conditions of subaerial exposure.

The fabrics of sediments reflect the geochemistry of the waters to which they are or were exposed. These waters may have been the waters in the depositional environment or more likely the pore-waters that passed through the geologic column. A constant stream of fluids passes through the interparticle pore space of sediments and rocks. The composition of this water varies. The purpose of the present paper is to show some relationship between the nature of the interparticle cement or carbonate matrix, i.e., its fabric, and the composition of the water which was responsible for the deposition of the cement or matrix. The objective of the present study is to determine from the fabric of the cement or matrix the composition of the water that was of importance in the conversion of the carbonate sediment into a limestone. We would like to tell by fabric examination under the petrographic microscope the environmental conditions to which the limestones were exposed in an ancient depositional basin.

For clarification, it should be pointed out that more than cement is involved in the diagenetic changes. As has already been indicated, moldic porosity may be developed in the diagenetic sequence; this moldic porosity may be ultimately destroyed by a calcite infilling or it may be preserved by oil which migrated through the geologic column.

B. Fabric and Mineralogy of Cement and Matrix and the Geochemistry of the Waters

I. Marine Cements

Cementation can occur in the marine environment. Mineralogically the cement is composed of aragonite or high-magnesian calcite. Each of these two minerals may leave a distinctive fabric. Aragonite commonly leaves a fibrous cement (Fig. 1), whereas high-magnesian calcite leaves a micritic matrix (Fig. 2). However in some sediments, such as in grapestones, aragonite also forms a micritic matrix and in Bermuda reefs high-magnesian calcite forms a fibrous cement (GINSBURG et al., 1967). Cementation in the marine environment is more common than has been generally admitted in the past.

1. Aragonite Cement

Fibrous cement is usually deposited in the form of aragonite under hypersaline conditions or in very shallow marine water in which CO_2 can be easily expelled. The latter may also be a slightly hypersaline environment. It has been recorded from deep water environments of the Red Sea (GEVIRTZ and FRIEDMAN, 1966), from beachrocks of many areas (NESTEROFF, 1954; STODDART and CANN, 1965) and from reefs (CULLIS,

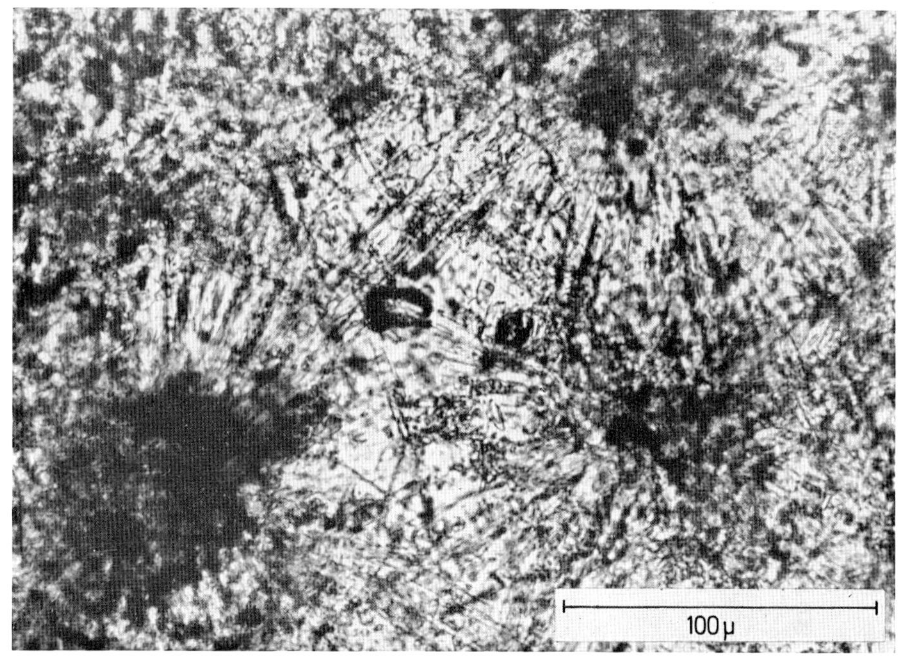

Fig. 1. Fibrous aragonite as cement in deep-sea carbonate sediment of the Red Sea (see GEVIRTZ and FRIEDMAN, 1966)

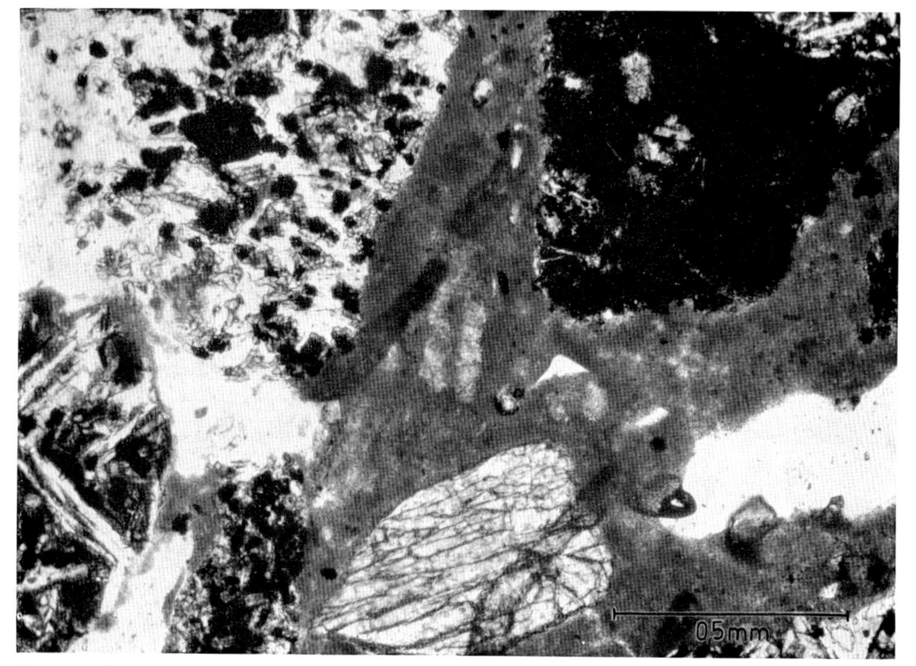

Fig. 2. High-magnesian calcite cementing detrital basalt and augite fragments in Recent beachrock, Punta di Agando near Gran Tarajal, Fuerteventura, Canary Islands. Beachrock was sampled by GERD TIETZ, University of Heidelberg

1904; Friedman, unpublished; Lees, 1964; and Pray, 1965). In reefs, however, this fibrous cement has been observed mostly in ancient environments and is now calcite, but by analogy, is interpreted as originally aragonite. During diagenesis this cement is paramorphically replaced by calcite (Friedman, 1964) and the original depositional fabric is retained. The cement of beachrocks is usually aragonite but as indicated in the next section it may be high-magnesian calcite.

Aragonite in the form of micrite serves as a cementing agent in grapestone. Aragonite micrite is indistinguishable from high- and low-magnesian calcite micrite. Hence after paramorphic replacement of this cement the original mineralogy can no longer be recognized. Inasmuch as low-magnesian calcite micrite is of freshwater origin and aragonite micrite of marine origin, no information on depositional environment nor on the salinity of water can be inferred from micrite-cemented carbonate sediments, unless diagnostic fossils are present.

2. High-Magnesian Calcite

Lithified carbonate sediments composed of high-magnesian calcite were described for the first time from the top of Atlantis Seamount, part of the Atlantic ridge from a depth of 150 fathoms (Friedman, 1964, p. 806—807). These sediments were dated at 9,000 to 12,000 years. The cement between the grains is composed entirely of high-magnesian calcite micrite. More recently John Milliman (1966) of Woods Hole Oceanographic Institute has discovered numerous examples of the same type of material from various depths in the ocean. He found that Quaternary examples were composed of high-magnesian calcite and Tertiary ones of low-magnesian calcite. Apparently high-magnesian calcite is removed and low-magnesian calcite formed on a microscale by solution-deposition without a textural change of the material affected (Friedman, 1964). The mechanism of formation of high-magnesian calcite as a cement under marine conditions is not yet understood, but the writer believes that a biological influence may exist.

High-magnesian calcite cement commonly lithifies skeletal and detrital silicate grains in beachrock. Along the Gulf of Aqaba near the city of Elat, Israel, this writer (Friedman, in press) has noted a cement in beachrock which was in part composed of high-magnesian calcite. Gerd Tietz of the University of Heidelberg has found a cement in beachrock of the Canary Islands which was for the most part composed of high-magnesian calcite (see Fig. 2). The cement of beachrocks is usually given as calcite or aragonite. Aragonite is said to be of marine origin and calcite of fresh water origin (Stoddart and Cann, 1965). The cement of beachrocks, as indicated by the Gulf of Aqaba and Canary Island examples, may be partly or largely composed of high-magnesian calcite of marine origin. Low-magnesian calcite has not been found by this writer as an important constituent of beachrock. Hence a freshwater influence in beachrock formation must at best be slight.

The high-magnesian calcite fabric, though diagnostic, is very likely to be confused in ancient marine limestones with an original aragonite micrite fabric. A high-magnesian calcite beachrock may, however, have some fibrous aragonite cement which is diagnostic of many beachrocks (Fig. 3). Ginsburg et al. (1967) discovered fibrous high-magnesian calcite cement in Bermuda reefs.

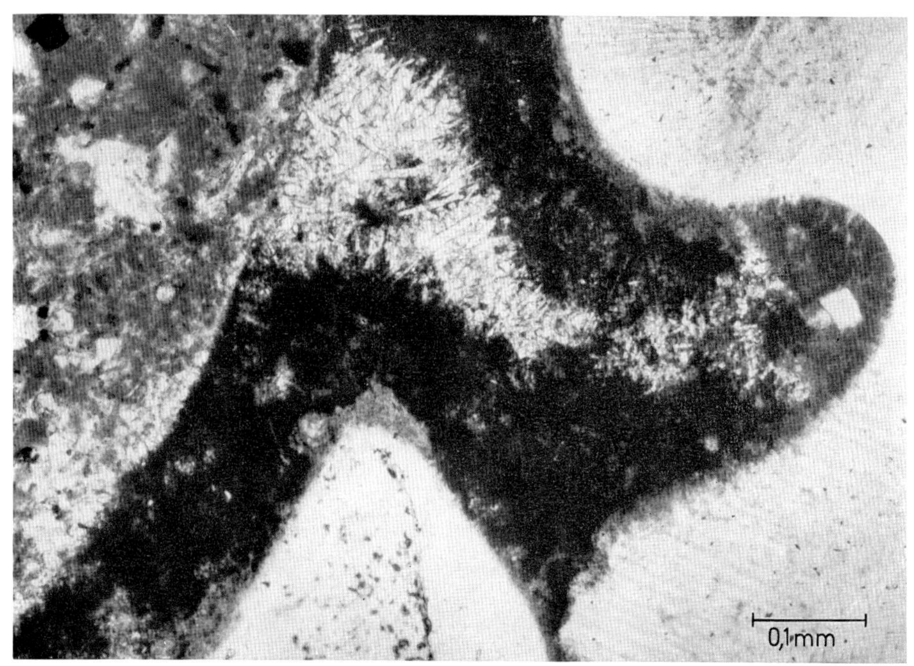

Fig. 3. Cement in beachrock consisting of both high-magnesian calcite and aragonite. Wide dark band in center of photograph is high-magnesian calcite cement; fibrous pore filling is aragonite. Gulf of Elat (Aqaba), Red Sea, 5 km south of City of Elat

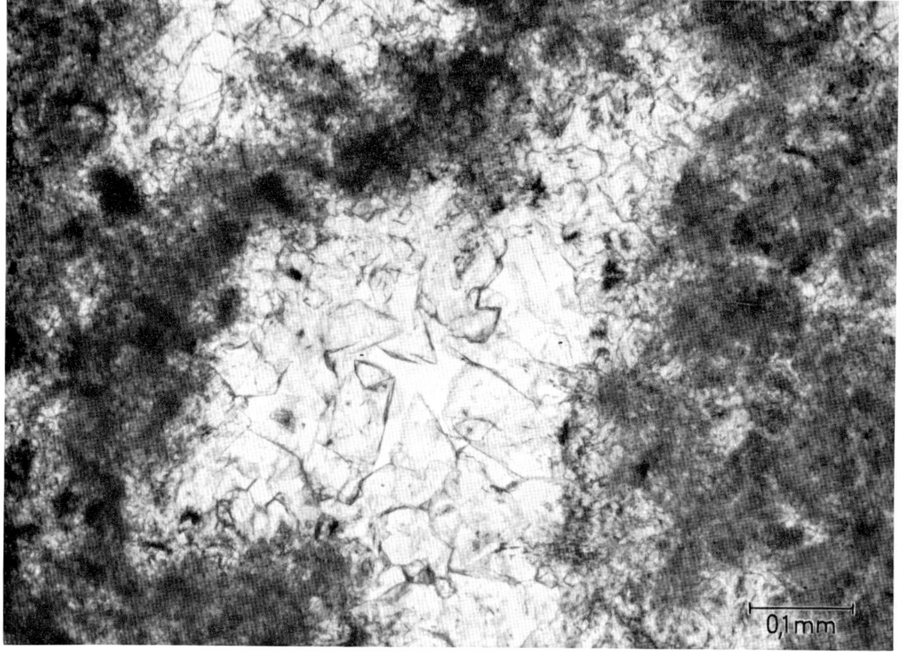

Fig. 4. Drusy calcite mosaic in freshwater tufa, Ensisheim, Schwäbische Alb, Germany. Tufa was sampled by GEORG IRION, University of Heidelberg

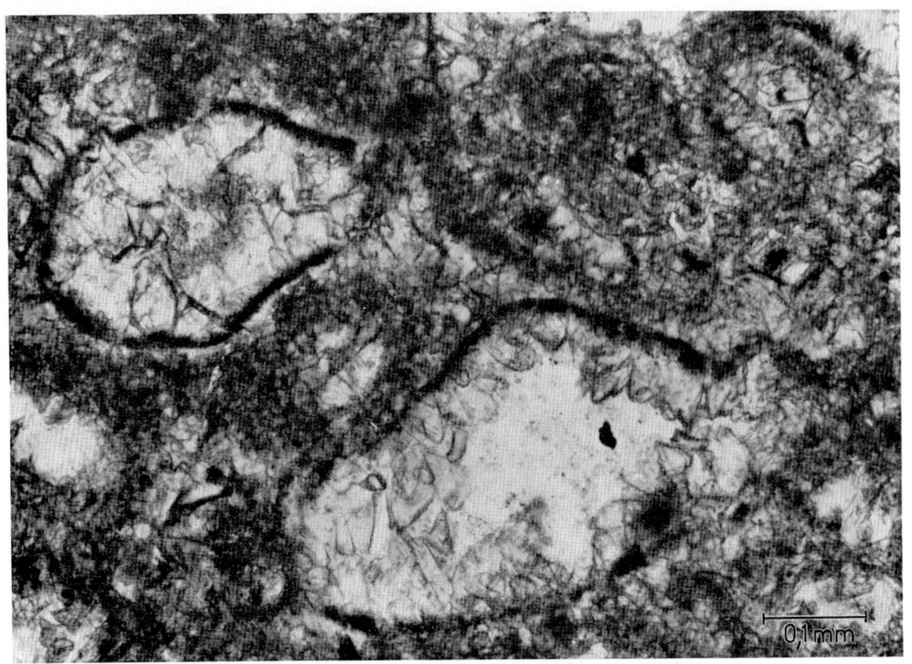

Fig. 5. Drusy calcite mosaic in freshwater tufa, Ensisheim, Schwäbische Alb, Germany. Tufa was sampled by GEORG IRION, University of Heidelberg

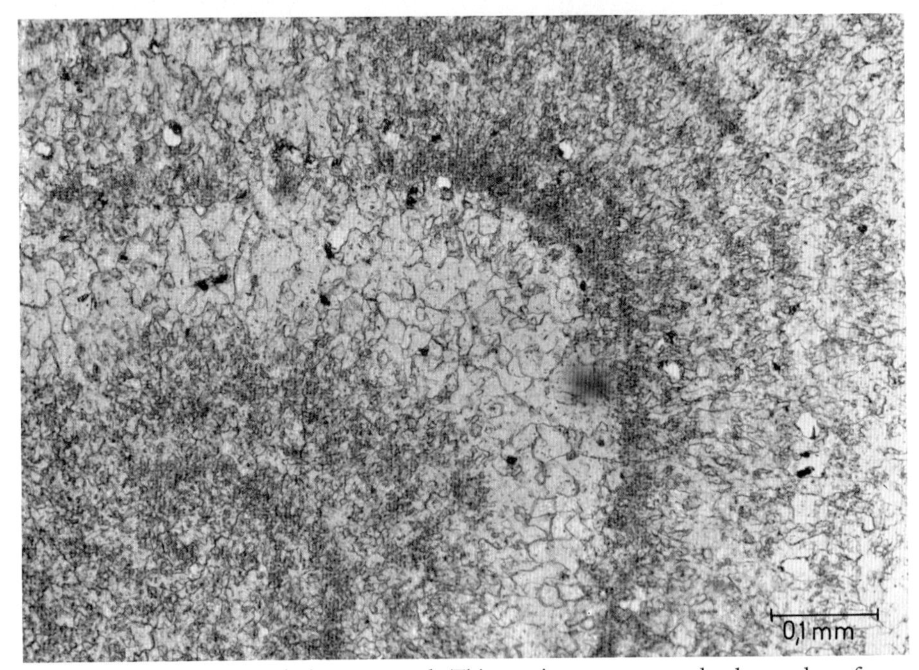

Fig. 6. Drusy calcite mosaic in cave pearl. Thin-section cut was made along edge of cave pearl. Guggenbach, Styria, Austria. Cave pearl was collected by MARTIN KIRCHMAYER, University of Heidelberg

II. Freshwater Cements

1. Drusy Mosaics, Wedge-Shaped Crystals, Rods

A drusy calcite mosaic which is so common in limestones has been deposited by fresh water (FRIEDMAN, 1964). This evidence has in the past been for the most part indirect and has been based on studies of carbon and oxygen isotopes. GROSS (1961,

Fig. 7. Rod and wedgeshaped crystals in cave pearl. Thin-section was cut along the largest diameter of cave pearl. This is the same pearl as that shown in Fig. 6

1964) showed that secondary calcite as cement and as alteration of fragments in carbonate sediments causes the C^{13}/C^{12} and O^{18}/O^{16} values of limestones to approach values observed in freshwater carbonates. This writer (1964) succeeded in correlating C^{13}/C^{12} and O^{18}/O^{16} values with mineralogical and fabric changes. He found that the approach towards freshwater values correlates with increase in the low-magnesian calcite content and with progressive development and ultimately more extensive spreading of drusy calcite mosaic. More recently BERNER (1965) using geochemical approaches confirmed the relationship between freshwater and these diagenetic changes. The drusy calcite mosaic is precipitated inorganically; the calcium bicarbonate is derived from the dissolution of aragonite (FRIEDMAN, 1964). Drusy calcite mosaic occurs in marine limestones in which it has been emplaced post-depositionally by fresh water.

A more direct approach has confirmed that drusy calcite mosaic forms through the action of fresh water. Thin-section cuts of fresh water tufa and cave pearls show that drusy calcite mosaic is truly a product of freshwater deposition (Figs. 4, 5, 6). Hence, direct evidence is now available that drusy mosaic is an important fabric of freshwater limestones. Previously this evidence depended on indirect geochemical approaches (FRIEDMAN, 1964). This mosaic is made up of calcite and fabricwise and mineralogically it remains unchanged during diagenesis.

Other freshwater carbonate fabrics include those of cave deposits which are composed of rods and wedgeshaped crystals. These have been reported by PETER KOLESAR, one of the writer's students at Rensselaer Polytechnic Institute and by GEORG IRION of the University of Heidelberg. KIRCHMAYER (1962) in Austrian cave pearls from mines found that wedge- and rod-shaped crystals (Fig. 7) in thin-sections cut approximately parallel to the largest possible diameter made up the rims of his pearls.

2. Micrite

Freshwater carbonate micrites, in the form of low-magnesian calcite, are deposited in lakes, such as in Green Lake, New York (KOLESAR, in preparation) or in Lake Constance (Fig. 8) (SCHÖTTLE, 1967). These carbonates are formed in Lake

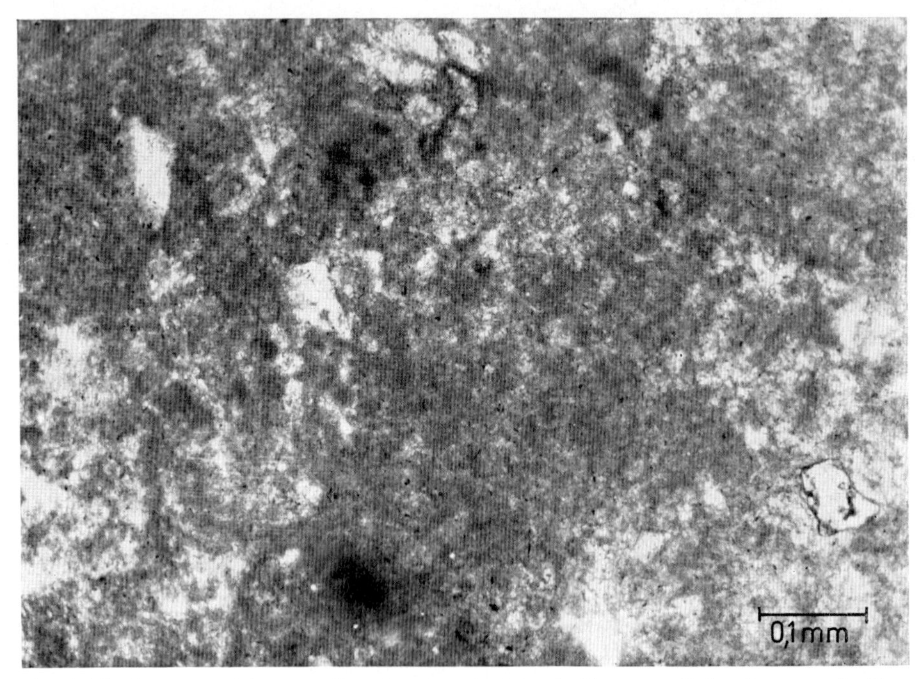

Fig. 8. Freshwater calcite oncolite, Langenrein Island, Untersee (Lake Constance), Germany. This sample was collected by MANFRED SCHÖTTLE, University of Heidelberg

Constance biologically either by blue-green algae or by large waterplants. The carbonates of Green Lake appear to be of inorganic origin. Texturally freshwater carbonate micrites cannot be distinguished from lithified lime-mud nor from marine micritic high-magnesian calcite cement or its low-magnesian diagenetic equivalent. However biological or fossil evidence should indicate whether a micrite is of fresh water or marine origin. The freshwater low-magnesian calcite micrites make up oncolites or massive limestones. Some of the latter, such as those in Green Lake, New York, are colloquially described as "reefs".

Conclusions

A relationship has been established between the fabric of carbonate cement and the salinity of the depositing water. Drusy mosaic, wedgeshaped crystals or rods are formed under fresh water conditions. If a drusy mosaic is found in marine limestones, these limestones must have been subjected to subaerial diagenetic changes or to fresh waters in the subsurface. Fibrous cements are the result of diagenetic changes in a marine environment. Carbonate micrite can form under marine or fresh water conditions and only the organisms contained in it can indicate which of these two was the environment of deposition. Mineralogically this micrite would be high-magnesian calcite or aragonite in the marine environment and low-magnesian calcite in the non-marine environment. However during diagenesis the end-product is invariably a low-magnesian calcite.

Acknowledgements

Grateful acknowledgement is extended to MARTIN KIRCHMAYER, GEORG IRION, MANFRED SCHÖTTLE, and GERD TIETZ of the University of Heidelberg for the loan of study material and for discussing with the writer problems of carbonate diagenesis. This study was financed in part by American Chemical Society grant No. PRF — 2393-A 2.

References

BERNER, R. A.: Chemical diagenesis of some modern carbonate sediments. Am. J. Sci. **264**, 1—36 (1966).

CULLIS, C. G.: The mineralogical changes observed in the cores of the Funafuti borings. In: The Atoll of Funafuti. Roy. Soc. London, Coral Reef Comm., Rept., p. 392—420 (1904).

FRIEDMAN, G. M.: Early diagenesis and lithification in carbonate sediments. J. Sediment. Petrol. **34**, 777—813 (1964).

— Carbonate sediments, reefs, and beachrocks, Gulf of Aqaba (Elat), Red Sea. J. Sediment. Petrol. (In press).

GEVIRTZ, J. L., and G. M. FRIEDMAN: Deep-sea carbonate sediments of the Red Sea and their implications on marine lithification. J. Sediment. Petrol. **36**, 143—151 (1966).

GROSS, M. G.: Carbonate sedimentation and diagenesis of Pleistocene limestones in the Bermuda islands. Unpublished Ph. D. thesis. California Institute of Technology, 244 p. 1961.

— Variations in the $0^{18}/0^{16}$ ratios of diagenetically altered limestones in the Bermuda Islands. J. Geol. **72**, 170—194 (1964).

KOLESAR, P.: In preparation. The fabrics of freshwater limestones. Unpubl. M. S. THESIS, Rensselaer Polytechnic Institute.

LEES, A.: The structure and origin of the Waulsortian (lower carboniferous) "reefs" of west-central Eire. Roy. Soc. London. Philos. Trans. Ser. B, **247**, 483—531 (1964).

MILLIMAN, J. D.: Submarine lithification of carbonate sediments. Science **153**, 994—997 (1966).

NESTEROFF, W. D.: Sur la formation des grès de plage ou "beach-rock" en Mer Rouge. C. R. Acad. Sci. Paris **238**, 2547—2548 (1954).

PRAY, L. C.: Limestone clastic dikes in Mississippian bioherms, New Mexico, p. 154—155. In: Geol. Soc. America, Abstracts for 1964. Geol. Soc. Am. Spec. Papers **82**, 400 p. (1965).

SCHOETTLE, M.: Die Sedimente des Gnadensees. Ein Beitrag zur Sedimentbildung im Bodensee. Unpublished Ph. D. thesis, University of Heidelberg, 104 p. 1967.

STODDART, D. R., and J. R. CANN: Nature and origin iof beachrock. J. Sediment. Petrol. **35**, 243—247 (1965).

Since this paper was written, the following papers have been published on this subject:

FISCHER, A. G., and R. E. GARRISON: Carbonate lithification on the sea-floor. J. Geol. **75**, 488—496 (1967).

GINSBURG, R. N., E. A. SHINN, and J. H. SCHROEDER: Submarine cementation and internal sedimentation within Bermuda reefs. Geol. Soc. America Program, Annual Meeting, 78—79 (1967).

The Formation of Dolomite in Sediments

Hans-Eberhard Usdowski*

With 4 Figures

Abstract

Dolomite forms in two ways: 1. The origin of dolomite on the surface of the earth, i.e., the early diagenetic formation of dolomite and 2. the origin of dolomite in sedimentary rocks, i.e., the late-diagenetic formation of dolomite. In both cases, pre-existing $CaCO_3$ reacts with solutions, resulting in the formation of dolomite. The reaction solution of the early diagenesis is modified, pre-concentrated sea water. The reaction solution of the late diagenesis consists of pore solutions. Both reactions proceed as follows: Metastable $CaCO_3$ which forms in place of stable dolomite is changed into dolomite by diagenetic solutions which generally contain abundant Mg. Different amounts of dolomite are thus formed from a definite amount of $CaCO_3$ depending on the anion content of the solutions. These processes are discussed with the aid of solubility equilibria of the Ca-Mg-carbonates, -sulfates and -chlorides.

A. Introduction

The sedimentary rocks of the earth's crust contain approximately 10^{17} tons of carbonate rocks. Of these, approximately $7 \cdot 10^{16}$ tons are limestones and $3 \cdot 10^{16}$ tons dolomitic rocks.

The origin of carbonate rocks in general is known for a long time. They are mostly precipitated from sea-water. The precipitation is purely inorganic, inorganic with the aid of organisms, and purely organic. Likewise, the basic concepts about the origin of dolomitic rocks are old. The geologic occurrence of dolomites is identical to that of limestones. Thus it was already assumed in the last century, that they are precipitated from sea water in a manner similar to inorganically formed limestones. However, since no Recent formation of dolomite was directly observed, it was assumed that dolomites were formed only during earlier periods of the earth's history. For about 10 years Recent dolomite is known to form during marine sedimentation. It is also known that the same processes which today lead to dolomitization were also operative in the geologic past. For example, the present formation of dolomite in the Persian Gulf may be compared with the origin of the German Zechstein dolomite. However, not all dolomitic rocks formed in the same way as Recent dolomite. Petrographic observations show that dolomites are also formed in buried sedimentary rocks through the action of Mg-bearing pore solutions on limestones. This concept, too, is known for some time (Twenhofel, 1932).

One must, therefore, distinguish two dolomitization processes: 1. The origin of dolomite on the earth's surface during sedimentation in a marine environment and 2. the formation of dolomite in buried sedimentary rocks by poresolutions acting on limestones. Both processes are diagenetic. The formation of dolomite on the earth's

* Sedimentpetrographisches Institut, Universität Göttingen, Germany.

surface generally starts shortly after sedimentation of the carbonate minerals; this dolomite is called early diagenetic. Late diagenetic dolomite, on the other hand, forms at a later stage, after burial of the carbonate sediments.

B. Systems

I. Representation

Solution always participate in the formation of dolomites. The most important solutions are sea water, modified sea water (sea water in various stages of evaporation), and pore solutions. All these solutions contain the principal components K^+, Na^+,

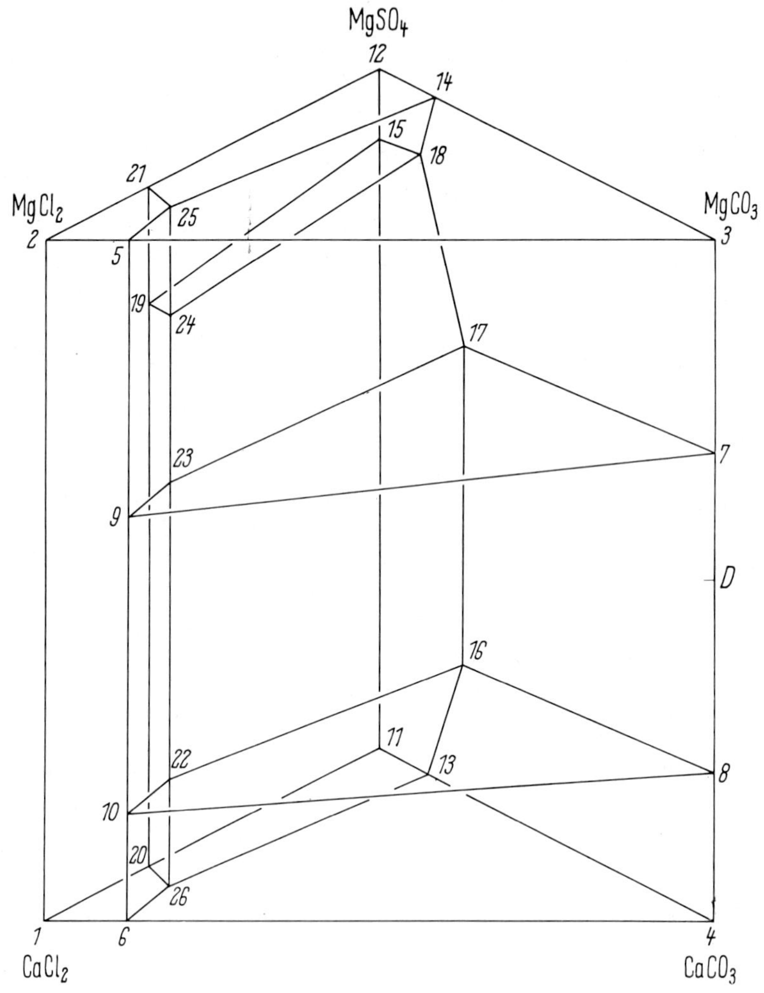

Fig. 1. Schematic representation of the phase-relationships in the 5-component-system $Ca^{2+}-Mg^{2+}-CO_3^{2-}-SO_4^{2-}-Cl_2^{2-}-H_2O$ by an equilateral prism. The stability field of the Ca-Mg-chloride is shown enlarged. It lies close to the edge 1 to 2. Also the stability field of the $MgSO_4$-hydrate is shown enlarged. D: dolomite. Point 11: $CaSO_4$ · Lengths of the edges of the prism: 100 mole-%. The representation also shows the phase-relationships of a part of the 6-component-system $Na_2^{2+}-Ca^{2+}-Mg^{2+}-CO_3^{2-}-SO_4^{2-}-Cl_2^{2-}-H_2O$

Ca^{2+}, Mg^{2+}, HCO_3^-, SO_4^{2-}, Cl^-, and H_2O. Disregarding K^+ which is much less abundant than Na^+, the remaining components form a 6-component-system. Since variations in the Na^+- and Cl^--content also occur in natural solutions besides fluctuations in the Ca^{2+}-, Mg^{2+}, HCO_3^--, and SO_4^{2-}-concentrations, conditions during the formation of dolomite are best considered using two extreme cases: 1. the NaCl-saturated, 6-component-system Na_2^{2+}-Ca^{2+}-Mg^{2+}-CO_3^{2-}-SO_4^{2-}-Cl_2^{2-}-H_2O and 2. the NaCl-free 5-component-system Ca^{2+}-Mg^{2+}-CO_3^{2-}-SO_4^{2-}-Cl_2^{2-}-H_2O.

The NaCl-saturated, 6-component-system is represented isobarically and isothermally by a tetrahedron, neglecting NaCl and H_2O (Usdowski, 1967). For dis-

Table 1. *The phase-relationships of the 5-component-system $Ca^{2+}-Mg^{2+}-CO_3^{2-}-SO_4^{2-}-Cl_2^{2-}-H_2O$ and the 6-component-system $Na_2^{2+}-Ca^{2+}-Mg^{2+}-SO_4^{2-}-Cl_2^{2-}-H_2O$. In the gaps between the solid phases solution and vapor must still be added to all solid phases, for both systems. With the 6-component-system, moreover, halite must still be included as an additional solid*

Points in Fig. 1	Solid phases
6–26–13–4–10–22–16–8	Calcite
10–22–16–8–9–23–17–7	Dolomite
9–23–17–7–24–18–5–25–14–3	Magnesite
26–13–22–16–23–17–24–18–19–15–20–11	Anhydrite
24–18–25–14–21–12–19–15	$MgSO_4$-hydrate
1–20–26–6–2–21–25–5	Ca-Mg-chloride
26–22–16–13	Calcite + anhydrite
10–22–16–8	Calcite + dolomite
22–23–17–16	Dolomite + anhydrite
9–23–17–7	Dolomite + magnesite
23–24–18–17	Magnesite + anhydrite
24–25–14–18	Magnesite + $MgSO_4$-hydrate
24–19–15–18	Anhydrite + $MgSO_4$-hydrate
22–16	Calcite + dolomite + anhydrite
23–17	Magnesite + dolomite + anhydrite
24–18	Magnesite + anhydrite + $MgSO_4$-hydrate
22	Calcite + dolomite + anhydrite + Ca-Mg-chloride
23	Magnesite + dolomite + anhydrite + Ca-Mg-chloride
24	Magnesite + anhydrite + $MgSO_4$-hydrate + Ca-Mg-chloride

cussing the solubility equilibria of the rock-forming carbonates, it is unnecessary to use the entire tetrahedron. It is sufficient to consider that part of the system which contains the composition of most of the natural solutions. This portion of the system can be represented by an equilateral prism (Fig. 1). Since the representation is isobaric and isothermal, those solution compositions are expressed by the prism which are in equilibrium with one or more solids. The compositions of the solutions are given in mole percent. Each side of the prism represents 100 mole percent. Absolute concentrations of the solutions must be considered separately. Also, the NaCl-content of the solutions is not represented.

Fig. 1 also represents the NaCl-free, 5-component-system. The difference of the two systems is, that, in the 6-component-system, halite must be added as additional phase to each solid component. In addition, NaCl is present in the solutions up to saturation concentration.

The phase relationships of the systems are compiled in Table 1. They are based on experimental studies made at 50°, 80°, 120° and 180°C below the vapor pressures of the solutions. The experiments were made in bombs of pyrex-glass. Equilibrium data were determined by solubility measurements an by reactions between solids and solutions (USDOWSKI, 1967). The data show that the positions of equilibria in both systems differ only slightly from one another. The following discussion, therefore, applies qualitatively to cases of NaCl-absence and also to the case of NaCl-saturation.

II. Reactions

The most important equilibrium concerning the formation of dolomite in sediments is that between calcite and dolomite (surface 8–10–22–16, Fig. 1) and the equilibrium between magnesite and dolomite (surface 7–9–23–17, Fig. 1). The equilibrium between calcite and dolomite in the 5-component-system can be obtained according to relationship (1), through the reaction of $CaCO_3$ with Mg-ions. Thus, dolomite is formed and Ca-ions are liberated.

$$2\,CaCO_3 + Mg^{2+} + (CO_3^{2-}, SO_4^{2-}, Cl_2^{2-}) \rightleftharpoons CaMg(CO_3)_2 + Ca^{2+} + (CO_3^{2-}, SO_4^{2-}, Cl_2^{2-})\,. \qquad (1)$$

The equilibrium can also be attained in that dolomite reacts with Ca-ions, whereby calcite is formed and Mg-ions are liberated. For the equilibrium between magnesite and dolomite relationship (2) applies correspondingly.

$$2\,MgCO_3 + Ca^{2+} + (CO_3^{2-}, SO_4^{2-}, Cl_2^{2-}) \rightleftharpoons CaMg(CO_3)_2 + Mg^{2+} + (CO_3^{2-}, SO_4^{2-}, Cl_2^{2-})\,. \qquad (2)$$

These equations show that, with a sufficient supply of cations, solids already present react and are transformed into new solids. These equations do not show, however, that, when calcite changes to dolomite, 2 moles of $CaCO_3$ are transformed into 1 mol of $CaMg(CO_3)_2$ by incorporating 1 mol of Mg-ions into the crystal lattice. This would suppose that the change results from diffusion in the solid state. Although experiments show that Mg-ions are adsorbed on crystal faces of Ca-carbonate (BERNER, 1966), the rate of diffusion of Mg^{2+} in the solid state is probably insufficient to cause on appreciable incorporation of Mg^{2+} in $CaCO_3$.

Dolomite forms from the reaction of $CaCO_3$ with Mg-bearing solutions by the dissociation of Ca-carbonate in the solution. Simultaneously, new seed crystals are precipitated from the solution. During subsequent growth, Ca^{2+} diffuses out of the surface of the seed crystals while Mg^{2+} diffuses into it. At the temperatures of the earth's surface the Ca/Mg-ratio on the surface is larger than one. The thickness of the surface layer is about 10^{-2} μ (order of magnitude). The growth proceeds slowly. With recent material, a growth rate of approximately 0.1 $\mu/10^3$ years was determined for the edges of the rhombohedra (PETERSON, BORCH, and BIEN, 1966). The following is true concerning the formation of dolomite from $CaCO_3$ and solutions in the 5-component-system: a definite amount, a, of $CaCO_3$ dissociates in a solution that contains defined amounts of Ca^{2+}, Mg^{2+}, CO_3^{2-}, SO_4^{2-}, and Cl_2^{2-} (3a). The CO_3^{2-} is essentially present as HCO_3^-. Since in the 5-component-system, however, the compositions of the solutions are given in formular units of the solid phases, the HCO_3^- is formulated as

CO_3^{2-}. From the solution formed after (3a) a definite amount, g, of $CaMg(CO_3)_2$ is precipitated (3b). The ions remaining in the solution are at equilibrium with the solid. Since reactions (3a) and (3b) proceed simultaneously, the summary equation (3c) results.

$$a\,CaCO_3 + b\,Ca^{2+} + c\,Mg^{2+} + d\,CO_3^{2-} + e\,SO_4^{2-} + f\,Cl_2^{2-} \rightarrow \quad (3a)$$
$$(a + b)\,Ca^{2+} + c\,Mg^{2+} + (a + d)\,CO_3^{2-} + e\,SO_4^{2-} + f\,Cl_2^{2-}\,.$$
$$(a + b)\,Ca^{2+} + c\,Mg^{2+} + (a + d)\,CO_3^{2-} + e\,SO_4^{2-} + f\,Cl_2^{2-} \rightarrow \quad (3b)$$
$$g\,CaMg(CO_3)_2 + (a + b - g)\,Ca^{2+} + (c - g)\,Mg^{2+} + (a + d - 2\,g)\,CO_3^{2-}$$
$$+ e\,SO_4^{2-} + f\,Cl_2^{2-}\,.$$
$$a\,CaCO_3 + b\,Ca^{2+} + c\,Mg^{2+} + d\,CO_3^{2-} + e\,SO_4^{2-} + f\,Cl_2^{2-} \rightarrow \quad (3c)$$
$$g\,CaMg(CO_3)_2 + (a + b - g)\,Ca^{2+} + (c - g)\,Mg^{2+}\,(a + d - 2\,g)\,CO_3^{2-}$$
$$+ e\,SO_4^{2-} + f\,Cl_2^{2-}\,.$$

These formulas show better than Eqs. (1) and (2), that the formation of dolomite from $CaCO_3$ and solutions does not take place in the solids states. They also show that 1 mole of $CaMg(CO_3)_2$ not necessarily results from 2 moles of $CaCO_3$ of the available solid and 1 mole Mg^{2+} of the solution.

Following the reversibel reactions of relationships (1) and (2), which should be interpreted following (3), calcite, dolomite, and magnesite were synthesized at the temperature interval from 120° to 180°C. The dolomite obtained was ordered dolomite with the symmetry R $\bar{3}$. Below 120°C, the synthesis of dolomite was no longer possible. Furthermore, the dolomite only formed in relatively concentrated solutions saturated with CO_3^{2-} and Cl_2^{2-} (points 9 and 10, Fig. 1) and saturated with CO_3^{2-}, SO_4^{2-} and Cl_2^{2-} (points 22 and 23, Fig. 1). In relatively low concentrations no reaction took place (points 7, 8, 16, and 17, Fig. 1).

The absence of reaction in dilute solutions an below 120°C is due to hydration of the Mg-ions present in solution. The number of non-, or only partially hydrated ions is larger at higher than at lower temperatures. Moreover, the amount of such ions increases with increasing concentration of the solution. The number of collisions between the elements suitable for the formation of dolomite within a unit of time therefore decreases with decreasing temperature and decreasing concentration of the solution. Directly below 120°, this quotient is already so small that no reaction was found during the time available for the experiment. Field observation, however, shows that in nature, where more time is available, the formation of dolomite also takes place in relatively weakly concentrated solutions and at relatively low temperatures. Therefore, the equilibrium data, which were not determined by reactions, but by solubility measurements, may also be applied to the relationships in nature.

C. The Early Diagenetic Origin of Dolomite

The recent formation of dolomite on the surface of the earth is a model of early diagenetic dolomite. Presently, dolomitic sediments form in S-Australia (ALDERMAN, 1959; ALDERMAN and VON DER BORCH, 1960, 1961, 1963; ALDERMAN and SKINNER, 1957; VON DER BORCH, 1965; SKINNER, 1963), in the Persian Gulf (CURTIS, EVANS, KINSMAN and SHEARMAN, 1963; KINSMAN, 1964; WELLS, 1962), on the Bahama Islands, on the S-coast of Florida (SHINN, 1964; SHINN, GINSBURG, and LLOYD, 1964), and on the Antilles (DEFFEYES, LUCIA, and WEYL, 1964). In these areas, dolomitization is caused by sea water, transported to the sediment surface by cappilar action through the pore space of a $CaCO_3$-sediment or by periodic flooding of this sediment, and concentrated by evaporation of H_2O at the sediment-surface or in lagoons. Dolomite does not form, however, by exceeding the solubility product of an initially relatively weakly concentrated solution. Dolomite forms by reaction of the preexisting Ca-carbonate with the more concentrated solutions of an advanced stage of evaporation.

A section through the prism of Fig. 1 (Fig. 2) illustrates the reactions during the early diagenetic formation of dolomite. This section proceeds from the $CaCO_3$-

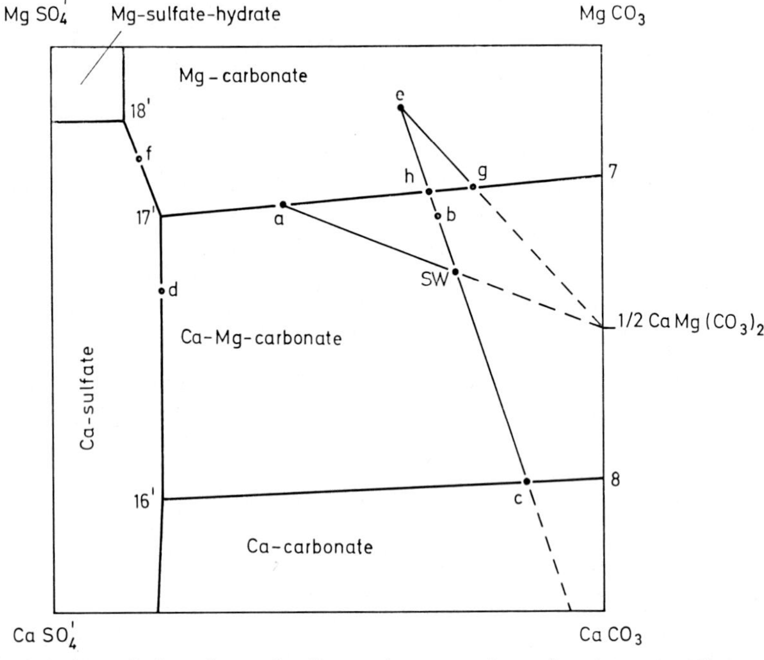

Fig. 2. Reactions during the early diagenetic conversion of carbonate sediments. The schematic representation shows a section through the prism of Fig. 1. The points $MgSO_4'$, 18', 17', 16', and $CaSO_4'$ are points of intersection on the lines $MgSO_4$ to 21, 18 to 24, 17 to 23, 16 to 22 and $CaSO_4$ to 20 of the three-dimensional diagram

$MgCO_3$-edge. It extends to the system along the edge, $CaCl_2$-$MgCl_2$-$MgSO_4$-$CaSO_4$. The stability relationships in this section do not differ qualitatively from those of the system along the edge, $CaCO_3$-$MgCO_3$-$MgSO_4$-$CaSO_4$. This section shows solutions which also contain $CaCl_2$ and $MgCl_2$ in solution or, in addition, also NaCl besides $CaCO_3$, $MgCO_3$, $MgSO_4$ and $CaSO_4$ in contrast to the system at the edge.

Solubility data indicate that the Ca-Mg-carbonate dolomite is in equilibrium with sea water (KRAMER, 1959; SASS, 1965). During isothermal evaporation of sea water, the Ca-Mg-carbonate should be precipitated first. In Fig. 2, SW represents the point for sea water. The actual position of the composition of sea water in the diagram is not taken into consideration because the diagram would become overcrowded. Precipitation of the Ca-Mg-carbonate would change composition of the solution SW along the extension of the line $^1/_2$ $CaMg(CO_3)_2$-SW from SW to a. At point a, Mg-carbonates should form. If both the phases are precipitated simultaneously, the composition of the solution should change along the line a-17'. Ca-sulfate should form in addition at point 17'. Further precipitation, during which halite forms at first, may be discussed using the equilibria of the remaining salts of sea water. Any precipitation under stable conditions should therefore show the sequence: Ca-Mg-carbonate, Mg-carbonate, Ca-sulfate.

During actual evaporation of sea water, however, the stable Ca-Mg-carbonate is not precipitated. Metastable Ca-carbonate forms instead in the stability range of the Ca-Mg-carbonate. The composition of the sea water hereby changes along the extension of the line $CaCO_3$–c–SW. If sea water so modified, e.g., with the composition at point b, now remains in contact with metastable or pre-existing $CaCO_3$, a reaction resulting in the formation of the Ca-Mg-carbonate must take place, because the composition of solution b is in the stability range of the Ca-Mg-carbonate. In this reaction, the composition of the solution is changed from b to c. When the solution has reached point c, the reaction stops, because here the Ca-carbonate and the Ca-Mg-carbonate are in equilibrium together with solution c. It can also occur that the composition of the modified sea water lies, e.g., along the phase boundary Ca-sulfate–Ca-Mg-carbonate (point d), because of a metastable precipitation of $CaCO_3$ plus $CaSO_4$. This solution, too, must react with available $CaCO_3$ leading to the formation of the Ca-Mg-carbonate. The composition of the solution during reaction changes along the line d-1 G because the solution is saturated with both carbonate- and sulfate ions. The reaction stops, when there is equilibrium at point 16' between the Ca-sulfate, the Ca-carbonate, and the Ca-Mg-carbonate.

The evaporating sea water can also change through a metastable precipitation of $CaCO_3$ and $CaSO_4$ so that the composition of the solution lies in the stability range of the Mg-carbonate (points e and f). Pre-existing Ca-Mg-carbonate must react with these solutions and form Mg-carbonate. The composition of solution e changes along the line e to g. At point g the reaction stops. In the case of $CaSO_4$-saturation, the composition of the solution changes along the line f to 17'. The solution is at equilibrium at point 17'. When modified sea water in the stability range of Mg-carbonate comes in contact with $CaCO_3$ it may react to form $MgCO_3$. Hereby the composition of the solution e is changed from e to h. Because $MgCO_3$ cannot be in equilibrium with $CaCO_3$, the $MgCO_3$ reacts at point h with the solution by forming Ca-Mg-carbonate. With subsequent formation of Ca-Mg-carbonate the equilibrium of point c is reached. In case of $CaSO_4$-saturation (point f), metastable Mg-carbonate forms with $CaCO_3$; at point 17' it is transformed into Ca-Mg-carbonate; this reaction stops at point 16'.

All these reactions, however, occur only at an advanced stage of evaporation of the solution. This is due to the hydration of the Mg-ions in the solution so that the amount of energy needed for the formation of Ca-Mg-carbonate seed crystals can be attained only in more contracted solutions. No reaction takes place with normal sea water. Moreover, the solutions must also remain in contact with the solids long enough, because reaction rates are very small.

Because of the reaction of $CaCO_3$ with modified sea water, equilibrium can be reached along the line 16' to 8. During the reaction of Ca-Mg-carbonate with modified sea water, an equilibrium on the line 17' to 7 can be attained. These processes which also include the condition of $CaSO_4$-saturation can be expressed in formulas by the equilibria of points 16′ and 17′. At point 16′, equilibrium exists between Ca-carbonate, Ca-Mg-carbonate, Ca-sulfate, and the solution 16′. Mg-carbonate, Ca-Mg-carbonate, Ca-sulfate, and the solution 17′ are stable at point 17′. The maximum number of phases in equilibrium obtainable can be expressed by the following equations:

$$x\ CaCO_3 + y\ CaMg(CO_3)_2 + z\ CaSO_4 \rightleftharpoons \text{solution } 16', \tag{4}$$

and $$u\ MgCO_3 + v\ CaMg(CO_3)_2 + w\ CaSO_4 \rightleftharpoons \text{solution } 17'. \tag{5}$$

The factors x, y, z, and u, v, w have the following meaning: if, e.g., no $CaSO_4$ is present, then z and w = 0. In place of solutions 16′ and 17′, solution must then be entered into the equation whose composition lies on the lines 16′ to 8, and 17′ to 7, e.g., the solutions c or g. Substitution of the stable and metastable minerals present in Recent dolomitic sediments for the formular units of Eqs. (4) and (5), result in a good correspondence between the observed phase assemblages and those derived from the system. Calcite, dolomite, magnesite, and anhydrite occur as stable minerals. Aragonite, Mg-calcite, protodolomite, hydromagnesite, and gypsum are metastable minerals. The origin of the mineral assemblages of Recent dolomitic sediments can, therefore, be interpreted with the aid of the stability relations of the 5-component- and 6-component-system. It should be taken into consideration, however, that the natural assemblages and solution compositions are not equilibrium assemblages and compositions. They merely represent metastable conditions of a long process that tends toward stable conditions.

D. The Origin of Late-Diagenetic Dolomite

Stable conditions are generally attained during late diagenesis, when the sediment is buried below a thick cover of rocks. Here the conditions necessary to attain stable equilibria are stable over longer periods than on the surface of the earth. Pore solutions are instrumental in the transformation of the unstable into stable minerals. Aragonite is transformed into calcite, gypsum generally into anhydrite. Mg-calcite and protodolomite are converted likewise by reaction with solutions.

The most important process of late diagenesis of carbonate rocks is the formation of dolomite from limestones and solutions that circulate in the pore space of sedimentary rocks. These solutions are generally — like sea water or modified sea water — solutions with 6 components. The origin of late diagenetic dolomite can therefore be discussed exactly like that of early diagenetic dolomite by using the 6-component- and 5-component-systems. There is no basic difference between both processes because they can be expressed by identical reactions in identical systems.

A reaction between $CaCO_3$ and solutions occurs only if the solutions have a suitable composition. The question of late-diagenetic formation of dolomite is therefore principally a question of the composition of the solutions circulating within the earth's crust. There are no restrictions regarding the concentrations of the solutions, as in the case of early diagenetic dolomite, because temperatures within the earth's crust are generally higher than on its surface. Besides, much time is available, so that the stable equilibria can be attained. The compositions of the pore solutions are best studied in two projections of the 5-component- and 6-component-system disregarding NaCl-content: one projection on the base of the prism and one of the $CaCl_2$-$MgCl_2$-edge on to the surface $CaCO_3$-$MgCO_3$-$CaSO_4$-$MgSO_4$. Thus the mixing

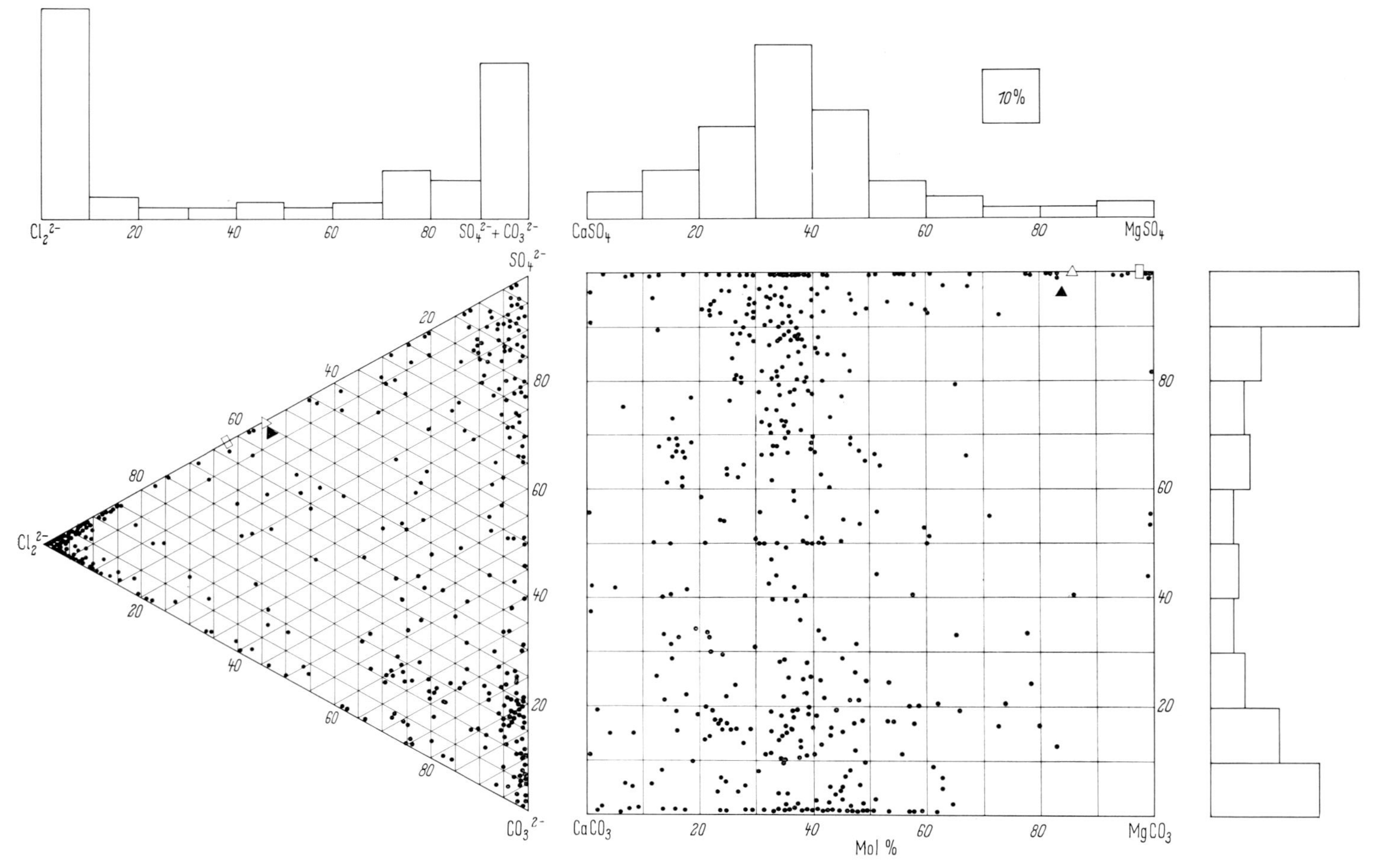

Fig. 3. Relative composition of pore solutions in two projections of the prism-representation of the 5-component- and 6-component-system. The HCO_3^--contents of the solutions are formulated as CO_3^{2-}. 85 additional points with a relative Cl_2^{2-}-content of more than 90 mole-% are not entered here. For comparison: ▲ sea water (SW), △ SW at the start of $CaSO_4$-precipitation, △ = SW at the start of NaCl-precipitation

ratios of the anions, CO_3^{2-}, SO_4^{2-}, Cl_2^{2-}, and the mixing ratios of the cations, Ca^{2+} and Mg^{2+} can be considered separately.

Fig. 3 shows an evaluation of approximately 450 analyses of pore waters. The variation diagram of anions shows that all mixtures of carbonate-, sulfate-, and chloride ions occur. Frequency maxima lie near the corners. The projection on the surface $CaCO_3$-$MgCO_3$-$CaSO_4$-$MgSO_4$ shows that pore solutions contain all mixtures of Ca^{2+} and Mg^{2+}. Particularly common are solutions between 20 and 50 mole-% Mg^{2+}.

Fig. 4 shows the average values of the equilibria which, in the 5-component-system, define the field of stability of dolomite against that of calcite and magnesite. Calcite, dolomite, and solutions are stable along the left curve of the diagram. The

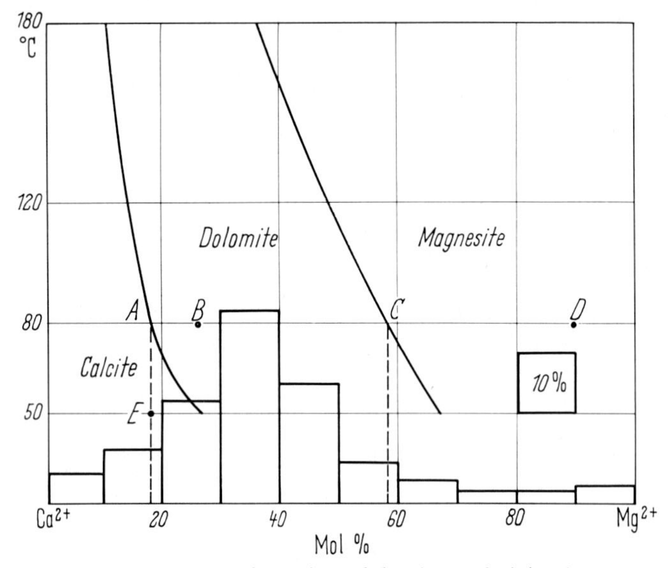

Fig. 4. The equilibrium positions for calcite-dolomite and dolomite-magnesite and the frequency of Ca^{2+} and Mg^{2+} in pore solutions. Both curves are intermediate values of the equilibrium data for points 8, 10, 22, and 16 of Fig. 1

right curve represents the equilibrium between magnesite, dolomite, and solutions. If, at a temperature of 80° C, a solution with the composition B comes in contact with a limestone, a reaction resulting in the formation of dolomite takes place, because the solution is in the dolomitic field. The solution is enriched in Ca and changes its composition from B to A. The reaction ceases at point A. If a solution of the composition D comes into contact with a limestone, magnesite forms at first, because the solution is in the magnesite field. The composition of the solution changes through the reaction from D to C. At point C, the magnesite reacts with the solution, forming dolomite. After magnesite has been used up, the equilibrium A is attained with continuous formation of dolomite. Thus dolomite can form from limestone by reaction with relatively Mg-rich, as well as with relatively Mg-poor pore solutions. The only prerequisite for the formation of dolomite is: the reaction solutions must have higher relative Mg-content than those solutions that are at equilibrium with calcite and dolomite. A comparison of the positions of equilibrium between calcite and dolomite with the frequency of Ca^{2+} and Mg^{2+} in pore solutions, shows that most pore solutions contain more Mg^{2+} than corresponds with the equilibrium. In most cases, in which pore solutions come into contact with limestone, a reaction resulting in the formation of dolomite must therefore take place.

The average position of equilibrium between calcite and dolomite is at about 18 mol-% Mg^{2+} in solution at a temperature of 80° C. The equilibrium between magnesite and dolomite is at approximately 58 mole-% Mg^{2+}. The frequency of pore solutions in the calcite field amounts to approximately 13%, at 80° C; those in the dolomite field have a frequency of about 76%, and those in the magnesite field of about 11%. At 80° C, therefore, dolomite is directly formed in 76% of all cases where pore solutions come in contact with limestones. In 11% of all cases, the formation of dolomite takes place via the intermediate formation of magnesite. The frequency of dolomite-formation amounts, therefore, to $76 + 11 = 87\%$. No reaction takes place in only 13% of all cases, in which pore solutions come into contact with limestones. If pore solutions come in contact with dolomitic rocks, then in 76% of all cases no reaction takes place. Only in 13% of all cases are dolomites altered to limestones. Considering the abundance of limestones (70%) and of dolomitic rocks (30%), late diagenetic formation of dolomite from limestones and pore solutions at a temperature of 80° C has a probability of approximately 61%. The reversed process, i.e., the formation of limestones from dolomite and pore solutions has a probability of about 4%. Limestones are not changed by solutions of the calcite field and dolomites by solutions of the dolomite field with probabilities of 9% and 23% respectively. Reactions at the phase-boundary dolomite-magnesite have a probability of 3%. These reactions were not discussed, however, because fraction of magnesite within the carbonate rocks is only small, probably not exceeding 1%. The probability of dolomite formation within the sediment mantle of the earth's crust is, therefore, large. This is the explanation that dolomite is more abundant in older than in younger carbonate rocks (Chilingar, 1956; Ronov, 1959). The probability, that a limestone comes in contact with pore solutions that are suitable to alter it into dolomite, increases with its age. Moreover, depth of burial and therefore temperature increase with age of sedimentary rocks, thereby increasing the probability of dolomite formation.

For example, pore solutions with a Mg-content of 18 mole-% circulating in a rock are stable with calcite at 50° C (point E, Fig. 4). If this rock is heated to 80° C, the phase boundary calcite-dolomite is reached at 80° C. If this temperature is exceeded, the solutions enter the dolomite field and are now able to transform limestones into dolomites.

Fig. 4 only shows the importance of cations with regard to dolomite formation. The late diagenetic formation of dolomite must, therefore, still be discussed with respect to anions. Table 2 lists the quantities of dolomite which form at 80° C by the reaction of 1 m^3 limestone with pore solutions. The relative Mg-content of the reaction-solutions is constant. It corresponds to the relative Mg-content of the most common pore solutions (35 mole-%). The principal anions of the solutions are either carbonate, sulfate or chloride corresponding to points 8, 16, 22, and 10 of Fig. 1. Table 2 shows, that a definite amount of limestone leads to variable amounts of dolomite depending on the anions of the reaction-solutions. It also shows, that positive as well as negative volume changes of the solids can occur during the formation of dolomite. A volume-reduction occurs during reaction with Cl_2^{2-}-rich solutions. Reaction with predominantly CO_3^{2-}- and SO_4^{2-}-bearing solutions leads to a volume-increase. The reason is evident when the conversions of units of volume are recalculated in moles. Table 3 shows, that predominantly CO_3^{2-}- and SO_4^{2-}-bearing solutions lead to formation of more dolomite than corresponds to the reaction-scheme of Eq. (1). In these cases, additional Ca^{2+}-, Mg^{2+}-, and CO_3^{2-}-ions are carried in solution and precipitate as dolomite. Only in Cl_2^{2-}-rich solutions does the formation of dolomite proceed in such a way that 2 moles of the available calcite react with 1 mole Mg^{2+} of the solution forming 1 mole of dolomite.

Table 2. *Recalculation of turnovers during the formation of dolomite from limestones and pore solutions with various anion-contents (80 °C). ( ) = data for the condition of NaCl-saturation. The remaining data apply in the case of NaCl-absence. The compositions of the reaction- and equilibrium solutions used for computation of the turnovers after* USDOWSKI *(1967)*

Turnover at point	Principal anion	m³ calcite	m³ dolomite	△
8	CO_3^{2-}	1	1.37 (1.09)	+ 0.37 (+ 0.09)
16	SO_4^{2-}	1	1.06 (1.03)	+ 0.06 (+ 0.03)
10.22	Cl_2^{2-}	1	0.86 (0.86)	— 0.14 (— 0.14)

Table 3. *Recalculation of the turnovers as the result of dolomite-formation from limestones and pore solutions from units of volume into moles (80 °C). ( ) = data in the case of NaCl-saturation. The remaining data apply in the case of absence of NaCl*

Turnover at point	Principal anion	moles calcite	moles dolomite
8	CO_3^{2-}	2	1.58 (1.26)
16	SO_4^{2-}	2	1.22 (1.18)
10.22	Cl_2^{2-}	2	1.0 (1.0)

Syntheses of $BaMg(CO_3)_2$ (Norsethite) at 20 °C and the Formation of Dolomite in Sediments

FRIEDRICH LIPPMANN*

With 3 Figures

Abstract

The synthesis of norsethite, $BaMg(CO_3)_2$, is the only known way to incorporate large stoichiometric amounts of magnesium into an anhydrous carbonate at earth-surface temperatures. The structure of norsethite is similar to that of dolomite, especially with respect to the bonding of the magnesium. Since the building of magnesite layers seems to be the main obstacle in the growth of dolomite it is inferred that dolomite may form at low temperatures according to the overall reaction:

$$CaCO_{3(solid)} + Mg^{++} + CO_3^{--} \rightarrow CaMg(CO_3)_{2(solid)}$$

which is analogous to the mechanism found in the preparation of norsethite.

The alkalinity required in addition to magnesium for this process of dolomitisation may be derived from the hydrolysis connected with silicate weathering, or it may be produced from magnesium sulphate and organic matter by sulphate-reducing bacteria in buried carbonate sediments.

Introduction

Many, if not most sedimentary dolomites have formed at temperatures not essentially different from those prevailing at the earth's surface. This may be deduced from the great number of field studies and petrographic examinations of dolomites which have been published since the classical works of VAN TUYL, 1916 and KROTOV, 1925. Frequent occurrences of dolomite patches in limestones and intimate interbedding of dolomite and limestone, which are found in many ancient carbonate rocks, as e.g. in the Muschelkalk of Southern Germany (RUDOLF, 1959), cannot be explained by hydrothermal dolomitisation. They constitute the main evidence in favor of low-temperature dolomitisation. That dolomite does indeed form at earth-surface temperatures has been established by a number of findings of subrecent dolomite during the last years (see e.g. papers in PRAY and MURRAY, 1965). So far, however, no well documented synthesis of ordered dolomite at normal temperature has become known. This contrast is characteristic of the present state of the dolomite problem.

Important steps towards a solution of the dolomite question were the determination and subsequent refinements of the crystal structure of dolomite (BRAGG, 1937; BRADLEY, BURST and GRAF, 1953; STEINFINK and SANS, 1959). The structure of dolomite may be roughly described as a regular interstratification of two kinds of layers, one of which is similar to the ultimate layer in calcite, the other to that in magnesite. Since it is possible to synthesise calcite at room-temperature in the laboratory, but not magnesite, all efforts to elucidate the conditions of low-temperature dolomite formation should focus on the magnesite layer.

* Mineralogisches Institut der Universität, 7400 Tübingen, Wilhelmstraße 56.

Quite recently it became possible to produce at approximately 20 °C a compound $BaMg(CO_3)_2$ [LIPPMANN, 1967 (1)] which is chemically analogous to dolomite, $CaMg(CO_3)_2$, and which was first described as the rare mineral norsethite by MROSE, CHAO, FAHEY and MILTON (1961). According to the crystal structure analysis of the

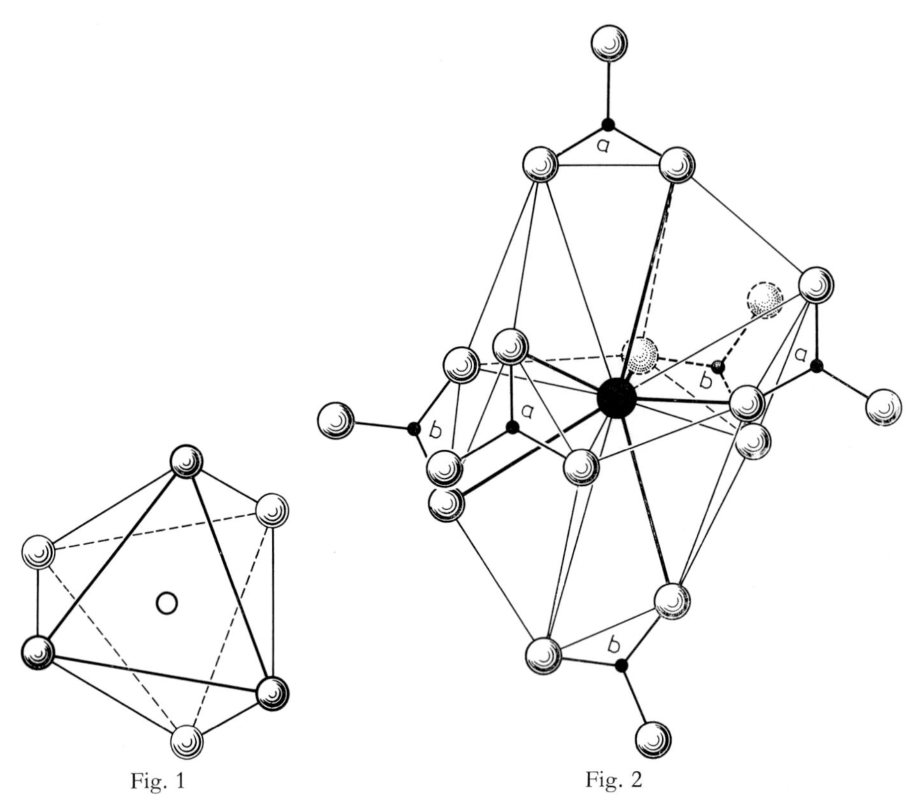

Fig. 1. Coordination octahedron around magnesium in the structure of norsethite, $BaMg(CO_3)_2$ [LIPPMANN 1967 (2)]. Projection in the direction of the c-axis on the basal plane. The magnesium (open sphere) is surrounded by three oxygens (heavy shaded spheres) from above and by three oxygens (light shaded spheres) from below. The oxygen triangle above (heavy lines) and the one below (dashed lines) are twisted according to the trapezohedral crystal symmetry. In dolomite one basal triangle is rotated by exactly 180° with respect to the other by a centre of symmetry in the magnesium

Fig. 2. Coordination polyhedron around barium in the structure of norsethite, $BaMg(CO_3)_2$ [LIPPMANN 1967 (2)]. View at 45° inclined to the basal planes. Ba: large black sphere; C: small black spheres; O: shaded spheres. The barium is surrounded octahedrally by six CO_3-groups. The three CO_3-groups in the basal level above the barium are marked by a; the ones below the barium by b. Since the gap between the oxygens of a CO_3-group points into the general direction of the barium the coordination of the latter by oxygen is twelve-fold

artificial material [LIPPMANN, 1967(2)] norsethite contains essentially the same magnesite layers as dolomite (Fig. 1) although the surrounding of the large barium cation by 12 oxygen (Fig. 2) is different from the octahedral coordination of the calcium in dolomite. Since the preparation of norsethite is the only known way to incorporate large stoichiometric amounts of magnesium into an anhydrous carbonate at normal temperatures conclusions from this synthesis concerning the mode of formation of dolomite in nature should be justified.

The "Norsethitisation" of Barium Carbonate

Because a direct precipitation of dolomite from sea-water had never been observed it became generally accepted that dolomite forms from calcareous sediments by reaction with magnesium-bearing solutions according to the reactions:

$$2\,CaCO_3 + MgSO_4 \rightarrow CaMg(CO_3)_2 + CaSO_4$$

(Haidinger-reaction after Krotov, 1925)

$$2\,CaCO_3 + MgCl_2 \rightarrow CaMg(CO_3)_2 + CaCl_2$$

(Marignac-reaction after Krotov, 1925)

which may be generalised as:

$$2\,CaCO_3 + Mg^{++} \rightarrow CaMg(CO_3)_2 + Ca^{++}.$$

If these reactions were of general validity at normal temperature norsethite $BaMg(CO_3)_2$ should form according to analogous equations. This, however, is not

Fig. 3. Crystals of norsethite, grown in a solution of 0.01 m $MgCl_2$ and 0.005 m $NaHCO_3$ by reaction with $BaCO_3$. Plain polarised light vibrating in the direction of the scale; scale length: 0.1 mm; refractive index of immersion medium: 1.66

the case: barium carbonate may be left in contact with magnesium chloride solutions of various concentrations for more than a year without appreciable changes taking place. If magnesium chloride solutions which contain carbonate ions at the same time are applied norsethite forms very quickly. In an aqueous solution of

$$0.01 \text{ m } MgCl_2 \text{ and } 0.005 \text{ m } NaHCO_3$$

norsethite may be identified after about a fortnight, and in

$$0.01 \text{ m } MgCl_2 \text{ and } 0.02 \text{ m } NaHCO_3$$

the first crystals may be observed even after two days of contact with barium carbonate at approximately 20 °C. In all preparations the crystal size is about 0.1 mm, so that in addition to the determination by X-ray diffraction the artificial norsethite can also be indentified with the petrographic microscope (Fig. 3). The analysis of the single reactions taking place during the synthesis of norsethite at normal temperature, which was reported in a previous paper [LIPPMANN, 1967 (1)], results in the following overall reaction:

$$BaCO_{3(solid)} + Mg^{++} + CO_3^{--} \rightarrow BaMg(CO_3)_{2(solid)}$$

It may be noted that a new compound $PbMg(CO_3)_2$ forms according to an analogous scheme, although at considerably slower rates than norsethite. The crystallographic study of $PbMg(CO_3)_2$ revealed that it is isotypic with norsethite (LIPPMANN, 1966). In the syntheses of both compounds not only Mg^{++} but also CO_3^{--} are taken from solution to form the double carbonate.

Conclusions Regarding the Low-Temperature Formation of Dolomite

From chemical analogy and structural similarities, especially with respect to magnesite layers of dolomite and norsethite [as well as of $PbMg(CO_3)_2$], one may infer that the formation of dolomite at near earth-surface temperatures proceeds after an overall reaction analogous to the one found for norsethite in the laboratory:

$$CaCO_{3(solid)} + Mg^{++} + CO_3^{--} \rightarrow CaMg(CO_3)_{2(solid)}$$

Although experiments along these lines are still under way and have not yielded any final results as yet, the consequences of this reaction shall be dicussed briefly. In contrast to the reactions so far assumed for the dolomitisation of calcareous sediments (see above) not only magnesium but also the carbonate ion, i.e. alkalinity, is required in stoichiometric amounts.

In nature there are two general ways to furnish the postulated alkalinity for the newly proposed mechanism of dolomite formation:

1. During weathering OH^--ions are released into solution by the hydrolysis of silicates. They react with the carbon dioxide of the air to form carbonate:

$$2\,OH^- + CO_2 \rightarrow CO_3^{--} + H_2O\,.$$

The weathering solutions may become concentrated to magnesium and/or alkali carbonate waters under semi-arid conditions. Their reaction with calcareous sediments in lacustrine and lagoonal environments may account e.g. for such dolomites as exhibit "primary" features.

2. In purely marine environments sulphate reducing bacteria (conf. WALLHÄUSER and PUCHELT, 1966) may produce alkalinity in the buried sediment according to the following reactions or to combinations of both:

$$2\,SO_4^{--} + 3\,C + H_2O + 2\,H^+ \rightarrow 3\,CO_3^{--} + 2\,H_2S \uparrow$$

$$SO_4^{--} + 2\,H_2 + 2\,H^+ \rightarrow 4\,OH^- + H_2S \uparrow$$

Under these conditions it is important that enough organic matter is available in the sediment as "food" for the sulphate-reducing bacteria. Magnesium sulphate is supposed to be introduced from percolating sea-water or by diffusion.

Both environments of dolomitisation should be distinct in their carbon isotope contents. Environment 1. should be characterised by a relative concentration of the heavy carbon isotope C_{13}, whereas environment 2. must be expected to be enriched in the light carbon C_{12}. The first example of the latter case is the "organic dolomite" described by SPOTTS and SILVERMAN (1966).

The proposed mechanisms allow explanation of the apparently erratic fashion in which dolomite formation seems to have proceeded in many instances. It becomes clear that even prolonged contact of calcium carbonates with sea-water or other magnesium-bearing solutions, even with concentrated ones, must not necessarily lead to dolomite in every case. The relatively low concentrations of magnesium which lead to norsethite in the laboratory cast serious doubt on the common opinion that evaporation, or more generally, that high concentrations of magnesium be required for dolomitisation.

References

BRADLEY, W. F., J. F. BURST, and D. L. GRAF: Crystal chemistry and differential thermal effects of dolomite. Am. Mineralogist **38**, 207—217 (1953).

BRAGG, W. L.: Atomic structure of minerals, pp. 115—116. New York, London: Ithaca 1937.

KROTOV, B. P.: Dolomite, ihre Bildung, Existenzbedingungen in der Erdkruste und Umwandlung im Zusammenhang mit dem Studium des oberen Teiles der Kazan-Stufe in der Umgebung von Kazan. Verhandl. naturforsch. Ges. Univ. Kazan **50** Lief. 6, 1—110 (1925) (in Russian with an extended German summary).

LIPPMANN, F.: $PbMg(CO_3)_2$, ein neues rhomboedrisches Doppelkarbonat. Naturwissenschaften **53**, 701 (1966).

— (1) Die Synthese des Norsethit bei etwa 20 °C und 1 at. Ein Modell zur Dolomitisierung. Neues Jahrb. Mineral., Monatsh. **1967**, 23—29 (1967).

— (2) Die Kristallstruktur des Norsethit, $BaMg(CO_3)_2$ im Vergleich zum Dolomit, $CaMg(CO_3)_2$. Naturwissenschaften **54**, 514 (1967).

— Die Kristallstruktur des Norsethit. Tschermaks Mineral. Petrogr. Mitt. **12**, 299—318 (1968).

MROSE, M. E., E. C. T. CHAO, J. J. FAHEY, and C. MILTON: Norsethite, $BaMg(CO_3)_2$, a new mineral from the Green River Formation, Wyoming. Am. Mineralogist **46**, 420—429 (1961).

PRAY, L. C., and R. C. MURRAY (editors): Dolomitization and limestone diagenesis. Soc. Econ. Paleontol. Mineral. Spec. Publ. No. 13 (1965).

RUDOLF, W. F.: Zur Dolomitisierung und Petrogenese im unteren Hauptmuschelkalk Württembergs. Diss., Tübingen 1959.

SPOTTS, J. H., and S. R. SILVERMAN: Organic dolomite from Point Fermin, California. Am. Mineralogist **51**, 1144—1155 (1966).

STEINFINK, H., and F. J. SANS: Refinement of the crystal structure of dolomite. Am. Mineralogist **44**, 679—682 (1959).

VAN TUYL, F. M.: The origin of dolomite. Iowa Geol. Survey Ann. Rep. 1914, **25**, 251—421 (1916).

WALLHÄUSER, K. H., u. H. PUCHELT: Sulfatreduzierende Bakterien in Schwefel- und Grubenwässern Deutschlands und Österreichs. Contrib. Mineral. Petrol. **13**, 12—30(1966).

Dolomitization of Biocalcarenites of Late-Tertiary Age from Northern Lanzarote (Canary Islands)

Peter Rothe*

With 8 Figures

Abstract

A calcarenite interbedded in a thick pile of basaltic rocks is described from Northern Lanzarote (Canary Islands). The limestone representing a former land surface has smoothed out a former relief of volcanic cones. The rock consisted originally of probably wind-transported biocalcarenite with fragments of mainly Lithothamnium, foraminifera, serpula, bryozoans, echinoderms and a little volcanic material which increases toward the top. Dolomitization has taken place. It is strongest in the uppermost parts of the sediment and decreases downward, and is restricted to the upper 3 to 4 m below the overlying basalt; lower in the section original calcite is preserved. Further downward sedimentary material is entirely calcite. Dolomitization is probably caused by leaching of Mg out of the overlying volcanics.

A. Introduction

Lanzarote is one of the Eastern Canary Islands, situated between 28° 50′ and 29° 15′ N Lat. approximately 200 km off the coast of Northwestern Africa. The Island is made up almost entirely of basalts of the alkali olivine group, ranging in age from the Tertiary to Historic times. We are only concerned with the older basalts in this paper. These flat lying basalts were erupted from fissures and/or flat shield volcanoes and occur mainly in the Northern part of the island. In Northern Lanzarote, near Orzola, a calcarenite bed is interlayered in the basal part of a basalt series about 300 to 400 m thick.

B. Geologic Setting

The underlying basalt-flows from sea level up to the sediments are mostly thick basalts; cinder and/or scoria cones are common. The overlying basalts are mostly thin "fresher looking" sheets with some layers of tuff and scoria. The whole section forms high escarpments (Fig. 1).

The limestone, containing land snails and ostrich eggs, is part of an ancient land-surface (Rothe, 1964, 1966). Foraminifera, probably reworked, indicate Upper-Miocene age, most likely corresponding to the Torton-Sarmat-stage. Thus the limestone seems Tertiary in age. Several lines of evidence indicate terrestrial deposition of the biocalcarenite:

I. The pre-sediment relief in Northern Lanzarote was formed by cones of tuff, scoria and basalt, partly weathered and eroded.

II. This relief has been filled and levelled with calcareous sand, probably derived from beaches. There is some evidence for wind-transport: outcrops along the eastern

* Laboratorium für Sedimentforschung, Universität Heidelberg, Germany.

coast NW of Orzola show considerable thickening of sediments south of a volcanic cone; the material seems to have been transported by winds blowing from northern directions; on top of a cone thickness of the sediment is zero.

III. The upper boundary of the limestone is knife-sharp, nearly horizontal, slightly dipping northward and about 30 m above sea level. The island was probably stable (concerning vertical movements) since the Quaternary; it is possible that the sediments were uplifted to their present position.

IV. The calcarenite components are entirely marine: coralline algae, foraminifera, serpula, bryozoans, echinoderms, generally well rounded. Volcanic material (basalt

Fig. 1. Calcarenites interbedded in basaltic series. Northern Lanzarote, looking towards Punta Fariones. Continuation of sedimentary layer is indicated by dotted line

fragments, feldspars), is intermingled with sediments and increases in amount toward the top of the bed.

V. From paleontologic evidence (land snails and ostrich eggs) the environment may have been desert-like and climatic conditions may have been similar to the present.

C. Description of Sections

Three sections were taken from outcrops NW of Orzola where the limestone is a few meters thick.

I. Section 1, Approximately 100 m S of Valle Chico

The limestone is quarried because of its purity 4 m below overlying basalt; the base of sediments is hidden by volcanic gravel. 10 samples (KLR 202 to 211).

II. Section 2, Valle Chico

Thickness 7 m. Underlain by violet-colored tuffs which are part of the south flank of a tuff-cone extending down to sea level. Overlain by olivine basalt, about 4 m thick. This flow extends south of Valle Chico to Punta Fariones, covering the ancient land surface. Horizontal stratification. Pure calcarenite at base; from the center of the section upwards increasing amounts of volcanic detritus. 13 samples (KLR 215 to 219, 228 to 235).

III. Section 3, 300 to 400 m N of Valle Chico

2.5 to 2.7 m thick. The upper part contains volcanic material. Some puzzling recrystallized limestone fragments may have been derived from another lower limestone layer. 7 samples (KLR 220 to 226)

D. Textures and Carbonate Mineralogy

The biogenic debris is composed of few fossils:

I. Mostly well rounded fragments of coralline algae, most probably of the genus Lithothamnium.

II. Foraminifera, both bentonic and pelagic species (ROTHE, 1966, p. 26 to 27).

III. Fragments of serpula, easily recognized by their fibrous structure, rings, and slightly curved walls with characteristic structure.

IV. Echinoderms, mostly spine fragments.

V. Landsnails (ROTHE, 1966, p. 23 to 24), considerably enriched in some layers mostly in the upper parts of the sections.

VI. Matrix material is now recrystallized, it may have been fine-grained carbonate dust.

X-ray analysis revealed low-magnesian calcite and dolomite distributed in the sections as follows: Section 1 consists entirely of dolomite from the basalt 3 m downward. Within 3 to 4 m there is a zone with both dolomite and a very small amount of calcite. Section 2 is all dolomite up to 4 m from top basalt, 4 to 7 m from above there is mostly calcite, two of the samples carrying some dolomite. In section 3 all carbonate is dolomite; however interbedded with the mostly cream-colored or slightly reddish dolomite there appears a lens of whitish-gray carbonate, 2×0.3 m in diameter, which is all calcite with only a very small amount of dolomite. The recrystallized limestone fragments consists entirely of dolomite.

Fossils with dark iron coatings are well preserved in the lower part of the Valle Chico section. The matrix, formerly calcilutite (?), shows grain growth of calcite crystals. Cavities within fossils are partly filled by drusy mosaic of calcite (Fig. 2).

Euhedral dolomite crystals occur in pore spaces in the transitional dolomite calcite zone of sections 1 and 2, fossil structures can still be determined (Fig. 3).

Dolomite rhombohedra are enlarged in the higher, purely dolomitic parts of the sections. No fossil-structures except Lithothamnium are preserved, although it is altered into cryptocrystalline dolomite similar to those described by MÜLLER and TIETZ (1966, p. 94). An increase in the degree of dolomitization approaching the top basalt is observed. KLR 208 (Fig. 4) is only 1.70 m from top basalt: the sample consists mostly of idiotopic dolomite (FRIEDMAN, 1965); some shadowish remains of former calcareous algae are still recognizable. In section 3 dolomite crystals are

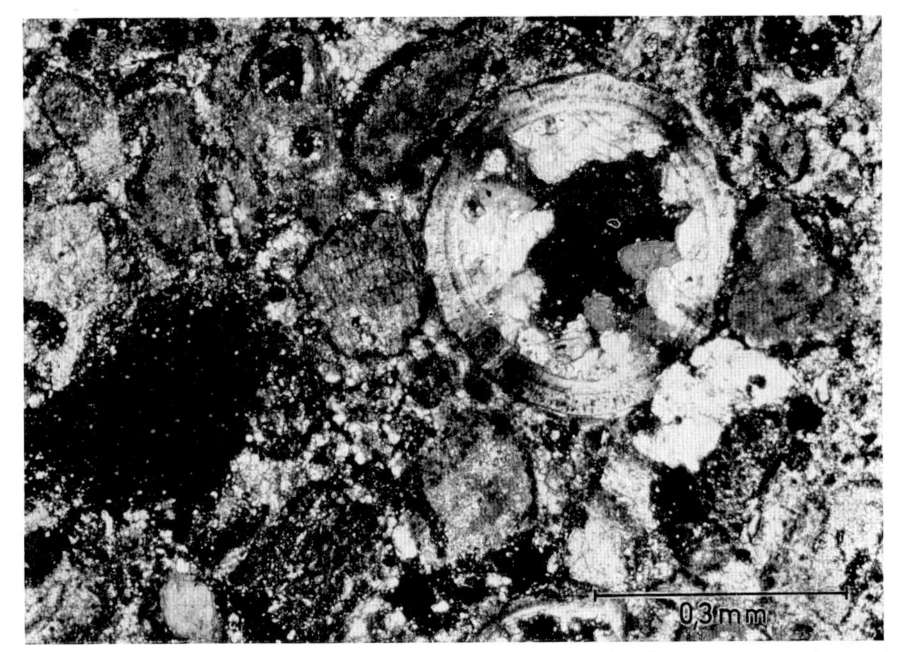

Fig. 2. KLR 217, section 2, about 5.30 m from overlying basalt. Biocalcarenite, carbonate is all calcite. Drusy mosaic within serpula. Crossed nicols

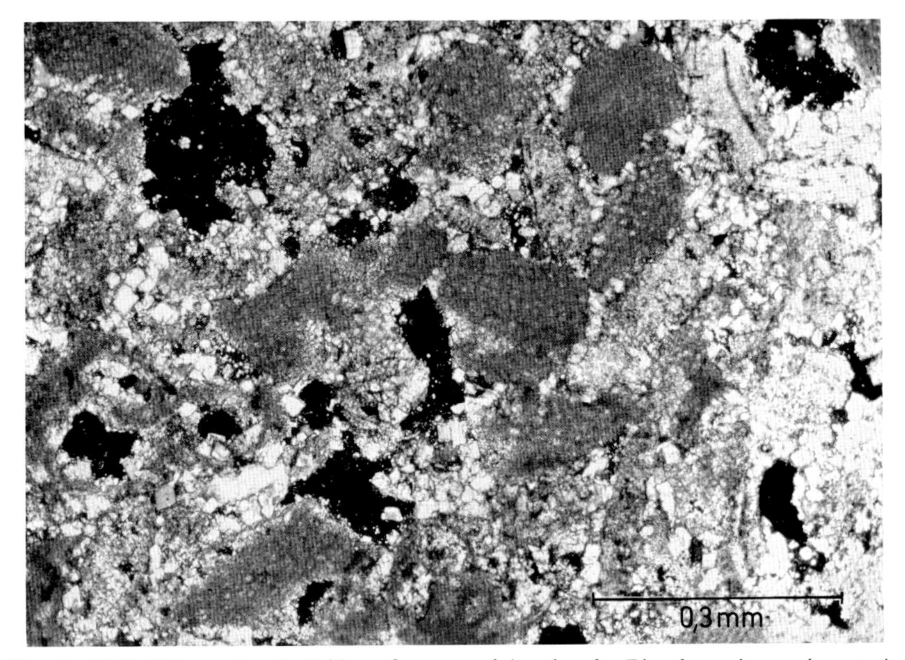

Fig. 3. KLR 202, section 1, 3.80 m from overlying basalt. Biocalcarenite, carbonate is calcite and dolomite. Euhedral dolomite crystals grown in pore spaces: first stage of dolomitization. Crossed nicols

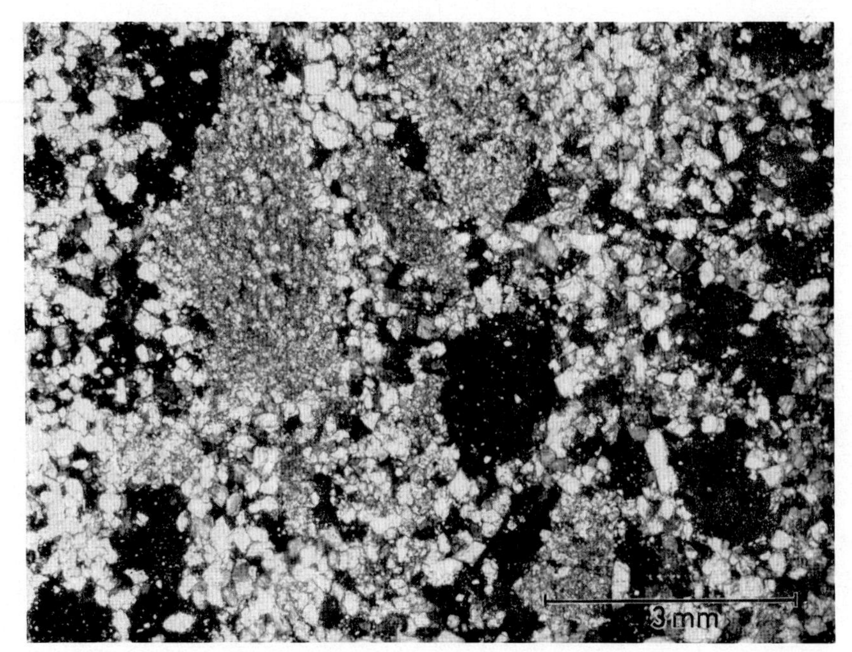

Fig. 4. KLR 208, section 1, 1.70 m from overlying basalt. Entirely dolomitized biocalcarenite. Shadowish structures of Lithothamnium between idiotopic dolomite. Crossed nicols

Fig. 5. KLR 220, section 3, 2.30 m from overlying basalt. Hypidiotopic-idiotopic dolomite, organic structures entirely recrystallized. Crossed nicols

Fig. 6. KLR 223, section 3, lens of whitish-gray biocalcarenite. Foraminiferan chamber is recrystallized microcrystalline calcite

enlarged. No organic structures can be recognized. Dolomite has grown into idiotopic-hypidiotopic dolomite (Fig. 5). The above cited lens of whitish-gray limestone (section 3) is exceptional. Fossils, even foraminifera, are well preserved in outline; cell structure of foraminiferan chambers are of recrystallized microcrystalline calcite (Fig. 6).

E. Geochemical Data

Ca, Mg, Fe and Sr were chemically determined from the HCl-soluble part of the samples. Ca was determined by the AeDTE-method (MÜLLER, 1956), Mg, Sr and Fe by Atomic Absorbtion Spectrophotometry with a Perkin-Elmer 303 model. The results were recalculated in relation to the sum of carbonates. Concerning dolomites Ca/Mg ratio is assumed to be 1. Lower Ca/Mg ratios, confirming higher Mg contents present in the samples indicate other minerals yielding Mg, too. This Mg may be derived from volcanic material, because both, the Mg-content of the bulk sediment and the amount of volcanic detritus increase towards the overlying basalt. Higher ratios occur in the carbonates made up of calcite and dolomite or calcite only. Diagram (Fig. 7) shows Ca/Mg ratios of the samples in relation to their distance from the overlying basalt.

Sr contents range between 0.02 and 0.075%, most of the samples carrying less than 0.03% Sr. Sr seems to increase slightly towards the overlying basalt (Fig. 8). Possibly, this higher Sr content is derived from former aragonite of land snails concentrated in the upper parts of the sections. A similar increase of Sr had been described by MÜLLER and TIETZ, but contents are higher in Fuerteventura. Sr content in the uppermost limestone layers of Lanzarote sections correspond to the average amount of Sr in limestone (TUREKIAN and KULP, 0.06%). The whitish-gray limestone lens is extremely low in Mg (0.75%) and extremely high in Sr (0.12%).

Iron contents in most of the samples are below 30 ppm, some range between 1000 and 9000 ppm. Such high amounts seem to be derived from weathered basaltic material present in the samples, because a few samples containing high iron contents are also low in carbonate content.

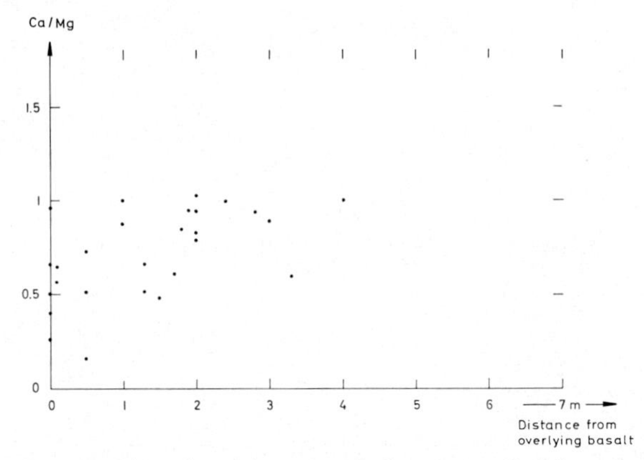

Fig. 7. Dolomitized biocalcarenites, Orzola/Lanzarote. Ca/Mg-ratio in relation to distance from overlying basalt

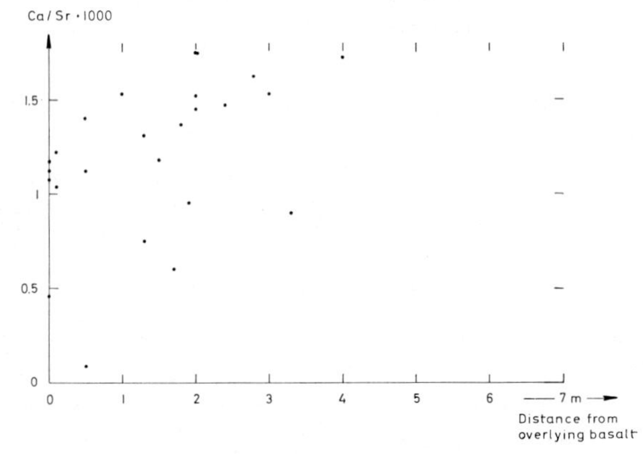

Fig. 8. Dolomitized biocalcarenites, Orzola/Lanzarote. Ca/Sr-ratio in relation to distance from overlying basalt

F. Possible Factors Causing Dolomitization

The biogenic calcarenite consisted originally of calcite (mainly high magnesian calcite: Lithothamnium) and some aragonite (serpula, land snails). Dolomitization decreases downwards and is restricted to the upper 3 to 4 m below the basalt (old land surface). Some possible factors causing dolomitization are discussed below.

I. Contemporaneous Dolomitization Caused by Ascending Waters

Climatic conditions were probably similar to those at present. Ascending capillary solutions would have been able to transport Mg leached from biogenic high magnesian calcite. Possibly there has been a lower stage of the island in Tertiary times and thus the sediments may have been in contact with sea water. Isochemical formation of dolomite within the sediment with Mg derived from algal calcite should be indicated by leaching of Mg in the lowermost parts of the sections. Section 3, however, has no zone of Mg leaching. Also, Mg contents within the sediments themselves would not be sufficient for dolomitization. This process is, therefore, discarded.

II. Dolomitization by Descending Waters Carrying Mg from Weathered Volcanic Rocks

Weathering in the thick pile of overlying volcanics might have liberated Mg. Percolating waters could have redeposited Mg within the sediments causing dolomitization. This would agree with the restriction of dolomite to the upper 3 to 4 m below the top basalt. No Mg was transported farther down than 3 to 4 m.

III. Dolomitization by Descending Brines

Dolomitization could have occurred during a possible higher sea level. In this case, brines percolating downward might have dolomitized the upper layer of the terrestrial calcarenite prior to its covering by the basalt flows in a similar way as described by Müller and Tietz.

G. Relation to other Occurences of Dolomitized Biocalcarenites in the Canary Islands

Müller and Tietz, and Tietz (in preparation) have pointed out dolomitization of biocalcarenites from Fuerteventura which are most probably of Quaternary age. Some of the outcrops are comparable with the Lanzarote sections; the material is more or less the same and so are geologic conditions including a top basalt layer. The Fuerteventura limestone is dolomitized to a lesser degree. Perhaps time is an important factor in dolomitization of the type described. In the case of ascending solutions the land surface may not have been exposed long enough. Alternatively, the over-lying volcanics have not weathered to the same degree as in Orzola. Compared with Fuerteventura the Orzola-sediments show an advanced stage of dolomitization.

Acknowledgements

Sincere thanks are offered to the Deutsche Forschungsgemeinschaft for kind financial assistance, Prof. Dr. G. Müller for discussion and Dr. H.-U. Schmincke for reviewing the english manuscript.

References

Friedman, G. M.: Terminology of crystallization textures and fabrics in sedimentary rocks, Fig. 1—11. J. Sediment. Petrol. **35,** No. 3, 643—655 (1965).

Müller, G.: Die Schnellbestimmung des $CaCO_3/MgCO_3$-Anteils in karbonatischen tonarmen Gesteinen mit dem Dinatriumsalz der Äthylendiamintetraessigsäure (AeDTE). Neues Jahrb. Geol. u. Paläontol. Monatsh. **1956,** 330—344.

— and G. Tietz: Recent dolomitization of Quaternary Biocalcarenites from Fuerteventura (Canary Islands). 8 Figs. Contrib. Mineral. Petrol. **13,** 89—96 (1966).

Rothe, P.: Fossile Straußeneier auf Lanzarote. 8 Abb. Natur u. Museum Frankfurt/Main **94**/5, 175—187 (1964).

— Zum Alter des Vulkanismus auf den östlichen Kanaren. 14 Abb., 5 Tab., 1 Karte. Soc. sci. Fennica Commentationes-Math. **XXXI/13** (1966).

Turekian, K. K., and J. L. Kulp: The geochemistry of strontium. Geochim. et Cosmochim. Acta **10,** 245 (1956).

Some Notes on the Insoluble Residues in Limestones

H. W. FLÜGEL*

With 2 Figures

Most carbonate rocks contain insoluble residue to various extent, whereby the clay-content forms the basis for a classification of the limestones (CORRENS, 1939, p. 200). Rocks with a clay mineral content of up to 25% are hereby designated as limestones. This classification is rarely come across in geologic literature, however; most of the time they are referred to as clayey limestones. Along with organic remains (conodonts, fossil teeth, spicules, Radiolaria, chitinous periderma of graptolithes, pyritized or silicified fossils), insoluble residues may consist of authigenic components (quartz, feldspar, pyrite, dolomite, zeolite, glauconite, etc.) on the one hand, and of detrital components supplied by the action of wind or water (illite, chlorite, kaolinite, quartz, feldspar, heavy minerals, etc.) on the other. The delineation between both of the latter components is particularly difficult within the realm of the clay minerals, since it may also be a matter here of synsedimentary, epigenetic or diagenetic modifications and authigenic products. The percentage of such authigenic products is, however, relatively small, since the clay content in present seas is for the most part detrital (BISCAYE, 1965). Nevertheless, this uncertainty makes a genetic interpretation of the mineral composition of the clayey residue more difficult, since certain lateral or vertical changes in the composition of clay residue do not necessarily have to be connected with a change in the source area, but might just as well be of a secondary nature. On the other hand, however, it are precisely such often short-lasting, qualitative and/or quantitative changes in the residue, which contribute to the clarification of definite geologic questions. The clarification of stratigraphic hiata represents such a case. These hiata are associated in part with a temporary local or regional regression and exposure to the air of an up till there marine realm of sedimentation. This may lead to soil formation as controlled by climate, in which case the soils which develop largely fall victim to the next-following transgression and are reworked, but in the process, however, can also change the insoluble residue picture within the transgression strata quantitatively and qualitatively over short distances.

Thus DALTON et al. (1958) were able to demonstrate, that the disconformable superposition of the Holdenville Shale (Desmoinesian) by the Hepler Sandstone (Missourian) in SE Kansas is reflected by a noticeable increase in the kaolinite content in the upper portions of the Holdenville Shale coupled with a simultaneous decrease in the feldspar content and the incipient decomposition of the chlorite, as well as the dominance of vermiculite and kaolinite in the basal strata of the Hepler Sandstone. They interpreted this as an indication of terrestrial soil-formation along the Desmoinesian/Missourian boundary.

---

* Institut für Geologie und Paläontologie, Universität Graz, Austria.

HAYES (1963) was able to substantiate a similar indication of fossil soil-formation and therewith proof of an unconformity and a stratigraphic hiatus in the Mississippian of SE Iowa. Here the mudstones of the Lower Warsaw Formation are overlain by the clayey biocalcarenites of the Upper Warsaw Formation. While the insoluble residue of both these formations normally consists predominantly of chlorite, illite, as well as of kaolinite in the higher parts also, in the crest of the Warsaw anticline, where the Lower Warsaw Formation noticeably is of no great thickness, vermiculite, as well as mixed-layer illite-montmorillonite instead of chlorite is found along the boundary between the Lower and Upper Warsaw Formations. HAYES connects both of these to weathering of the uppermost mudstone of the Lower Warsaw Formation, whereby — as in the example given above — the derived detritus after having transgressed beyond the apex of the anticline, brought about a local, short-lasting disruption in the normal insoluble residue picture of the limestones of the Upper Warsaw Formation as well.

That such soil-formations and redepositions as the result of a transgression may not only give rise to a change in the mineralogical composition of the clay minerals of the insoluble residue, but that quantitative changes are also connected therewith, is shown in a study of the Palaozoic around Graz (Austria) carried out by KODSI (1967). In connection with an investigation of the Devonian/Carboniferous boundary he was able to determine in a boundary profile, that the biogenic-rich micrites of the lower Upper Devonian (to I—III) are concordantly overlain by micrites, completely identical with respect to microfacies, of the upper Lower Carboniferous (anchoralis-zone, cu II β/γ). They start with an 80 cm thick succession of limestone, which contains a mixed conodont-fauna of Upper Devonian and Lower Carboniferous elements, whereby the number of Devonian forms clearly decreases upwards. Traces of reworking, redeposition, the formation of necks, etc. can't be determined; nevertheless very clear differences in the character of the insoluble residue are found (Table 1):

Table 1

	Upper Devonian tol-III	Lower Carboniferous Mixed-fauna Zone	cu II β/γ
Amount of insoluble residue			
average value	4.9%	3.6%	4.9%
median value	1.5—4.8 my	5.2— 9.6 my	1.6—3.4 my
mean value	4.5—9.5 my	11.1—16.6 my	5.0—7.8 my
mineral composition	predominantly illite	illite/kaolin-minerals in the coarser fractions, predominantly illite in the finer fractions	predominantly illite

These qualitative and quantitative differences pertain only to the lowermost horizons of the upper Lower Carboniferous, to the immediate realm of transgression therefore after a stratigraphic hiatus, in which the entire upper Upper Devonian was demonstrably eroded away. This realm is characterized by reworked Upper Devonian conodonts, a reduction in the amount of insoluble residue of about 25%, as well as a distinct coarsening of the residue, and possibly also a slight change in the mineralogy of the coarser fractions. One could attribute the coarsening coupled with a simultaneous reduction in the amount of insoluble residue to a stronger turbulence of the water at the start of the transgression and therewith to the removal of the fine-grained material, in which case the insoluble residue itself would represent the accumulation of a lower Upper Carboniferous soil-formation.

Such secondary changes of the insoluble residue may also have other causes, however, as an investigation by Pölsler (1967) of the Upper Devonian of the Karnic Alps shows. Here, the Pipeline-tunnels through the Karnic Alps locally disclosed a rhythmic layering of approximately 6 cm thick layers of gray and red Upper Devonian clayey micrites. Both can't only be differentiated on the basis of their color, but also very clearly on the basis of their lithology, as well as their incorporated fauna. Whereas the red limestones show an insoluble residue content of 12 to 16%, composed of quartz, illite, chlorite (or kaolinite) and numerous, well-preserved conodonts, in the only 0.6% insoluble residue derived from the gray limestones pyrite, feldspar and quartz are found as authigenic products along with illite; thereby is the number of conodonts small and the individuals are badly preserved. This observation can only be explained by means of a secondary change in the insoluble residue, which is of interest to the paleontologists only insofar, that it resulted in a qualitative and quantitative reduction of the conodonts, which up until elimination already led in a primary stage to species which individually were poorly represented.

Whereas the mineralogical composition of the clayey residue should hardly change more distinctly, however, within a geologic-stratigraphic layered complex that belongs together, investigations up to now (Valeton, 1954; Boger, 1966) show, that very clear quantitative differences in the insoluble residue content exist from layer to layer. Thereby, as E. Seibold (1952) has shown and as personal investigations have been able to substantiate up to now, the carbonate content within the individual beds remains fairly constant from the foot-wall to the hanging-wall and the transitions from layer to layer are very clear. This speaks for the fact, that each layer represents a genetic unit, which originated under homogeneous conditions of sedimentation. Lithification of the layers, before deposition of the next-following layer probably did not always take place in the process, since cases of worm burrows, which run straight through two or more layers across the boundaries of the layers are described in the literature. The individual layers clearly differ in their carbonate — clay ratio (Böger, 1966; Freiberg, 1966). If a comparison is made between the quantitative insoluble residue investigations of longer, closed profiles published up to now, then ever recurring types of insoluble residue become apparent. They can be explained, if we assume, that in the formation of clayey layered limestones three different factors, namely the detrital insoluble residue, the carbonate and the time of formation, play a role. In this case, the following possibilities are conceivable.

1. The duration of sedimentation changes with the thickness of the bedding.

a) The sedimentation of detritus is continuous and constant over a certain duration of time. The sedimentation of carbonate changes continuously or discontinuously during the time of deposition. A diagram, which shows the absolute insoluble residue content versus the thickness of the bedding, will show a continuous increase in the insoluble residue coupled with an increase in the thickness of the bedding, whereas the percentage of insoluble residue will fluctuate. This, for example, is demonstrated by the Upper Jurassic Oberalmer beds analyzed by H. Flügel and Fenninger (1966), and Garrison (1967) (Fig. 1, No. B). They were interpreted as pelagic-bathyal formations. The absolute insoluble residue content increases clearly with increasing

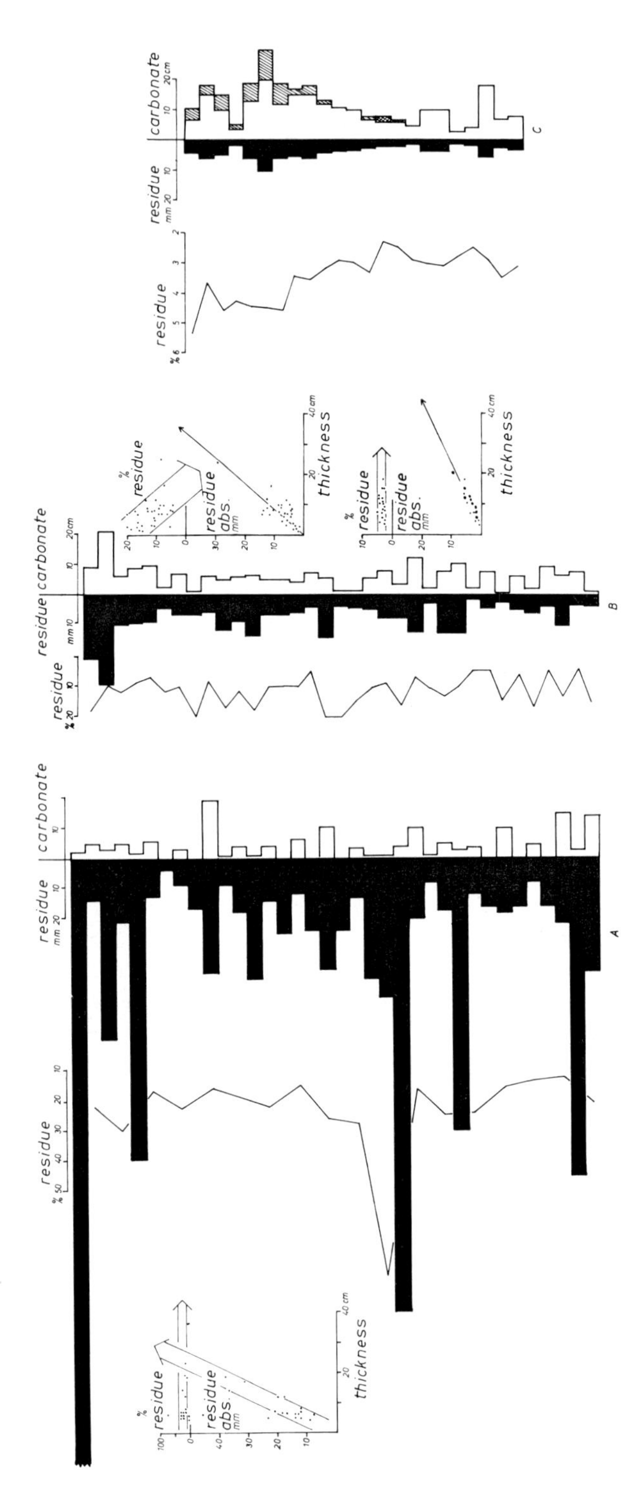

Fig. 1. Insoluble residue-diagrams: A) Amphiclinen-beds (Triassic) of Juda Huzna (Slovenia); B) Oberalm-beds (Upper Jurassic) of Oberalm (Salzburg); C) Titon-Limestone of the Pechgraben (Upper-Austria). Hachured: carbonate deficit, chequered: carbonate surplus at 3% residue

thickness of the bedding, the percentage of insoluble residue decreases. In profile this is expressed by a strongly fluctuating of the insoluble residue percentage curve due to the fluctuating thickness of the individual beds. This increase in the carbonate content with increasing duration of deposition (thickness) of the individual beds could geologically be explained in this manner, that a continuous carbonate-clay-sedimentation is here overlain discordantly by a discontinuous one; a process, which becomes the more frequent, the longer the duration of sedimentation lasts.

An exactly opposite case was reported by E. SEIBOLD (1952, p. 360) from the Lower Malm in Schwaben. Here also, a continuous increase in the absolute clay content in stages was found; in this case, however, the amplitude of the insoluble residue-percentage curve increases successively; i.e., the carbonate content does not increase correspondingly with the clay content, but somewhat less clearly. Here, interpreting geologically, one could perhaps think of the superimposed activity of limestone solution.

b) The sedimentation of detritus, as well as carbonate is continuous and constant with respect to the unit of time. Here a diagram of the absolute insoluble residue content will show a continuous increase of the same with an increase in the thickness of the bedding, while the percentage of insoluble residue will only fluctuate a little, independent of the thickness of the bedding, i.e., the duration of sedimentation.

FLÜGEL (1967) described such a situation from the Upper Steinmühl-Limestones and HOLZER (unpubl.) respectively from the Upper Jurassic limestones of the Pechgraben. In the former (possibly also in the second case) it is a question of sedimentation on submarine mounds within the pelagic realm, which is characterized by very thin bedding. At their type-locality (Arracher quarry near Weidhofen a. d. Y.), the upper portions of these can be broken down into two parts: the Saccocoma- and the Calpionellen-Limestones on the basis of microfacies. Both are micrites bearing biogenic components, which can only be differentiated by means of their faunal content. The absolute clay content (Fig. 2) fluctuates strongly from layer to layer; it does, however, show a distinct increase with an increase in the thickness of the bedding. The insoluble residue percentage curve, on the other hand, is remarkably constant. It fluctuates around about 7.5% in lower portions, and around about 4.5% in upper portions, whereby no relationship to the thickness of bedding at all can be observed. The sudden transition between both realms also becomes evident in the overall pictures: The lower part of the section is considerably more irregular as far as the thickness of the bedding is concerned, and with respect to the absolute insoluble residue content as well, than the upper portion, the bedding of which is less thick. The somewhat stronger variations in the percentage of insoluble residue in the lower portion of the approximately 15 m thick section might be explained by a changing supply of carbonate. If under this condition the clay content is kept constant at about 7.5%, then the deficit or the surplus of carbonate in the individual beds can be calculated. Compare Fig. 2. If one compares the values obtained therewith with the microfacies findings, it can be seen, however, that one does not get far with such calculations. Thus, within the microfacies of horizons 26 and 21 a very pronounced increase in echinoderm remains is found. In both cases the beds consist up to more than 80% of biogenic remains ("encrinites"). One would, therefore, expect a definite carbonate surplus, which however, is only the case with horizon 24. Horizon 21, however, shows a carbonate deficit. It is suspected, that the variations of the insoluble residue of the lower part of the section are connected with primary variations in the clay content.

The section of the Pechgraben quarry (H. HOLZER), which is comparable with regard to the increasing absolute insoluble residue content at a constant content expressed in percentage, differs from the one of the Steinmühl-Limestone at first sight in that its bedding is less thick (thickness of bedding about 10 cm). Here also we are dealing with fossiliferous micrites to biomicrites. The insoluble residue content for the most part lies below 5%. The cut of the section represented here is of interest insofar, that it shows in stages the condition of a continuous increase in the insoluble residue content with a corresponding equivalent increase in the thickness of the

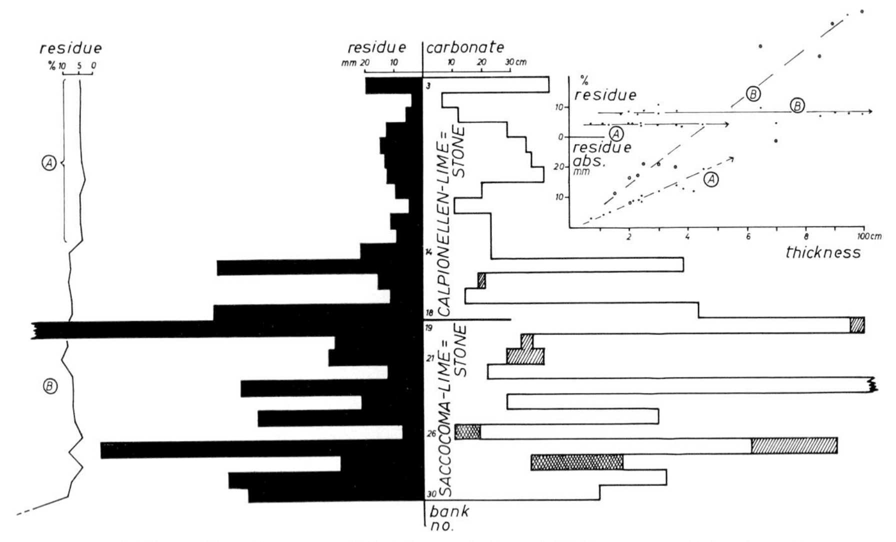

Fig. 2. Insoluble residue-diagram of the Upper Steinmühl-Limestone of the Arracher quarry (Weidhofen a. d. Y.). Hachured: carbonate deficit, chequered: carbonate surplus at 7.5% residue

bedding or the carbonate content. Accordingly, a small increase in the percentage of insoluble residue becomes evident. It shows, that within a certain type (1 b) a different development can be incorporated in time (1 a). In the case on hand the explanation could be found in a reduction of the carbonate content through solution with time, which is supported by the fact that the deficit, which is calculated by assuming an insoluble residue content of 3%, increases with an increase in the thickness of bedding, i.e., with an increase in the duration of time.

c) Detritus- and carbonate sedimentation vary with the thickness of bedding (i.e., with the duration of sedimentation). This is the most common condition encountered and consequently under the most extreme assumptions, no connection between the absolute insoluble residue content or the insoluble residue percentage, and the thickness of bedding can be demonstrated. As the analysis of the Triassic Amphiclinen strata of Huda Juzna in Slovenia shows[1], however, a certain relationship

[1] For the disposition of funds for this investigation, I thank the Österr. Akademie der Wissenschaften (The Austrian Academy of Sciences). For the helpful assistance rendered in the field, I wish to thank Prof. Dr. RAMOVS, Llubljana.

to the thickness of the bedding might be determinable. With these strata, we are dealing with a very turbulent alternation of thin beds of fossiliferous micrites (Echinodermata, ammonites, conodonts, Radiolaria), clay-shales and sandy shales up to sandstones. Their lithogenic analysis has been completed not yet at this time. The percentage of insoluble residue, the absolute quantity of residue, as well as the carbonate content vary clearly from bed to bed. This applies also then, when one does not take the interbedded layers of clay-shale (likewise with a strongly variable carbonate content) into consideration. A comparison between the thickness of the bedding and the absolute insoluble residue content of the micrites shows a tendency towards an increase in the residue content with increasing thickness of the bedding with a relatively large range of variation (Fig. 1, No. A), whereas indeed the percentage of insoluble residue likewise clearly varies, but without clearly showing a tendency in this case, however, whether it be ascending or descending. The insoluble residue content versus thickness of the bedding-diagram is similar in appearance to that of the Tithon-Limestones of the Pechgraben (Fig. 1, No. C). A reference made to the percentage clearly shows the marked difference, however: whereas the range of variation in the Pechgraben amounts to 2 to 3%, within the Amphiclinenstrata it lies between 6 and 33% without taking the clayey-marley and clayey beds into consideration and with an average insoluble residue content of 19%; i.e., the range of variation amounts here to approximately 27%.

From this example we can conclude, that in the case of a strongly variable insoluble residue supply, along with variable carbonate sedimentation, the duration of sedimentation as reflected by the varying thickness of the bedding also shows itself in the increase of the insoluble residue content as is to be expected, although this is not a must theoretically. Percentage-wise the insoluble residue content is variable to a great extent, however, and no relationship to the thickness of the bedding can be demonstrated.

2. A second possibility would be, that without detriment to its thickness, each bed was deposited within the same time span. Such a case was illustrated by E. SEIBOLD (1952) with examples from the Lower Malm of Schwaben and is characterized by the fact, that here the absolute insoluble residue content remains constant unaffected by the thickness of the bedding. If in this case the percentage of insoluble residue present does not change, then the thickness of the individual beds remains constant; if the percentage does vary, then the carbonate content increases (with decreasing insoluble residue content) or decrease (with an increase in the insoluble residue content) and the thickness of the bedding therewith likewise. Such cases were confirmed by HOLZER in stages in the Pechgraben-section.

As the work by E. SEIBOLD (1952) and our own investigations have shown, most of the time in nature several of the cases presented here appear to occur side by side or superimposed in one profile. Investigations dealing with the lateral variations of the insoluble residue content within a single bed, or within sequences of sections equivalent in time are unknown to me.

During the investigation of the Red Eagle Cyclothem (Lower Permian) by MCCRONE (1963) it was found, that the percentage of insoluble residue in both limestone members of this series (Glenrock Limestone, Howe Limestone), both no more than 1 m thick, varies distinctly over the more than 250 mile long stretch. To judge in accordance with the data supplied by MCCRONE (1963), no relationships to the thickness of the bedding appear to exist in this case. The change of the insoluble residue content from bed to bed within the stratigraphic sequence was not investigated, however.

Concerning the distribution and ordering of the insoluble residue in carbonate rocks, hardly any investigations have been published up to now. The reasons for this, on the one hand, might be the wide-spread assumption, that the insoluble residue occurs distributed in an accidental and irregular manner and on the other, that the largely small quantities and small grain size of the residue make investigation difficult.

After the study of terrigenous rocks with radiography had found application, with the result that the orientation of clayey matter was explained, it was natural to test this method also with carbonate rocks. Whereas K. H. Wolf et al. (1967) could find no structures in very pure limestones as was to be expected, H. Flügel and Fenninger (1966) produced photographs of clayey limestones, which permitted the recognition of an irregular distribution of dark spots of various densities (compare Fig. 8 in Table 29 in H. Flügel and Fenninger, 1966). Possibly, this is indicative of a local agglomeration of the insoluble residue. Electron-microscope investigations often reveal a more or less accidental distribution of the clay minerals. All these studies up to now were carried out on rocks that were poor in insoluble residue content, however; while marl, clayey marl and other clay-rich rocks are still unstudied. Most recently an attempt was made therefore to add some to the clarification of the degree of ordering of the clayey residue particularly with regard to these rocks by making use of the X-ray diffractometer (H. Flügel and Walitzi, 1967). The same method was used here as the one utilized by Meade (1961) to study the ordering within clay-shales, and by H. Flügel and Walitzi (1966) to investigate the ordering of micritic limestones.

The degree of ordering $s\,\frac{I\,002}{I\,020}\,s\,\frac{I\,002}{I\,020}$ (E. M. Walitzi) of illite, the carbonate content, the porosity, the distribution of the grain size and the density of a series of clay marls, composed from drilling profiles of the Neogene from the Vienna Basin, was investigated. The thickness of the series studied amounted to 2880 m. The carbonate-content varies from 14 to 24%. The degree of ordering amounts to 1.3 in the uppermost sample (depth 225 m). It gradually increases up to 5.0, in the course of which distinct variations could be ascertained, however. Thus a sample from a depth of 1641 m showed a degree of ordering of 5.6, while a sample from 2010 m depth only showed one of 2.4. With the carbonate content a weakly negative correlation (—0.116) is prevalent; i.e., the degree of ordering decreases with increasing carbonate content. This influence of the carbonate upon the degree of ordering is also shown by the fact, that the positive correlation increases from + 0.599 to + 0.650 when calculating the partial correlation between the degree of ordering and depth with exclusion of the carbonate content; i.e., the degree of ordering becomes more distinct with increasing depth.

These observations show two facts: First, rocks with a high clay content obtain only a weak ordering of their clay minerals (illite) with increasing depth of burial. If on the other hand the result is generalized taking the carbonate content into consideration, it could be deduced, that an increase in the carbonate content inhibits an ordering. Clayey limestones, but marly limestones as well, have not revealed of any ordering, which perhaps is connected with the small illite content, which here no longer permits such an investigation.

On the other hand, a clayey marl sample with a carbonate content of 15% from the Amphiclinen strata shows an excellent ordering of the illite, with a degree of ordering of 12. The layers dip steeply. The ordering here is probably not subject to diagenetic-sedimentary conditions, but is the result of tectonism.

References

BISCAYE, P. E.: Mineralogy and sedimentation of recent deep-sea clay in the Atlantic Ocean and adjacent seas and oceans. Bull. Geol. Soc. Am. **76**, 803—832 (1965).

BÖGER, H.: Paläökologische Untersuchungen an gebankten Kalken. Am Beispiel des Ooser Plattenkalkes, Oberdevon I der Eifel. Geol. Fören. i. Stockholm Förh. **88**, 307—326 (1966).

DALTON, J. A., A. SWINEFORD, and J. M. JEWETT: Clay minerals at a Pennsylvanian disconformity. Clay and Clay Minerals **5**, 242—251 (1958).

FLÜGEL, H. W.: Die Lithogenese der Steinmühl-Kalke des Arracher Steinbruches (Jura, Österreich). Sedimentology **9**, 23—53 (1967).

—, u. A. FENNINGER: Die Lithogenese der Oberalmer Schichten und der mikritischen Plassen-Kalke (Tithonium, Nördliche Kalkalpen). Neues Jahrb. Geol. u. Paläontol. Abhandl. **123**, 249—280 (1966).

—, u. E. M. WALITZI: Untersuchungen über Calzit-Regelung in Kalken mit Hilfe des Diffraktometer-Verfahrens. Anz. Akad. Wiss. Wien, Math.-naturw. Kl. 67—72 (1966).

— — Regelung und Porosität in Tonmergeln des Wiener Beckens. Neues Jahrb. Geol. u. Paläontol. Monatsh. **1968**, 1—10 (1968).

FREYBERG, B. v.: Der Faziesverband im Unteren Malm Frankens. Ergebnisse der Stromatometrie. Erlanger Geol. Abh. **62**, 112 S. (1966).

GARRISON, R.: Pelagic limestones of the Oberalm beds (upper-Jurassic-lower cretaceous), Austrian Alps. Bull. Canadian Petrol. Geol. **15**, 21—49 (1967).

HAYES, J. B.: Clay mineralogy of Mississippian strata of southeast Iowa. Clay and Clay Minerals **10**, 413—425 (1963).

KODSI, G. M.: Zur Kenntnis der Devon/Karbon-Grenze im Paläozoikum von Graz. Neues Jahrb. Geol. u. Paläontol. Monatsh. **1967**, 415—427 (1967).

MCCRONE, A. W.: Paleoecology and biostratigraphy of the red eagle cyclothem (lower permian) in Kansas. Bull. State Geol. Survey Kansans **164**, 114 S. (1963).

MEADE, R. H.: X-ray diffractometer method for measuring preferred orientation in clays. Profess. Papers Geol. Survey **424-B**, 273 (1961).

PÖLSLER, P.: Geologie des Plöckentunnels der Ölleitung Triest-Ingolstadt (Karnische Alpen, Österreich/Italien). Karinthia II **77**, 37—58 (1967).

SEIBOLD, E.: Chemische Untersuchungen zur Bankung im unteren Malm Schwabens. Neues Jahrb. Geol. u. Paläontol. Abhandl. **95**, 337—370 (1952).

VALETON, I.: Beitrag zur Petrographie des mittleren Muschelkalkes Süd-Deutschlands. Heidelb. Beitr. Mineral., Petrogr. **4**, 207—216 (1954).

WOLF, K. H., A. J. EASTON, and S. WARNE: Techniques of examining and analyzing carbonate skeletons, minerals, and rocks. In: Carbonate rocks. Physical and chemical Aspects. Developments in Sedimentology **9 B**, 253—341. Amsterdam-London-New York: Elsevier Publ. comp. 1967.

Relationship Between Carbonate Grain Size and Non-Carbonate Content in Carbonate Sedimentary Rocks

HANNELORE MARSCHNER*

With 2 Figures

Abstract

Grain size and fabric of 40 samples of Triassic carbonate rocks were studied. The grain size of early diagenetic carbonate minerals is inversely proportional to the clay mineral content. Probably, envelopes of clay minerals prevented formation of larger carbonate crystals during diagenetic recrystallization.

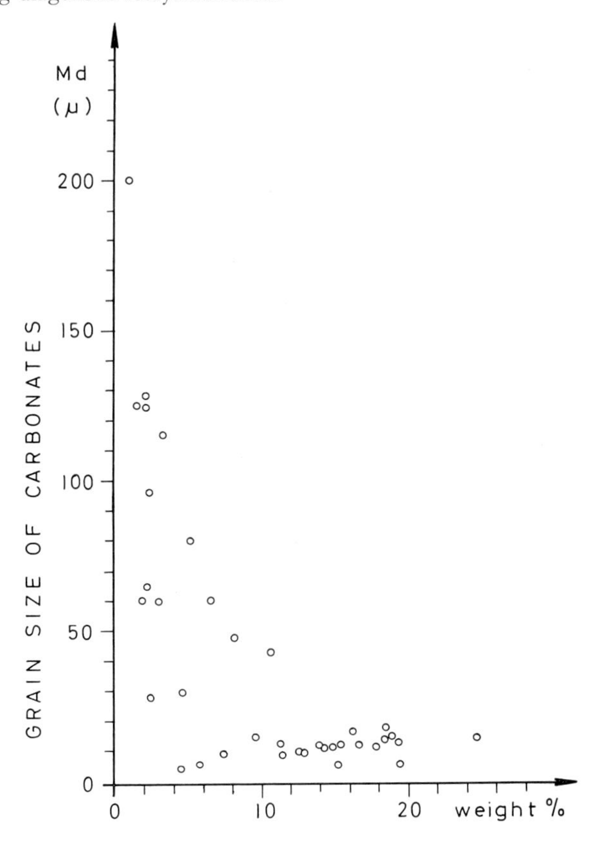

Fig. 1. Dependence of the carbonate grain-size on the content of the non-carbonate fraction of the samples

* Laboratorium für Sedimentforschung, Universität Heidelberg, Germany.

The grain size distribution of carbonate minerals is often used to characterize carbonate sediments and their origin (WEBER, 1964). However, the origin of a particular grain size distribution can be due to different factors. One of these, the influence of non-carbonate content on grain size of carbonate minerals, is discussed below.

Grain size and fabric of 40 samples of several carbonatic beds of the Lower Keuper (Triassic) from NW Germany were studied. These rocks are shallow-water deposits. They differ much petrographically with regard to their carbonate/non-

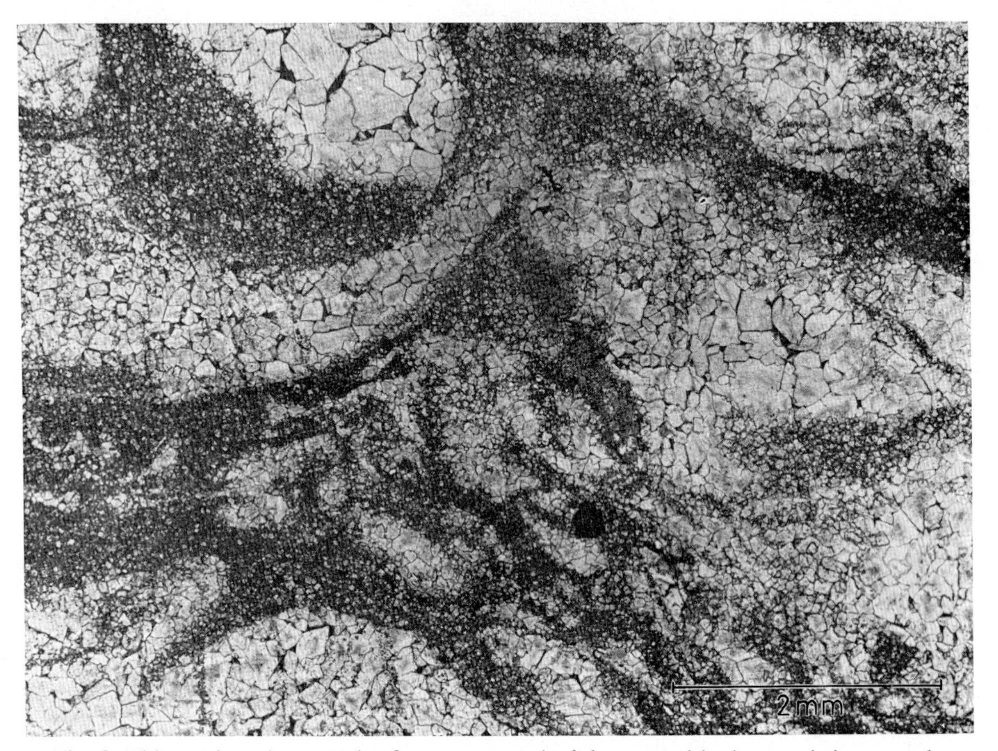

Fig. 2. Thin-section photograph of an organogenic dolostone with characteristic, strongly varying grain size distribution. Note that the carbonate grain size decreases with increasing clay mineral fraction. Parallel polarizers

carbonate ratio, carbonate grain size, and texture (with and without shell fragments; bedded and massive). The carbonate minerals are dominantly of early diagenetic origin (MARSCHNER, 1966).

Average carbonate grain size of all samples is plotted against non-carbonate content (fraction $< 6.3\,\mu$) in Fig. 1. A decrease of carbonate grain size with increasing non-carbonate material is obvious.

A possible explanation is, from thin section evidence:

Clay minerals always occur along carbonate grain boundaries and envelope these. The envelopes are the thicker (in comparison to the diameter of the carbonate grains) the greater the ratio clay material/carbonate. The relationship carbonate grain-size/non-carbonate matter is thus probably due to the "protective" envelopes of the clay minerals during recrystallization. Organic substances and envelopes of metallic hydroxides have a similar effect. This applies only to the early diagenetic carbonate minerals.

A biogenic dolostone (Fig. 2) is a good illustration of the above relationship: The carbonate crystals are largest in shell fragments or in spaces closed off by shells from clay sedimentation. In the interstitial spaces, however, grain-size changes depend on the content in foreign matter. This interstitial distribution of non-carbonate substances occurred during sedimentation. It is probably the reason why organogenic forms are still recognizable as such after diagenetic recrystallization. This is the explanation for the pronounced grain-size differences that characterize organogenic dolomite. In principle this is true for all carbonate sediments, which contain sedimented carbonate intraclasts with smaller or larger quantities of non-carbonate matter as their surrounding matrix. In chemically precipitated carbonate rocks the differences in grain-size are much smaller. The carbonate grain-size is then approximately fixed by the sedimentation rates carbonate/clay minerals whose ratio is more or less constant for a given bed. This is true only if no general or locally restricted carbonate dissolution and reprecipitation has taken place.

References

MARSCHNER, H.: Mineralogisch-petrographische Untersuchungen an karbonatreichen Gesteinen aus dem Unteren Keuper des Weserberglandes. 137 p. Diss., Univ. Hamburg, priv. circ. 1966.

WEBER, J. N.: Trace element composition of dolostones and dolomites and its bearing on the dolomite problem. Geochim. et Cosmochim. Acta **28**, 1817—1868 (1964).

Experimental Compaction of Carbonate Sediments

GÖTZ EBHARDT*

With 8 Figures

Abstract

Recent carbonate sediments of different grain size and with different carbonate content were compacted under pressures of up to 650 atm. The reduction of porosity with increasing pressure as well as the progress of compaction with time at constant pressure and their dependance on grain size, temperature and chemical nature of the pore medium were evaluated and presented graphically. Of special importance for a comparison with natural processes is the differentiated investigation of the various interdependent mechanisms which produce compaction: reorientation of the individual grains, plastic and shear deformation of the components and chemical processes such as pressure solution and solution transport.

A. Introduction

Compaction experiments on clays have been performed for some time, starting from investigations in soil mechanics. Their results show good correspondence with the porosity of natural burried claystones. Limestones, however, rarely show direct indication of compaction, although Recent carbonate sediments have high porosities of up to 70%, and limestones, especially micrites, normally show very little pore space. Hence there must have been compaction, if not all of the cement introduced by pore solution was from outside the carbonate system.

Surely the compaction of lime mud and sand is more complex than that of clastic sediments, mainly because of the high solubility of calcium carbonate, the thermodynamic metastability of some of the primary minerals and of the wide variability in grain shape. Therefore compaction experiments are useful, only if the different interdepending mechanisms — reorientation of the grains, shear and plastic deformation of the individual components and chemical processes (e.g., pressure solution) — are separated as well as possible. The time factor also has to be taken into consideration, because at low temperatures chemical processes need more time than is available in the laboratory. This is also of great consequence to grain deformation relationships to reorientation of the grains.

Due to these difficulties only relatively little work on the experimental compaction of lime sediments has been done in the last years: R. TERZAGHI (1940), ROBERTSON, SYKES and NEWALL (1962), HATHAWAY and ROBERTSON (1961), FRUTH, ORME and DONATH (1966).

B. Apparatus and Procedure

The compaction apparatus consists of a pressure cell, an oil-hydraulic press, a heating mantle and a control unit for pressure and temperature. The sediment sample is contained in a steel cylinder, 40 mm in diameter, pressure is applied by a

* Institut für Geologie, Universität Würzburg, Germany

steel piston. In the experiments described here, pore solution was allowed to escape freely at the base of the cell through a paper filter and a porous steel plate. The movement of the piston was measured with an accuracy of 0.01 mm, pressure varied from 20 to 650 atm, with a precision of $\pm$ 2 atm below 80 atm and about 10 atm at higher pressures; temperatures could be elevated up to 200°C.

The sediment samples weighing about 30 g (dry weight) were dispersed in sea water or, if sandy, directly poured into the cell; than the cell was evacuated to remove air bubbles. The pressure was increased slowly in steps of $\sim$ 5 min. After each step we waited, until compaction was completed. This normally required 2 or 3h. At the end of each run the porosity was determined by drying the compressed sample. Porosity is expressed in terms of the void ratio E:

$$E = \frac{\varepsilon}{1-\varepsilon} = \frac{V_v}{V_s} = \frac{m_w}{V_b - m_w} = \frac{V_b \varrho_s - m_s}{m_s}.$$

(ε = porosity, V_v- and V_s = volume of voids and of solid grains, V_b = bulk volume, m_w and m_s = mass of interstitial water and of solid grains, ϱ_s = solid grain density). In cases, where ϱ_s is known, it is possible to control the other values.

To get some information about the mechanism of compaction, the progress of compaction with time was determined after rapid increasing the pressures from 20 to 70 atm and related to the total compaction of the sample in this step.

C. Samples

The sediment samples and their properties are listed in Table 1. In the text and in the diagrams they are always designated with the abbrevations given in column 1. Grain size analysis was made by sieving and by pipette-analysis with the Andreasen-cylinder.

Table 1. *Sediment Samples*

sample	provenance	type of sediment	carbonate %	sand %	silt %	clay %	Md μ	So
Ad IV	Adria	sand	86	96	3	1	450	1,6
Rm	Red Sea	sandy silt	75	20	80	—	20	1,3
P	Persian Gulf	clayey silt	65	17	56	27	6	3,7
F II	Florida Bay	clayey silt	86	8	61	31	7	3,8
Ad I	Adria	clayey silt	41	4	78	18	11	3,3
Hz	Persian Gulf	silty clay	36	—	48	62	2	—
$CaCO_3$	—	silt	100	—	100	—	22	1,5
Bd	Lake of Constance	sandy silt	91	24	70	6	17	3,5

Md = Median diameter; So = $\sqrt{Q_3/Q_1}$ Sorting coefficient

Ad IV consists mainly of skeletal debris with some well rounded particles of rather hard pelletal micrite. Rm contains within the sand fraction Foraminifera and irregular aggregates of up to 2 mm in diameter. Samples P and F II contain some badly rounded quartz sand. F II and Ad I contain much organic substance in disseminated and in particulate form. Hz was ground before the experiments and consists of clay and carbonate debris. The sand fraction of Bd is largely composed of oncoids. For comparison, experiments were also run with chemically precipitated $CaCO_3$ (calcite) of silt size with sharpedged, partly rhombohedral grains.

D. Experimental Results

1. *Relation of void ratio to pressure, different sediments* (Fig. 1). For the compaction of clays Terzaghi (1925) has establisged the relation $E_p = E_1 - b \lg p$, where E_1 = void ratio at unit pressure, b = factor of compressibility, p = pressure, and E_1 and b are material constants. This relation is not really valid for carbonate sedi-

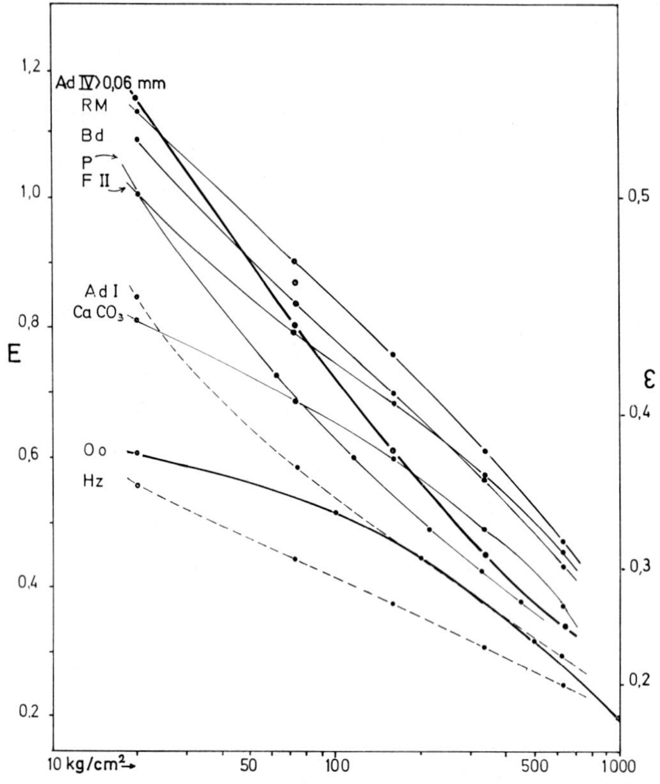

Fig. 1. Void ratio versus confining pressure, different sediments. Thick lines: sand size sediments. Dashed lines: clayey sediment. Oo oolithic sediment, after Fruth et al. (1966)

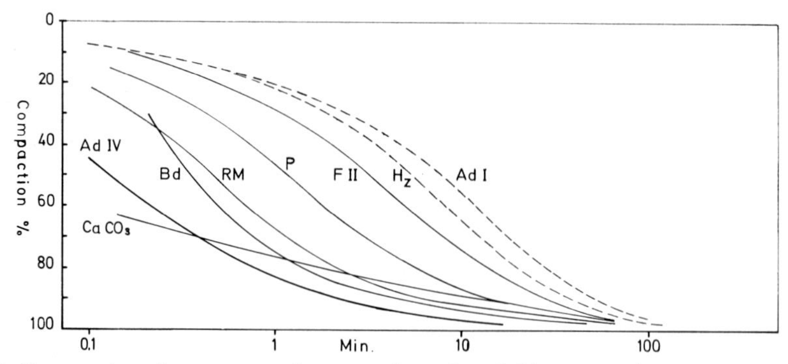

Fig. 2. Compaction (in percent of compaction after 2 h) versus time at constant pressure, different samples

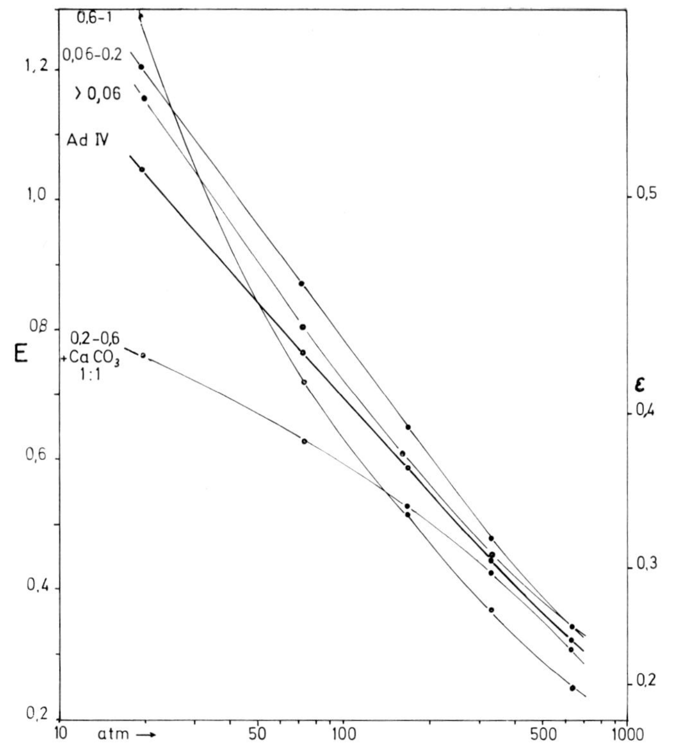

Fig. 3. Void ratio versus pressure, different sieve fractions of sample Ad IV

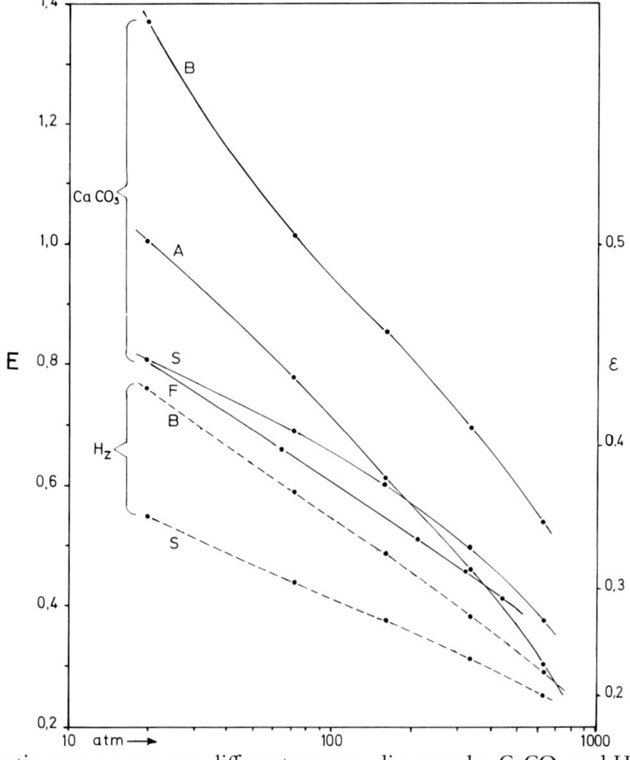

Fig. 4. Void ratio versus pressure, different pore media, samples $CaCO_3$ and Hz. *B* benzene, *A* air, *S*, *F* sea and fresh water

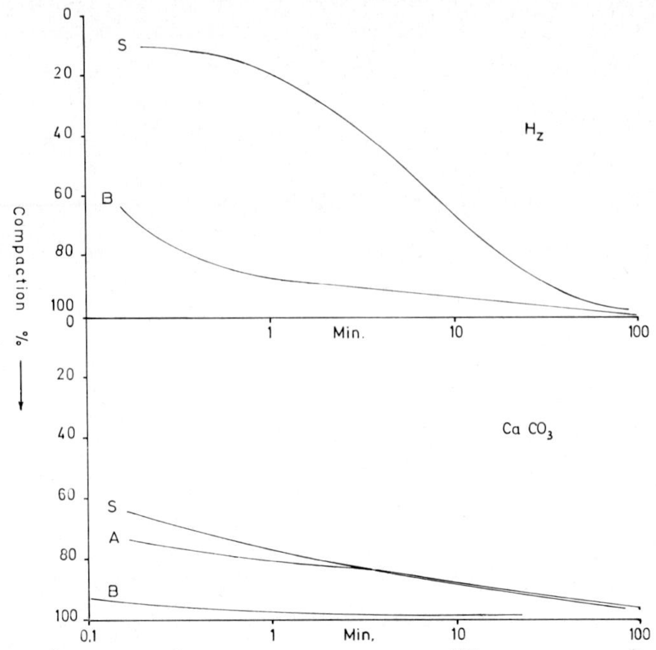

Fig. 5. Compaction versus time at constant pressure, different pore media. *B* benzene, *A* air, *S* sea water

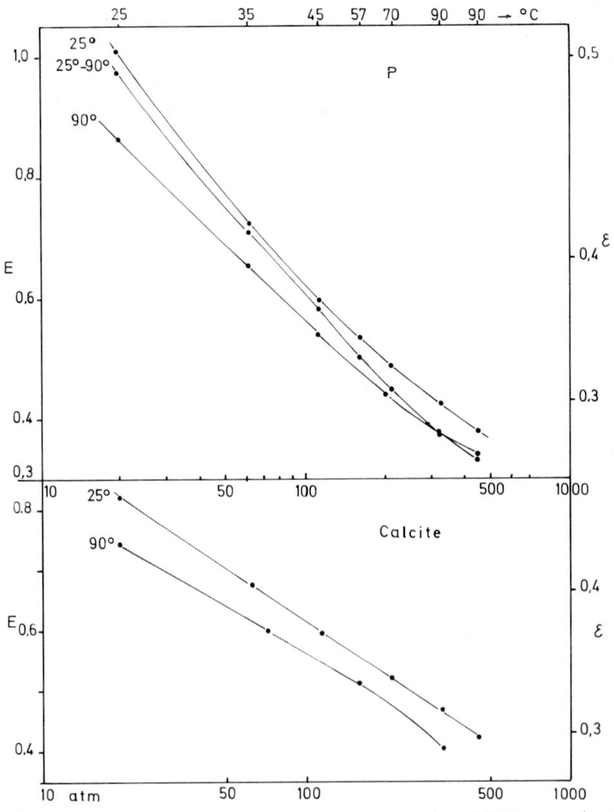

Fig. 6. Void ratio versus pressure at room temperature and at 90 °C, samples P and $CaCO_3$. One curve with temperature elevated stepwise (scale at the Top of the upper graph)

ments, but deviations may serve to recognize details of the compaction mechanism. Therefore the pressure is plotted logarithmically.

Curves Ad IV and Oo belong to sand size sediments, where crushing and penetration account for most of the possible compaction, as can be seen from thin sections (Fig. 7). Curve Oo is taken from FRUTH, ORME and DONATH (1966). In Ad IV, deformation begins at relatively low pressures, depending on the high primary porosity, whereas the oolite Oo starts to be crushed remarkably only at pressures higher than 100 atm.

Fig. 7. Microphoto of sample Ad IV, compressed with 650 atm. Fractured shell debris, plastically deformed micritic pebbles (centre).

Curves Bd, Rm and F II belong to predominantly silty sediments with some sand size components. The compaction roughly follows the Terzaghi rule, but at higher pressures compaction is more intense than expected. Influences of redistribution or deformation of single grains and may be of pressure solution are difficult to separate.

The curve of $CaCO_3$ is convex, similar to that of the oolite, though the grain size is much smaller. Probably there is an increase of crushing at higher pressures due to the low sorting coefficient.

Samples P, Ad I and Hz are characterized by an increasing clay content and give curves similar to those of pure clays (CHILINGAR and KNIGHT, 1960; v. ENGELHARDT and GAIDA, 1963). The coarse fraction seems to play a rather passive role.

2. *The Compression versus time diagram* at constant pressure (Fig. 2) shows dependance mainly on the grain size: the finer the sediment, the longer time is needed for

compaction. With the exception of Bd and $CaCO_3$ all curves are of a similar type. In the case of $CaCO_3$ the angular shape and the uniform grain size seem to lower the compaction rate in the later stages of compaction.

3. *Dependance of compaction on grain size* (Fig. 3). Experiments were run with different sieve fractions of sample Ad IV: Removal of the silt and clay fraction causes a somewhat higher initial porosity, but no fundamental difference (their share in the original sediment is rather low). The isolated coarse fraction shows intensive

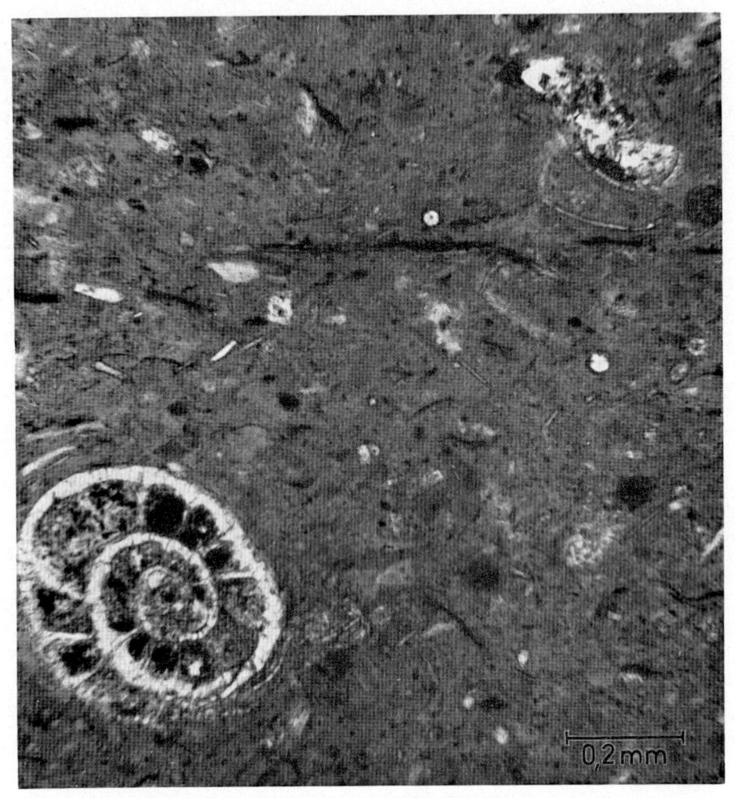

Fig. 8. Thin section of F II, compressed with 650 atm. Fine-grained carbonate mud with compacted organic substance and shells, one collapsed (above right) and another one, though partly hollow, nearly undeformed. Direction of stress: vertical.

compaction at low pressures and very low porosity at high pressure. The curve of the fraction 0.2 to 0.6, diluted 1:1 with $CaCO_3$ (silt) is nearly the same as that of pure calcite. Here, too, the fine fraction is responsible for the type of compaction.

4. *Compaction with different pore media* (Fig. 4, 5). Two samples were compacted with fresh and salt water, air, and benzene as an unpolar medium. Both the pure $CaCO_3$ and the clayey Hz preserve a much higher porosity when immerged into benzene, especially at lower pressures. On the other hand compaction is much quicker in benzene than in water. Supposedly in this case benzene lowers the friction between the grains and prevents much of the crushing of single grains. In the case of clayey material also electrochemical forces play an important role. With the

pores filled with air, friction seems to be elevated, which causes more crushing thus leading to rather low porosity at high pressure.

5. *Influence of temperature* (Fig. 6). Elevated temperatures of 90°C cause more intense compaction; the void ratio is about 10% lower than for runs at room temperature. In one experiment, where the temperature was increased stepwise, the void ratio approaches about the same value as in the run, where the temperature was elevated from the beginning.

Conclusions

Two micrographs show the structure of the compressed samples. It is not yet possible to state any connections to natural diagenetic compaction. Additional work has to be done concerning the extent of crushing in finer grained sediments and the influence of elevated temperature, which might compensate for the lack of time in laboratory experiments.

Acknowledgements

The author is indebted to Prof. Dr. G. Knetsch and Dr. D. H. Welte, Geological Institute of Würzburg University, who provided the impulse for this work and much helpful advice. The experiments were supported by the Deutsche Forschungsgemeinschaft. The author also thanks Dr. H. Genser, Freiburg, Dr. D. Meischner, Göttingen, Prof. Dr. G. Müller, Heidelberg, Prof. A. C. Neumann, Miami, and Prof. Dr. E. Seibold, Kiel, for supplying the samples.

References

Chilingar, G. V., and L. Knight: Relationship between pressure and moisture content of kaolinite, illite and montmorillonite clays. Bull. Am. Assoc. Petrol. Geologists **44**, 101—106 (1960).

Engelhardt, W. v.: Der Porenraum der Sedimente, 207 p. Berlin-Göttingen-Heidelberg: Springer 1960.

—, and K. H. Gaida: Concentration changes of pore solutions during the compaction of clay sediments. J. Sediment. Petrol. **33**, 919—920 (1963).

Fruth, L. S., Jr., G. R. Orme, and F. A. Donath: Experimental compaction effects in carbonate sediments. J. Sediment. Petrol. **36**, 747—754 (1966).

Hathaway, J. C., and E. C. Robertson: Microtexture of artificially consolidated aragonitic mud. U.S. Geol. Survey, Profess. Papers **424-C**, 301—304 (1961).

Parasnis, D. S.: The compaction of sediments and its bearing on some geophysical problems. Geophys. J. **3**, 1—28 (1960).

Robertson, E. C.: Laboratory consolidation of carbonate sediment. In: Marine Geotechnique, a symposium, 118—127 (Ed.: A. F. Richards). Urbana (Ill.) 1967.

—, L. R. Sykes, and M. Newall: Experimental consolidation of calcium carbonate sediment. U.S. Geol. Survey Profess. Papers **350**, 82—83 (1962).

Rodgers, J. J. W., and W. B. Head: Relationships between porosity, median size and sorting coefficients of synthetic sands. J. Sediment. Petrol. **31**, 467—470 (1961).

Terzaghi, K.: Erdbaumechanik. Wien 1925.

Terzaghi, R. D.: Compaction of lime mud as a cause of secondary structure. J. Sediment. Petrol. **10**, 78—90 (1940).

Weller, J. M.: Compaction of sediments. Bull. Am. Assoc. Petrol. Geologists **43**, 273—310 (1959).

Stylolitization in Carbonate Rocks

Won-choon Park and Erik H. Schot*

With 2 Figures

Abstract

Stylolite features are defined in terms of rock grain fabric relationships: intergranular and aggregate stylolites are differentiated. Geometric and genetic classifications of stylolites are offered. Megascopic and microscopic observations on the stylolites occuring in carbonate rocks of the Uppper Mississippian System in southern Illinois fluorspar mines and in the Lower Ordovician Jefferson City Formation near Rolla, Missouri are listed and their interpretation is briefly discussed.

Difficulties in explaining the origin of stylolite formation by means of "contraction-pressure" and "solution-pressure" theories are brought out. An attempt has been made to determine, within the present knowledge of carbonate petrology and diagenesis, when stylolites begin and cease to form by comparative means using the various diagenetic fabric relationships. Stylolitization in carbonate rocks usually begins during the deposition of "dog-tooth" type cement (radiaxial or early drusy mosaic stage) and ends concordantly with the virtually complete cementation of pore space by late drusy mosaic calcite; thus stylolites are of a diagenetic pre-complete cementation origin.

A. Introduction

Stylolites are recognized as irregular planes of discontinuity, along which two rock units (i.e., lithology and rock- or mineral grains) appear to be interlocked or mutually interpenetrating. These planes are usually, but not always, characterized by the accumulation of relatively insoluble residues or authigenic minerals along them, which form the stylolite seam. The composition of this residual material is in effect a function of the host rock. The following seam materials have been reported: clay minerals (Thomson, 1959), organic material (Ramsden, 1952), coal (Stockdale, 1945), silica (Brown, 1959; Park, 1962), pyrite (Park and Schot, 1968), sphalerite (Amstutz and Park, 1967), molybdenite (observation by W. C. P.), löllingite (personal communication, Dr. N. Markham), limonite (Park and Schot, 1968), fluorite (Park, 1962), carbonates (Brown, 1959; Freeman, 1965), phosphates, serpentine, V-U minerals, sericite and hydromicas (Bushinskiy, 1961).

Stylolites are most frequently found in limestones and dolomites (Wagner, 1913; Stockdale, 1922; a.o.), but have also been observed in marbles (Gordon, 1918; Bushinskiy, 1961; a.o.), sandstones and quartzites (Tarr, 1916; Sloss and Feray, 1948; Conybeare, 1949; Heald, 1955; a.o.), cherts (Hunt, 1863; Bastin, 1933; Trefethen, 1947; a.o.), novaculites (Park, 1962), sedimentary barite (Puchelt and Müller, 1963; Zimmermann, 1964; Gilluly and Gates, 1965), clay shales and siliceous slates (Bushinskiy, 1961), agates and petrified wood (Shaub, 1955), tectonic breccias (Herbert and Young, 1957), conglomerates (Bastin, 1940; Bushinskiy,

* Mineralog.-Petrographisches Institut, Universität Heidelberg, Germany.

1961), phosphorites (BUSHINSKIY, 1961), evaporites (HALL, 1843; MEYER, 1862; BUSHINSKIY, 1961), rhythmic magnesite layers (LLARENA, 1965), "coon-tail" fluorite layers (AMSTUTZ and PARK, 1967), porphyries (BLOSS, 1954; GOLDING and CONNOLLY, 1960), pegmatites and quartz lenses (BAILLY, 1954), hydrothermal molybdenite-quartz veins, gold-quartz veins and ribbon quartz bodies (MCKINSTRY and OHLE, 1949; CHACE, 1949), and asbestos veins (DU TOIT, 1948; MALES and GOLDING, 1961).

In this paper, discussion is restricted to stylolites in carbonate rocks only. At the present, two theories on the origin of stylolites are adhered to: the "solution-pressure" theory advocated by WAGNER (1913) and STOCKDALE (1922) and the "contraction-pressure" theory proposed by MARSH (1868) and SHAUB (1939). Certain difficulties are inherent in explaining the origin of stylolites by these two theories based on the present knowledge of carbonate petrology. The presence of relatively insoluble residues and of partly eliminated rock- and mineral grains in and along stylolite seams supports the origin of stylolites through solution processes, however.

The problem of stylolitization is not so much a matter of how stylolites originate, but more a matter of the time at which they were formed. With regard to the time of formation, two schools of thought exist; one proposes a post-diagenetic (*post-complete cementation) origin, whereas the other assumes a diagenetic (*pre-complete cementation) origin. The first school of thought includes the majority of the solution-pressure theory adherents, whereas the latter tends to be supported in part by advocates of the contraction-pressure theory. It is the opinion of the authors that a solution to the stylolite problem may be found through detailed paragenetic studies of fabric and chemical aspects of the host rock throughout its diagenetic history. Recently, AMSTUTZ and PARK (1967) pointed out the importance of stylolite features in interpreting the origin of certain Mississippi-Valley-type deposits. Stylolitization and related solution processes may play a rôle in explaining certain features of ore accumulation in stratiform deposits; it may also play an important róle in oil-migration and the subsequent development of oil-reservoirs (RAMSDEN, 1952; DUNNINGTON, 1967). More detailed information concerning the points brought forward in this paper as well as virtually all pertinent literature may be found in PARK (1962), SCHOT (1963), TRURNIT (1967), and PARK and SCHOT (1968).

B. Definition and Classification

I. Definition

Stylolite features can be defined in terms of their relationship to the fabric of the host rock; intergranular stylolites and aggregate stylolites may be differentiated.

1. *Intergranular stylolites*. Here the amplitude of the seam must be smaller than the grain size of the host rock. This term includes those features which in the literature

* Post-complete cementation and pre-complete cementation are used here in place of the former and rather ambiguous terms, post(pre)-lithification or -induration. Oolites, pellets or carbonate clasts held together by dog-tooth-type cement or radiaxial cement may retain a porosity of 30 to 40%; at this stage the rock may already be indurated. In the literature, lithification and induration in part usually signify virtually complete cementation. In carbonate rocks, those having a porosity of less than 12% may be regarded as completely cemented, and vice versa. Terms like post-induration (post-lithification) or pre-induration (pre-lithification) are ambiguous and must be eliminated not only from the discussion on the time of formation of stylolites, but elsewhere as well. These terms are not satisfactory unless one can specify quantitatively how much porosity could have been present.

have formerly been referred to as pressure-solution, concave-convex-sutured contacts, and pitted pebbles, oolites or rock- and mineral grains. With intergranular stylolites, (crystallographic) orientation (of minerals) and contact-relationship (amount of contact) between the two grains involved, relative solubility, stress-strain relationships and surface tension relationships between the two grains play a rôle in determining their geometry.

2. *Aggregate stylolites:* Any seam which departs from the bedding plane and with an amplitude greater than the width of the individual columns that make up the seam is here defined as an aggregate stylolite. "Residual seams or grooves" with amplitudes less than the width of their individual columns may also be included here, particularly so when in the outcrop they can be observed to converge into seams with amplitudes greater than the width of their individual columns, but then only when they can definitely be established as being solution seams. The geometric features of aggregate stylolites are dependent on the mutual fabric relationship, pressure and solution effects, porosity and permeability, rate and duration of the stylolitization, in addition to the crystallographic orientation and contact-relationship between the individual grains.

II. Geometric Classification

Stylolites may be classified geometrically in two ways: first by means of the two-dimensional geometry of the stylolite seam itself, and secondly by means of the congruency of the stylolite(s) in relation to the bedding plane. Fig. 1 shows the first

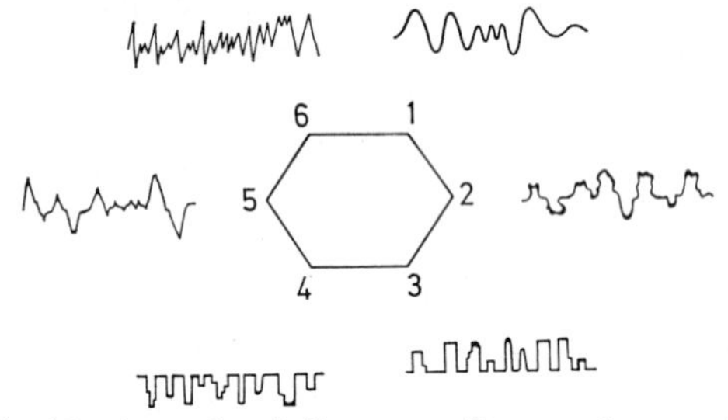

Fig. 1. Megascopic classification of stylolites according to the two-dimensional geometry of the stylolite seam itself. Amplitude and thickness of the seam are not expressed, because they can vary in a short distance. Gradational transitions between all of the different types are possible. Types: 1. Simple or primitive wave-like type; 2. Sutured type; 3. Up-peak type (Rectangular type); 4. Down-peak type (Rectangular type); 5. Sharp-peak type (tapered and pointed); 6. "Seismogram" type.

classification and Fig. 2 the latter. A detailed account of both classifications is given by PARK and SCHOT (1968). With reference to Fig. 1, it may be added here that type 1, 5 and 6 stylolites are most frequently observed in micron to decimicron-sized carbonates, whereas types 2, 3 and 4 were more frequently observed in bio-oosparitic carbonates. Type 1 stylolites frequently converge into "residual seams or grooves", which are usually congruent to the bedding. Highly inclined to vertical stylolites tend to be of types 5 and 6. Concerning Fig. 2, STOCKDALE (1922) also recognized different types of stylolites in relation to the bedding plane; inclined stylolites were accepted by him and many other investigators as evidence for the post-complete cementation origin of stylolites. This interpretation is not justified, because during diagenesis many directions of stress and strain may be operative.

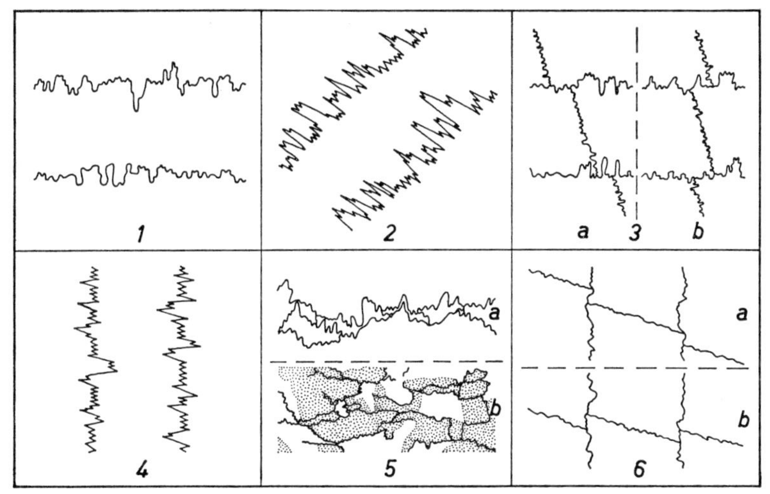

Fig. 2. Megascopic classification of stylolites in relation to the bedding. The bedding is here expressed by the horizontal center-line (bedding plane) of the figure. Scale and regional extend of the stylolite types shown are not expressed. Types: 1. Horizontal stylolites; 2. Inclined stylolites; 3. Horizontal-inclined (vertical) crosscutting stylolites; 4. Vertical stylolites; 5. Interconnected network stylolites; 6. Vertical-inclined (horizontal) crosscutting stylolites.

III. Genetic Classification

Two types of stylolites have been differentiated in terms of time of formation: diagenetic and tectonic stylolites. For the diagenetic stylolites an early diagenetic or late diagenetic origin may be differentiated by means of detailed studies of fabric changes within the host rock. Diagenetic stylolites may include virtually all the stylolites represented in Fig. 2. Many tectonic stylolites may already be recognized in the field and may include types 4, 5 and 6 of Fig. 2. It appears that tectonic stylolites are more the result of pressure-solution and recrystallization phenomena due to the physical parameters (pressure and temperature) involved than to chemical factors.

C. Observations on Stylolites

Detailed field observations were made on stylolites in carbonate rocks of the Upper Mississippian System in southern Illinois fluorite mines and in the Lower Ordovician Jefferson City Formation near Rolla, Missouri. A detailed account on the stylolites and the lithology of the host rocks are found in Park (1962), Schot (1963) and Park and Schot (ibid.).

I. Megascopic observations

A few of their characteristic megascopic features are listed below:

— Stylolites are largely of types 1, 2, 3, and 4 of Fig. 1 and type 1 of Fig. 2; types 2, 3 and 5a, Fig. 2 occur subordinately.

— Stylolite seams are essentially parallel to the bedding; cross-cutting stylolites are rare.

— Stylolites may converge into residual clay seams or grooves.

— Consecutive horizontal displacement of inclined stylolites by horizontal stylolites (Fig. 7, PARK and SCHOT, ibid.) occurs.

— Displacement of diagenetic calcite veinlets by stylolitization and slickolitization occurs.

— Stylolites are present in intraformational breccia clasts and not in the matrix; stylolitization preceded brecciation (Fig. 6, PARK and SCHOT, ibid.).

— Stylolite seams trend downward when approaching "cut-and-fill" channel walls, with the thickness of the seam material increasing and the amplitude decreasing in the same direction; stylolitization initiated prior to the formation of the "cut-and-fill" (Fig. 8 and 9, PARK and AMSTUTZ, 1968).

— Seam material is generally thicker at the crests and valleys of the stylolites than along the vertical stylolite walls (portions); solution was greatest at the points of protrusion.

— Stylolites are associated with quartz and pyrite-lined miarolitic-type or geodic cavities, quartz and pyrite-bearing concretions and occasionally open vugs or cavities; the frequent association of all four structures suggests a relatively early diagenetic origin for them (PARK and SCHOT, ibid.).

— Frequently ore minerals (sulphides) are enriched along stylolites and residual clay seams or grooves.

II. Microscopic Observations

Microscopic observations revealed the following characteristics:

— Seam material usually consists of relatively insoluble residues; the following minerals were observed within the seam: clay minerals, organic material, silica, dolomite, fluorite, sphalerite, pyrite and limonite.

— Clay minerals within the seam show a linear orientation congruent with the nearest side of the stylolite wall.

— Silica usually occurs as an insoluble residue along the inner portions of the stylolite fingers; quartz, cryptocrystalline and fibrous chalcedony, and opal have been observed. Features of directional growth of euhedral, doubly terminated quartz during stylolitization were observed (TRENNER's α-rule).

— Carbonates (dolomite, strontianite, witherite [alstonite]) may occasionally be found either as insoluble residues or as authigenic products due to stylolitization in the seams not only within carbonate rocks, but in those within sandstones and fluorite ore-rocks as well. Coarse, sparry, replacement rimmed calcite occasionally occurs at the inner portions of the stylolite fingers; this may indicate that grain growth sensu stricto may have played a rôle in, and be enhanced by stylolitization.

— Stylolites between fluorite ore-rocks and micrites indicate that the fluorite was deposited before stylolitization; here the micrites were relatively more soluble.

— Sphalerite and pyrite are more abundant within the stylolite seams than within the enclosing host rock; generally also the grain size of these sulfides tend to be larger and the shape more euhedral. This may reflect a local reducing environment along the stylolite seams.

— Limonite appears to be strictly a secondary oxidation product of the pyrite.

— Pressure-twinning of calcite, pressure shadows of sulphides, microfaulting and shearing were observed within and along the stylolite seams.

— Geopetal features within and cavity filling along stylolite seams have been observed (PARK and SCHOT, ibid.).

— When a micrite grain and a monocrystalline calcite grain occur pitted, the micrite is always dissolved and the monocrystalline calcite appears almost intact. This may be attributed to relative surface tension effects and to improved intergranular and possibly also intragranular capillarity of the micrite.

— Where two pitted grains are of identical composition and fabric, the smaller grain tends to pit the larger one as reported elsewhere (KAHLE, 1966; SCHIDLOWSKI and TRURNIT, 1966); this is caused by the higher surface tension per unit area of the smaller grain at the periphery of the contact.

— Pitting and pressure-solution between oolite grains as an initiation of the stylolitization occurred during or after the "dog-tooth" type cementation, but early within the period of late drusy mosaic calcite cementation.

— Stylolitization may not neccessarily be one continuous process, but may take place periodically at certain intervals. The expression of such periodic intervals of stylolitization (i.e., periodicity) during diagenesis may be observed in stylolites with a rather thick accumulation of seam material and relatively higher amplitudes. The stylolite seams are thus composed of curved sets of laminae of seam material (PARK and SCHOT, ibid.).

D. Discussion

As for the present-date interpretation on the formation of stylolites, two schools of thought exist. One is the "solution-pressure" theory advocated by WAGNER (1913) and STOCKDALE (1922), which states in brief that stylolites originate through differential solution and pressure along a fracture or mechanical plane after the hardening (i.e., after complete cementation) of the carbonate host rock. The other is the "contraction-pressure" theory advocated by MARSH (1868), ROTHPLETZ (1900) and SHAUB (1939) which proposes the formation of stylolites through the plastic flow of a thin clay layer over a lime mud when both are in a plastic state.

Many difficulties are inherent in both theories in explaining stylolitization in view of the present knowledge of carbonate rocks. The great difficulty with the "contraction-pressure" theory is that it fails to account for and explain the presence of a film of relatively insoluble residual material in the stylolite seams and for the pressure-solution features between rock and mineral grains observed nearby and along stylolite seams. It is these features per se, that leave little doubt that stylolites are the result of solution and not simple plastic flow due to overburden-pressure as postulated by advocates of the "contraction-pressure" theory. Difficulties with the "solution-pressure" theory include: the advocation of the presence of a jointed surface, fracture, or mechanical plane along which stylolitization could have initiated (many such surfaces or planes do not indicate solution processes!) and differential pressures on such surfaces or planes, the lack of microtectonic adjustment planes due to volume reduction during stylolite formation (these adjustment planes should continue into the host-rock), the occurrence of generally flat grain boundary surfaces between late drusy mosaic calcite cement in indurated carbonates (grains cemented in this manner, i.e., with relatively flat grain boundaries, would hardly allow the development of differential pressure and the subsequent formation of a stylolite seam), and the lack or scarcity of observations of stylolites in folded carbonate rocks in the field and in the literature.

When both explanations for the origin of stylolites were formulated during the late 18hundreds and early 19hundreds, the petrology and diagenetic aspects of carbonate rocks were virtually unknown and relatively little understood. Also, curiously enough, no extensive microscopic work was employed by students of the stylolite problem, except for a few, to explain the formation of stylolites.

Based on ample field, and thin-section observations, discussed already in the previous section, it is concluded that the majority of stylolite features in carbonate rocks are of a *diagenetic pre-complete cementation origin*. In the literature, stylolitization and pitting between two rock- or mineral grains have been interpreted as having resulted prior to complete cementation by late drusy calcite by Sorby (1908), Heald (1959), Kahle (1966), and Amstutz and Park (1967).

Mechanical treatment of stylolite formation in virtually completely cemented carbonate rocks fails to account for the volume reduction which accompanies stylolitization. From the literature and in the field through personal observation volume reductions of up to 35 to 42% having resulted from stylolitization have become known. Provided that such appreciable amounts of dissolved carbonate fill in and cement the still available pore space in virtually completely cemented rock units affected by stylolitization, certain features normally associated with the processes of rock removal, such as open spaces or vugs, brecciation, and tension fractures due to settling of the overlying strata, would have to be present there. The lack of faulting at the margins of stylolite seams (these should continue into the host-rock beyond the stylolite seam), which one normally would expect to find as the result of subsidence of the rock column accompanying the stylolitization — particularly along the shorter, discontinuous seams — in completely cemented rocks points against the time of formation of stylolites as being post-complete cementation (or using conventional terminology as being post-induration or post-lithification), and indicates rather that the pressure-solution and stylolitization occurred while the host rock was still relatively plastic and not completely cemented, in other words, before the complete elimination of pore space by late drusy mosaic calcite.

With respect to the process of cementation, it is thought that pressure-solution and stylolitization may act as one of the important factors in supplying the cement, particularly in carbonate rocks (Bathurst, 1958; Heald, 1959).

The most heatedly debated and still not completely solved problem surrounding the stylolite controversy is the time of formation, as to when stylolites begin and cease to form during the various and long processes of diagenesis. The time factor is most probably variable depending upon the manner in which the elimination of pore space accompanying the conversion of the host from a sediment to a rock proceeds. For example, the elimination of pore space in carbonates is thought to be largely the result of cementation — not compaction, in contrast to pelitic rocks where the reverse is thought to be true; in sandstones both factors appear to be equally important.

Stylolitization appears to be a long, drawn-out process, which usually begins before cementation by late drusy mosaic calcite and ends concordantly with the complete cementation of pore space by late drusy mosaic calcite.

Studies of various fabric and chemical changes throughout the diagenetic history of the host rock must be incorporated to gain an understanding of pressure-solution and stylolitization processes.

Acknowledgement

For acquainting us with the stylolite problem, for helpful suggestions and discussions throughout our study, and for critically reading the manuscript, we would like to thank Prof. G. C. Amstutz of the University of Heidelberg and Prof. A. C. Spreng of the University of Missouri at Rolla. Discussions with Drs. R. A. Zimmermann, F. El Baz and P. Trurnit also were stimulating.

References

Amstutz, G. C., and Won C. Park: Stylolites of diagenetic age and their role in the interpretation of the southern Illinois fluorspar deposits. Mineralium Deposita **2,** 44—53 (1967).
Bailly, P. A.: Présence de microstylolites dans des pegmatites et des lentilles de quartz. Bull. géol. France **3,** 299—301 (1954).
Bastin, E. S.: Relations of chert to stylolites at Carthage, Missouri. J. Geol. **41,** 371—381 (1933).
— A note on pressure stylolites. J. Geol. **48,** 214—216 (1940).
Bathurst, R. G. C.: Diagenetic fabrics in some British Dinantian Limestones. Liverpool Manchester Geol. J. **2,** 11—36 (1958).
Bloss, F. D.: Microstylolites in a rhyolite porphyry. J. Sediment. Petrol. **24,** 252—254 (1954).
Brown, C. W.: Diagenesis of late Cambrian oolitic limestone, Maurice Formation, Montana and Wyoming. J. Sediment. Petrol. **29,** 260—266 (1959).
Brown, W. W. M.: The origin of stylolites in the light of a petrofabric study. J. Sediment. Petrol. **29,** 254—259 (1959).
Bushinskiy, G. I.: Stylolites. Jzv. Akad. Nauk S.S.S.R., Ser. Geol. **8,** 31—46 (1961) (Engl. transl.).
Chace, F. M.: Origin of the Bendigo saddle reefs with comments on the formation of ribbon quartz. Econ. Geol. **44,** 561—597 (1949).
Conybeare, C. E. B.: Stylolites in Pre-Cambrian quartzite. J. Geol. **57,** 83—85 (1949).
Dunnington, H. V.: Aspects of diagenesis and shape change in stylolitic limestone reservoirs. Proc. VIIth World Petrol. Congress, Vol. II, no. 3, Mexico City, 339—352 (1967).
Du Toit, A. L.: Origin of the amphibole asbestos deposits of South Africa. Trans. Geol. Soc. S. Africa **48,** 161—206 (1945).
Freeman, T.: Post-lithification dolomite in the Joachim and Plattin Formations (Ordovician), northern Arkansas. Abstract, G. S. A. Program 1965 Ann. Meetings, p. 58 (1965).
Gilluly, J., and O. Gates: Tectonic and igneous geology of the northern Shoshone Range, Nevada. U. S. Geol. Survey Profess. Paper **465,** 153 p. (1965).
Golding, H. G., and J. R. Connoly: Stylolitic lava from Goobang Creek, New South Wales. Australian J. Sci. **23,** 129 (1960).
Gordon, C. H.: On the origin and nature of the stylolitic structure in Tennessee marble. J. Geol. **26,** 561—568 (1918).
Hall, J.: Geology of New York, Part IV, Comprising the Survey of the Fourth Geol. District, p. 95—131 (1843).
Heald, M. T.: Stylolites in sandstones. J. Geol. **63,** 101—114 (1955).
— Significance of stylolites in permeable sandstones. J. Sediment. Petrol. **29,** 251—253 (1959).
Herbert, P., Jr., and R. S. Young: Late stylolites. J. Geol. **65,** 107 (1957).
Hunt, T. S.: Geology of Canada. Geol. Survey Canada, Rept. Progr. 631—634 (1863).
Kahle, C. F.: Some observations on compaction and consolidation in ancient oölites. Compass **44,** 19—29 (1966).
Llarena, J. C. de: Aportaciones gráficas al estudio de la magnesita sedimentaria de Asturreta (Navarra). Estud. Geol. **20,** 315—337 (1965).
Males, P., and H. G. Golding: Stylolitic structures in asbestos veins. Australian J. Sci. **24,** 197—198 (1961).

MARSH, O. C.: On the origin of the so-called Lignites or Epsomites. Proc. Am. Ass. Adv. Sci. **16,** 135—143 (1868).
MCKINSTRY, H. E., and E. L. OHLE, JR.: Ribbon structures in gold-quartz veins. Econ. Geol. **44,** 87—109 (1949).
MEYER, H. VON: Mitteilung. Neues Jahrb. Mineral., p. 590 (1862).
PARK, WON C.: Stylolites and sedimentary structures in the Cave-In-Rock fluorspar district, southern Illinois. Thesis, Univ. of Mo. at Rolla, 264 p. (1962).
—, and G. C. AMSTUTZ: Primary "cut-and-fill" slump channels and gravitational diagenetic features within the CaF_2-ZnS ore horizons, Cave-In-Rock fluorspar district, Illinois. Mineralium Deposita **3**, 66—80 (1968).
—, and E. H. SCHOT: Stylolites: their nature and origin. J. Sediment. Petrol. **38**, 175—191 (1968).
PUCHELT, H., u. G. MÜLLER: Mineralogisch-geochemische Untersuchungen an Coelestobaryt mit sedimentärem Gefüge. In: AMSTUTZ, G. C. (ed.), Developments in Sedimentology, Vol. II, Sedimentology and Ore Genesis, p. 143—156. Amsterdam: Elsevier 1964.
RAMSDEN, R. M.: Stylolites and oil migration. Bull. Am. Ass. Petrol. Geol. **36**, 2185—2186 (1952).
ROTHPLETZ, A.: Über eigentümliche Deformationen jurassischer Ammoniten durch Drucksuturen und deren Beziehungen zu den Stylolithen. Sitz.-Ber. math.-physik. Kl. bayer. Akad. Wiss. München **30,** 3—32 (1900).
SCHIDLOWSKI, M., and P. TRURNIT: Drucklösungserscheinungen an Geröllpyriten aus den Witwatersrand-Konglomeraten. Ein Beitrag zur Frage des diagenetischen Verhaltens von Sulfiden. Schweiz. Mineral. Petrog. Mitt. **46,** 337—351 (1966).
SCHOT, E. H.: The diagenetic origin of the dolomites, chert, and pyrite in the Jefferson City Formation. Thesis, Univ. of Mo. at Rolla, 264 p. (1963).
SHAUB, B. M.: The origin of stylolites. J. Sediment. Petrol. **9,** 47—61 (1939).
— Do stylolites develop before or after the hardening of the enclosing rock? J. Sediment. Petrol. **19,** 26—36 (1949).
— Notes on the origin of some agates and their bearing on a stylolite seam in petrified wood. Am. J. Sci. **253,** 117—120 (1955).
SLOSS, L. L., and D. E. FERAY: Microstylolites in sandstone. J. Sediment. Petrol. **18,** 3—13 (1948).
SORBY, H. C.: The application of quantitative methods to the study of rocks. Quart. J. Geol. Soc. London, 171—233, especially 224—227 (1908).
STOCKDALE, P. B.: Stylolites: their nature and origin. Indiana University Studies, 1—97 (1922).
— Stylolites with films of coal. J. Geol. **53,** 133—136 (1945).
TARR, W. A.: Stylolites in quartzite. Science **43,** 819—820 (1916).
THOMSON, A.: Pressure solution and porosity. Soc. Econ. Paleontol. Mineral. Special Pub. No. 7, "Silica in Sediments", 92—110 (1959).
TREFETHEN, J. M.: Some features of the cherts in the vicinity of Columbia, Missouri. Am. J. Sci. **245,** 56—58 (1947).
TRURNIT, P.: Morphologie und Entstehung von Druck-Lösungserscheinungen während der Diagenese. Diss., Univ. of Heidelberg, 498 p. (1967).
WAGNER, G.: Stylolithen und Drucksuturen. Geol. Paläontol. Abhandl. **11,** 101—128 (1913).
ZIMMERMANN, R. A.: The origin of the bedded Arkansas barite deposits (with special reference to the genetic value of sedimentary features in the ore). Diss., Uni. of Mo. at Rolla, 367 p. (1964).

Analysis of Pressure-Solution Contacts and Classification of Pressure-Solution Phenomena

PETER TRURNIT*

With 7 Figures

Abstract

Pressure-solution contacts may be 1. plain or curved, 2. smooth or differentiated. The geometry of the contact surface depends a) on the relative pressure-solubility between the partners in contact along the direction of stress, and b) on the relationship of the radii of curvature of the partners in contact. While point 1. depends on a) and b), point 2. is influenced only by a). 14 pressure-solution contacts exist. They can be expressed by contact formulas. Pressure-solution phenomena are classified on a geometric and genetic basis. Partings in carbonate rocks and quartzites in many cases consist of smooth pressure-solution surfaces.

A. Introduction

Pressure-solution phenomena have been referred to as pitted pebbles, indentations or impressions in pebbles and sandgrains, interlocking grains, microstylolites between pebbles and sandgrains, stylolites and sutures. Detailed studies concerning sutures and stylolites have been published by WAGNER (1913) and STOCKDALE (1922). New studies by BUSHINSKIY (1961), MANTEN (1966) and DUNNINGTON (1967) are of notable mention. SHAUB (1939, 1949) was one of the last ones to defend the theory of differential compaction in explaining the genesis of stylolite surfaces; this theory was first proposed by QUENSTEDT [1837 (1, 2)]. Pitted pebbles were discussed in detail by KUMM (1919), KUENEN (1942/43) and MORAWIETZ (1958). Interlocking quartz grains, microstylolites in sandstones and impressions in sandgrains were treated by WALDSCHMIDT (1941), TAYLOR (1950), GAITHER (1953), HEALD (1955), CAROZZI (1960), VON ENGELHARDT (1960) and SCHIDLOWSKI and TRURNIT (1966). The genetic connection between the black seam of certain partings and the pressure-solution residue along stylolite surfaces was discussed by SEMPER (1916), SCHOO (1922), WEPFER (1926), YOUNG (1945) and others. A complete discussion concerning the whole problem of pressure-solution during diagenesis with 700 references was given by TRURNIT (1967).

B. The 14 Pressure-Solution Contacts

Pressure-solution phenomena are composed of two partners (beds, layers, pebbles, sandgrains, fossils) and a contact surface (pressure-solution plane) along which solution residue may be concentrated (Fig. 1 A).

The contact surface may be 1. plain or curved (Fig. 1, B 1), 2. smooth or differentiated (Fig. 1, B 2).

* Mineralogisch-Petrographisches Institut, Universität Heidelberg, Germany.
or: RST Technical Services, Ltd., Kalulushi, Rep. Zambia, Africa.

The partners may have a) equivalent or different relative pressure-solubility along the direction of stress (Fig. 1, Ba), b) equivalent or different radii of curvature at the contact (Fig. 1, Bb). The geometry of the contact surface depends on the properties of the partners in contact. While point 1. is influenced by a) and b), point 2. is determined only by a) (Fig. 1 B). Equivalent relative pressure-solubility of the partners in contact along the direction of stress as well as equivalent radii of curvature of the partners at the contact result in a plain contact surface. Different relative pressure-solubility in the direction of stress and different radii of curvature of the partners at the contact result in a curved contact surface. Equivalent relative pressure-solubility of the partners along the direction of stress results in a differentiated (sutured, stylolitic) contact surface; different relative pressure-solubility results in a smooth contact surface. The relationship between the crystal-, or grain size of the partners also plays a rôle; difference

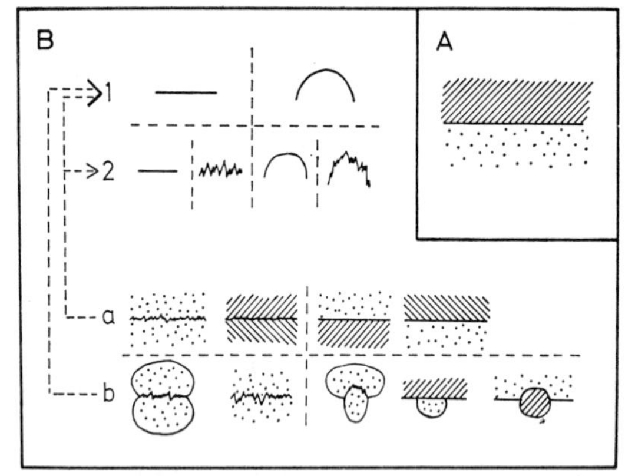

Fig. 1. A) Pressure-solution contacts are composed of a contact surface and two partners; B) the contact surface may be plain or curved (1), smooth or differentiated (2); the partners may have equivalent or different pressure-solubility along the direction of stress (a), equivalent or different radii of curvature at the contact (b); while point one is influenced by a) and b), point 2 is only determined by a)

in grains size of the partners as well as a large accumulation of solution residue prevent the differentiation of the contact surface.

A series of minerals with decreasing relative pressure-solubility was formulated by HEALD (1955), MORAWIETZ (1958), BUSHINSKIY (1961, 40), SCHIDLOWSKI and TRURNIT (1966) and TRURNIT (1967) through the study of a) the minerals found in the solution residue of pressure-solution planes, b) the heads of stylolites, c) the impressions in pebbles and sand grains, d) smooth and differentiated pressure-solution surfaces, and e) the tendency of minerals in contact and of equivalent pressure-solubility to be affected by stylolitization. This series is as follows: 1. halite and potassium salts, 2. calcite, 3. dolomite, 4. anhydrite, 5. gypsum, 6. amphibole and pyroxene, 7. chert, 8. quartzite, 9. quartz, glauconite, rutile, hematite, 10. feldspars, cassiterite, 11. mica, clay minerals, 12. arsenopyrite, 13. tourmaline, sphene, 14. pyrite, 15. zircon, 16. chromite. All minerals in the series which come after calcite can become a part of the solution residue along a pressure-solution surface in limestone or the head of limestones stylolites; all minerals which follow after quartz can become a part of the solution residue of a pressure-solution surface in a sandstone (quartzite) or the head of a stylolite of quartz. The first minerals in the series can be pitted, forming a smooth contact plane, by those minerals which occur further back in the series. Partners in an equivalent position in the series — that means partners of equivalent pressure-solubility — can develop differentiated pressure-solution contacts. Equivalent pressure-solubility very often, but not always, means the same composition. In a certain way, the positions in the series are interchangeable. In rarely reported cases, quartz and chert may have sutured or stylolitic contacts with calcite or dolomite; the same was observed between quartz and feldspar. The tendency to be affected by stylolitization decreases from 1. to 16. between partners of equivalent pressure-solubility.

Pressure-solution contacts were discussed or analysed by TAYLOR (1950), GAITHER (1953), LOWRY (1956), MORAWIETZ (1958), SIEVER (1959), VON ENGELHARDT (1960), SCHIDLOWSKI and TRURNIT (1966) and TRURNIT (1967).

Fig. 2 shows the 14 pressure-solution contacts, which were derived utilizing the above mentioned considerations. The first division is based on the relative pressure-solubility between the partners along the direction of stress, the subdivision is based on the relationship of the radii of curvature at the contact. The smooth contact-planes 5. to 8. develop with equivalent relative pressure-solubility at the start of the pressure-solution activity and at an advanced stage if much solution residue has been concentrated. A thick layer of solution residue behaves as an insoluble partner; different grain size of the partners inhibits differentiation.

With equivalent pressure-solubilities of the partners along the direction of stress, one can distinguish 8, with different pressure-solubilities 6, and altogether 14 pressure-solution contacts. The contact formulas are explained sufficiently in Fig. 2; the 14 pressure-solution contact types can be expressed by the numbers 1 through 14, in the form of a drawing or a formula.

C. Classification

Classifications of pressure-solution phenomena by BUSHINSKIY (1961, Fig. 1), PARK (1962, Fig. 38), AMSTUTZ and PARK (1964, 215; 1967, 46, Fig. 2) and PARK and SCHOT (in print: Fig. 1, 2) were confined to stylolite seams and surfaces. The classification, which is offered here (Fig. 4), assumes a pressure-solutive genesis for the mentioned phenomena. The development of a differentiated pressure-solution surface between partners of equivalent or similar pressure-solubility starts from a smooth plane. Certain points of each partner will be particularly resistant, others particularly soluble. By connecting the resistant and the soluble points of the partners, one obtains stylolites in an advanced stage of the pressure-solution activity (Fig. 3 A). The stylolites develop from flat to steep cones (Fig. 3 B). The base remains constant, the height increases parallel to the stress, the sides become steeper and strive to attain a parallel position, the angle between the mantle of the cone and the axis of the cone becomes smaller. Cone-shaped stylolites are stylolites of class 1, 1st order (Fig. 3 B, 1.1). The stylolites also may be affected by solution. The solution acts upon the sides of the cones. Solution residue can be, but is not necessarily concentrated; without solution residue, cone shaped stylolites of class 1, 1st order are preserved. When much solution residue is accumulated, the top of the cone will be cut by solution; obtuse cones arise (DUNNINGTON, 1967, Fig. 1 and 2) where solution acts in addition both above and below the forehead. Obtuse cones can develop to columnar-shaped stylo-

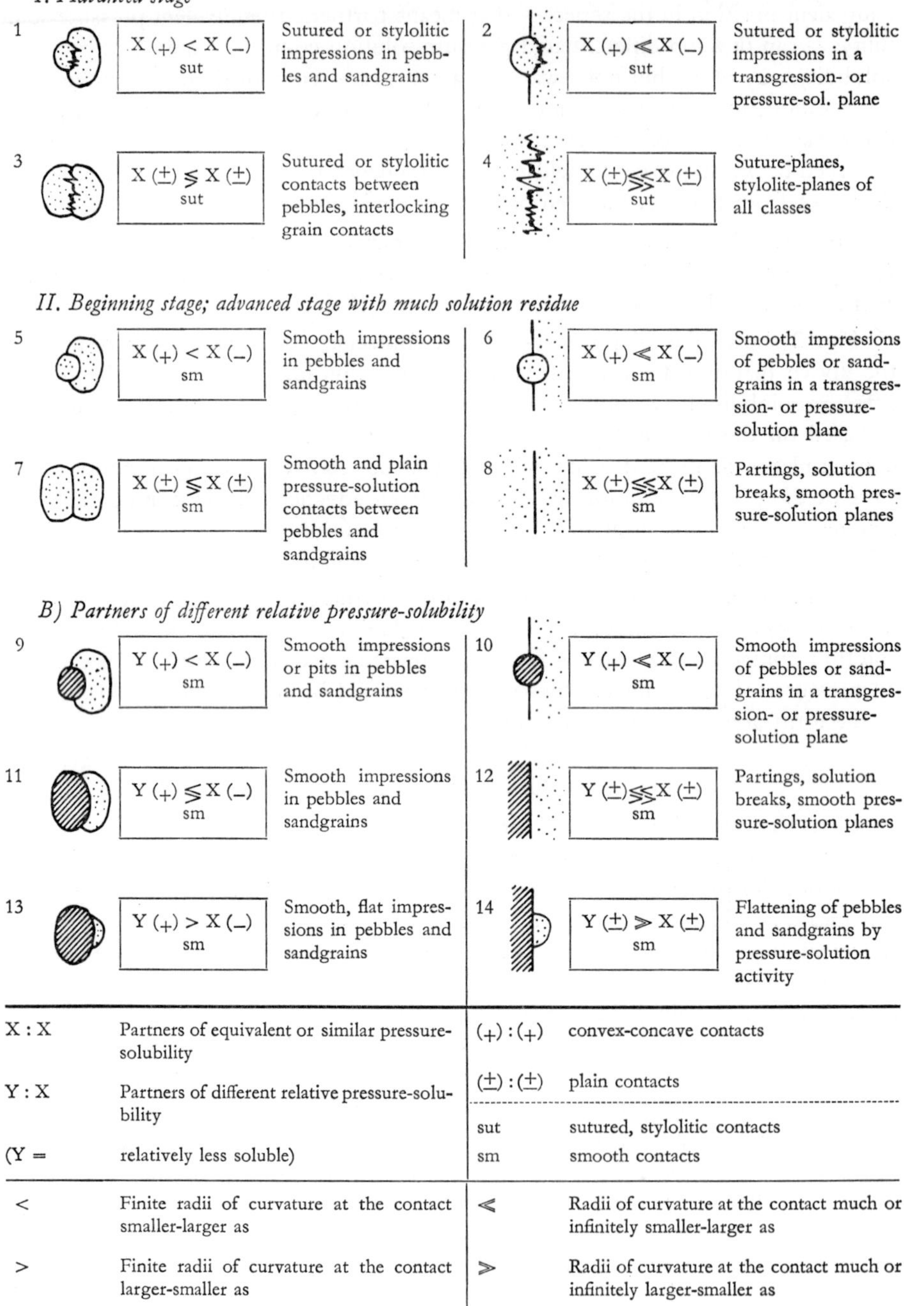

Fig. 2. The 14 pressure-solution contacts with contact formulas

lites with almost parallel sides, where solution acts only below and above the foreheads. Obtuse cones and columns are stylolites of class 1, 2nd order (Fig. 3, B 1.2).

A particle, for instance a quartz grain or a fossil in limestone, which is relatively less soluble in comparison to the surrounding host rock, can react in two ways, as soon as it arrives at a pressure-solution surface by solutive activity along that plane: 1. the solvent penetrates along the boundary between the limestone and the particle — the particle becomes a part of the solution residue, 2. the solvent does not penetrate along the boundary — the particle becomes the head of a columnar-shaped stylolite with parallel sides, a stylolite of class 2 (Fig. 3 C, a—d, 2). That stylolite is composed of a socket, a trunk, a mantle (two sides), a head, a smooth forehead and a cap consisting of solution residue.

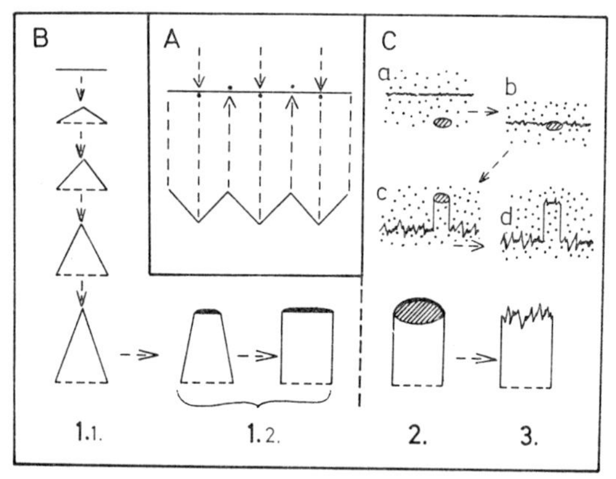

Fig. 3. A) The development of a differentiated pressure-solution surface by connection of the most resistant or the most soluble points of the partners; B) The development of cone-shaped stylolites of class 1, 1st order and their metamorphosis to obtuse cones and columnar shaped stylolites of class 1, 2nd order with the concentration of a large amount of solution residue; C) The development of columnar-shaped stylolites of class 2 with a head and the origin of columnar-shaped stylolites of class 3 after the solution of the head

If the head is not absolutely but only relatively less soluble, partners of equivalent relative pressure-solubility will be in contact with each other, after the complete solution of the head. The forehead of the stylolite, which was smooth until that time, can become differentiated, if solution residue has not accumulated above a certain amount. A stylolite of class 3, with a differentiated forehead made up of stylolites of one of the classes of a smaller order originates in this manner (Fig. 3, C 3). If too much solution residue has been concentrated, a stylolite of class 1, 2nd order with a smooth forehead originates. Stylolites of class 1, 1st and 2nd order, and of class 2 and 3 are the basic elements of pressure-solution phenomena.

Smooth impressions in pebbles and sandgrains represent the molds of stylolites of class 2 (casts), which are only composed of a head (Fig. 5, A 1 to 3). Smooth impressions develop only with partners of different pressure-solubility.

Differentiated impressions in pebbles and sandgrains or sutured stylolitic and microstylolitic contacts between pebbles and sandgrains originate if the partners are of equivalent pressure-solubility; they can be compared to and thought of in terms of the foreheads from stylolites of class 3 (Fig. 5, B 1 to 3).

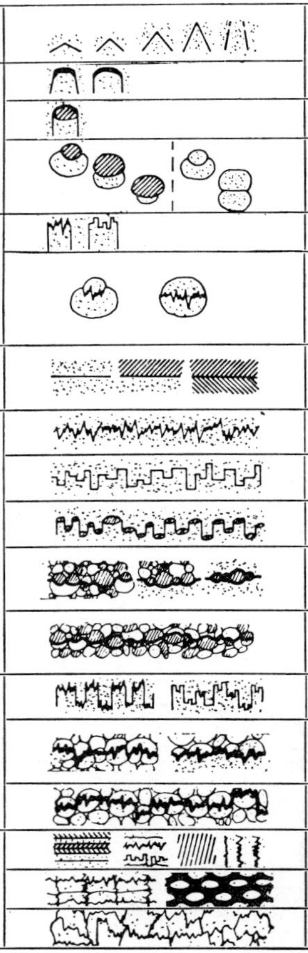

Activity	Planes	Surface type	Phenomenon	Subtype	Development	Head developed	Head	Mold/cast	
Pointed and linearly „advancing" pressure-solution			Stylolites of class 1	1st order	development from flat to steep cones			cast	transitions
Plain pressure-solution activity	Limited planes	Smooth, plain or curved pressure-solution surfaces	Stylolites of class 1	2nd order	development from obtuse cones to pillars		insoluble head	cast	transitions
			Stylolites of class 2		a pillar exists under the head from the beginning	only the head developed	insoluble head	mold	transitions
			Smooth impressions	on pebbles	a pillar exists under the head from the beginning	only the head developed	insoluble head	mold	transitions
			Smooth impressions	on sandgrains	a pillar exists under the head from the beginning	only the head developed	insoluble head	mold	transitions
		Differentiated, plain or curved pressure-solution surfaces	Stylolites of class 3		a pillar exists under the head from the beginning		soluble head	cast	transitions
			Differentiated impressions	on pebbles	a pillar exists under the head from the beginning	only head developed	soluble head	mold	transitions
			Differentiated impressions	on sandgrains	a pillar exists under the head from the beginning	only head developed	soluble head	mold	transitions
			Differentiated impressions	interlocking grains	only forehead without pillar developed		soluble head	mold	transitions
	Unlimited planes	smooth pressure-solution surfaces	Partings, solution breaks					mold and cast equivalent	transitions
		Differentiated pressure solution surfaces	Stylolite planes of class 1, 1st order, Suture planes					mold and cast equivalent	transitions
			Stylolite planes of class 1, 2nd order					mold and cast equivalent	transitions
			Stylolite planes of class 2					mold and cast equivalent	transitions
			Connected planes of smooth impressions in conglomerates; smooth pits of pebbles in a transgression-plane or pressure-solution plane					mold and cast equivalent	transitions
			Connected planes of smooth impressions in sandstones; pits of sandgrains in a pressure-solution plane					mold and cast equivalent	transitions
		higher order differentiated pressure-solution surfaces	Stylolite planes of class 3					mold and cast equivalent	transitions
			Connected planes of differentiated (sutured, stylolitic) impressions in conglomerates; differential impressions of pebbles in a transgression plane					mold and cast equivalent	transitions
			Connected planes of differentiated impressions in sandstones; suture planes					mold and cast equivalent	transitions
Pressure-solution in space by the combination of pressure-solution surfaces			Planeparallel fabric and structure						
			Interconnecting fabric and structure						
			Diffuse network fabric						

Fig. 4. Classification of pressure-solution phenomena

By planar lining up of stylolites or of their foreheads, unlimited, extended pressure-solution surfaces may result. Extended pressure-solution surfaces develop approximately normal to the pressure of the overlying sediments and approximately parallel to the bedding along preexisting channelways for the solution. Pressure-solution activity starts along limited contacts between components which already have been consolidated before or after sedimentation. With advancing pressure-solution activity, the limited pressure-solution contacts are connected either through the cement simultaneously with the cementation of the pore space (MORAWIETZ, 1958; TRURNIT, 1967) or the connection may also be attained by flattening of the detrital grains, the transfer of the segments to the center of the circle and the simultaneous lateral

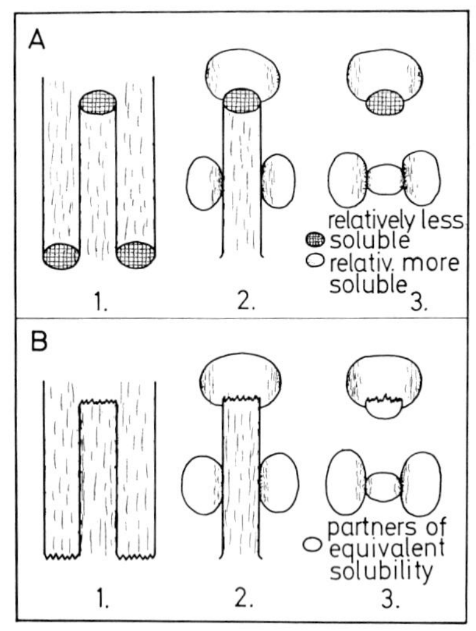

Fig. 5. The genetic relation between stylolites and impressions in pebbles and sandgrains (pitted pebbles); A) Smooth foreheads of stylolites of class 2 with a head are identical to the casts of smooth impressions (molds) in pebbles; they develop between partners of different pressure-solubility; B) Differentiated foreheads of stylolites of class 3 are identical with the casts of differentiated impressions (molds) in pebbles and with interlocking grain-contacts; they develop between partners of equivalent or similar pressure-solubility

approach of the limited contacts (Fig. 6 A, 1 to 3). Vertical pressure-solution planes (with horizontal stylolites) are established approximately normal to the tectonic stress along structural planes; they are extended from the beginning.

Smooth, unlimited pressure-solution planes may represent the first stage of differentiated surfaces in a relative homogeneous sedimentary rock. In most cases, however, they develop in the manner sketched in Fig. 7 through the successions of 1a, 1b and 2. Stylolites from a differentiated surface are removed from the forehead from one or both sides when the forehead meets a relatively less soluble bed. The insoluble bed may be a solution residue or a primary sedimentary layer that has undergone various diagenetic changes. The stylolites from one side will all have the same altitude. Type 3 and 4 ("up-peak and down-peak type = rectangular type") by PARK (1962, Fig. 38), AMSTUTZ and PARK (1964, 215; 1967, 46, Fig. 2) and PARK and SCHOT (in print: Fig. 1) originate in that manner. The meeting of the forehead of a stylolite with a relatively insoluble bed will in many cases take place as penetration of stylolites in a bed which is less soluble from mm to mm. At the beginning of the pressure-solution activity along an unlimited, extended pressure-solution surface, partners of equivalent pressure-solubility are in contact with each other. With advancing pressure-solution activity the composition and the relative pressure-solubility of the partners may change because of the layered inhomogeneity of a sediment in the vertical direction and the mutual approach of layers which originally were further apart. Because of the rule, that partners of different pressure-solubility have smooth contacts, the "growth" of the stylolites will decrease gradually and will lead

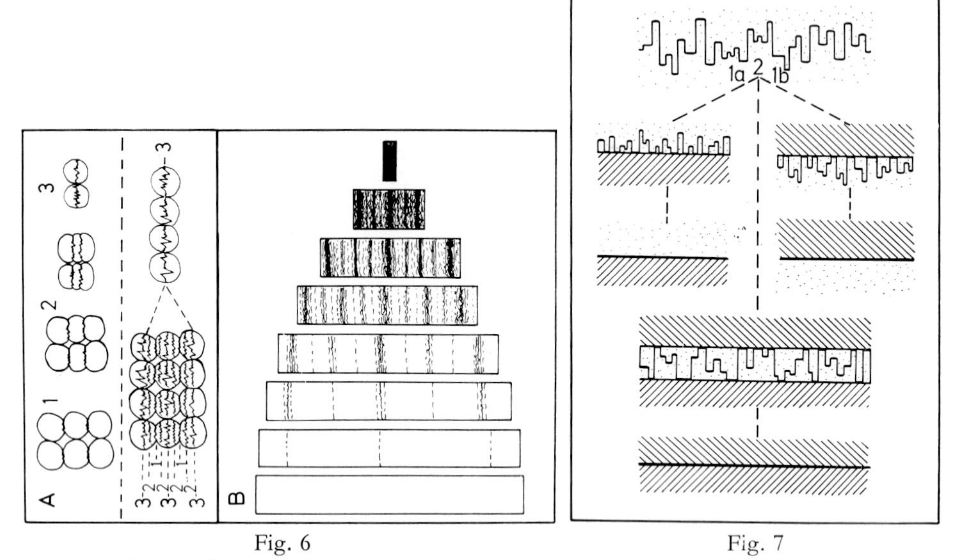

Fig. 6 Fig. 7

Fig. 6. A) Development of extended, unlimited pressure-solution surfaces by the flattening of the grains and the transfer of limited pressure-solution contacts from the border of the detrital grains to the center of the grains; B) The fusion of plane-parallel pressure-solution surfaces of a zone by solution of the soluble beds in between; origin of residual beds

Fig. 7. Origin of smooth partings (extended pressure-solution contacts) by the removal of the stylolites; 1a and 1b: Development of smooth residual layers between a relatively more and a relatively less soluble partner; 2: Development of a smooth residual layer between two relatively insoluble beds by the solution of the partner in between

from a stagnation of the "growth" to the removal of the stylolites from the foreheads. The concentration of a large amount of solution residue leads to the same result. Extreme differences in composition from one bed to another in older sedimentary rocks are not necessarily primary, but could have been developed gradually by diagenetic changes, where pressure-solution and transport of the dissolved material along pressure-solution planes plays a great rôle. The solution of complete beds between pressure-solution planes and the formation of residual beds (STOCKDALE, 1930) by the fusion of thin plane-parallel residual layers of a pressure-solution zone may create extreme contrasts. Extended, differentiated pressure-solution surfaces will appear in large quantities before the smooth, unlimited solution surfaces and the residual beds with smooth contacts to the adjoining soluble beds above and below. In a later stage, the smooth pressure-solution planes (partings) and residual beds will increase in number at the expense of the differentiated pressure-solution planes by the fusion of residual layers of differentiated pressure-solution surfaces, increase of solution residue along a single plane and the removal of the stylolites. Limited pressure-solution surfaces between single grains or particles, extended unlimited stylolite surfaces and extended unlimited smooth pressure-solution planes or residual beds can be found at the same time over a short distance of a stratigraphic succession.

Pressure-solution activity along the bedding starts from limited grain contacts, leads to extended, little differentiated and later on to unlimited stylolite surfaces, which develop into smooth pressure-solution partings and residual beds and end with residual rocks of secondary origin by solution of the relatively more soluble components. Partings and solution breaks remain at the final stage of development, caused by pressure-solution activity parallel to the bedding; at the very end of a similar developmental succession a certain type of cleavage (PLESSMANN, 1964; 1966) remains. Displacement is caused by solution and not gliding.

Pressure-solution zones parallel to the bedding, and others vertical to structural pressure may be interconnected. A special case is presented in connection with the genesis of the "Kramenzelkalke" (TRURNIT, 1967), which can not be discussed here in detail. Diffuse pressure-solution fabrics can be developed by changing directions of pressure and in between the fragments which make up tectonic breccias.

Acknowledgements

The author wishes to thank Prof. Dr. G. C. AMSTUTZ and Prof. Dr. G. MÜLLER for their interest in this work and Dipl. Geol. E. SCHOT for his assistance in translating the text.

References

AMSTUTZ, G. C., and W. C. PARK: New Mineralogic Observations of Stylolite Seams, Abstracts of A.A.P.—G.A.C.—M.A.C. Papers, 515—516, 18.—21. May 1964, Toronto (Kanada) 1964.

— — Stylolites of Diagenetic Age and their Role in the Interpretation of the Southern Illinois Fluorspar Deposits. Mineralium Deposita **2,** 44—53 (1967).

BUSHINSKIY, G. I.: Stylolites. Izvest. Akad. Nauk. S.S.S.R., Ser. Geol. (serija geologiceskaja) 31—46 (engl. transl.) (1961).

CAROZZI, A. V.: Microscopic Sedimentary Petrography, 1st ed. 485 p. New York and London: John Wiley & Sons 1960.

DUNNINGTON, H. V.: Aspects of Diagenesis and Shape Change in Stylolitic Limestone Reservoirs. Proc. VII World Petrol. Congr. Mexico, **2,** Panel Discuss. No. 3 (1967).

ENGELHARDT, W. VON: Der Porenraum der Sedimente, 207 p. Berlin-Göttingen-Heidelberg: Springer 1960.

GAITHER, A.: A Study of Porosity and Grain Relationships in Experimental Sands. J. Sediment. Petrol. **23,** 180—195 (1953).

HEALD, M. T.: Stylolites in Sandstones. J. Geol. **63,** 101—114 (1955).

KUENEN, P. H.: Pitted Pebbles. Leidsche Geol. Mededeel. **13,** 189—201 (1942/43).

KUMM, A.: Die Entstehung der Eindrücke in Geröllen. Geol. Rundschau **10,** 183—233 (1919).

LOWRY, W. D.: Factors in Loss of Porosity by Quartzose Sandstones of Virginia. Bull. Am. Assoc. Petrol. Geologists **40,** 489—500 (1956).

MANTEN, A. A.: Note on the Formation of Stylolites. Geol. en Mijnbouw **45,** 269—274 (1966).

MORAWIETZ, F. H.: Die Anlösungserscheinungen in der Juranagelfluh und ihre Bedeutung für die Diagenese, 132 p. PhD.-Thesis, Tübingen, Germany, 1958.

PARK, W. C.: Stylolites and Sedimentary Structures in the Cave-in-Rock Fluorspar District/ Southern Illinois. M. Sc.-Thesis, Rolla, Missouri, 1962.

—, and E. H. SCHOT: Stylolites: Their Nature and Origin. J. Sediment. Petrol. (In print).

Plessmann, W.: Gesteinslösung, ein Hauptfaktor beim Schieferungsprozeß. Geol. Mitt. **4** 69—82 (1964).
— Lösung, Verformung, Transport und Gefüge (Beiträge zur Gesteinsverformung im nordöstlichen Rheinischen Schiefergebirge). Z. deut. Geol. Ges. **115,** 650—663 (1966).
Quenstedt, F. A.: (1) Die Stylolithen sind anorganische Absonderungen. Wiegmann's Archiv Naturg. **3,** 137—142 (1837).
— (2) Auszüge. Neues Jahrb. Mineral. 496—497 (1837).
Schidlowski, M., u. P. Trurnit: Druck-Lösungserscheinungen an Geröllpyriten aus den Witwatersrand-Konglomeraten. Ein Beitrag zur Frage des diagenetischen Verhaltens von Sulfiden. Schweiz. min. petrog. Mitt. **46,** 337—351 (1966).
Schoo, J. H.: Zur Diagenese der alpinen Kreide, 92 p. Koblenz: Breuer (PhD.-Thesis, Zürich) 1922.
Semper, M.: Schichtung und Bankung. Geol. Rundschau **7,** 53—56 (1916).
Shaub, B. M.: The Origin of Stylolites. J. Sediment. Petrol. **9,** 47—61 (1939).
— Do Stylolites Develop before or after Hardening of the Enclosing Rock? J. Sediment. Petrol. **19,** 26—36 (1949).
Siever, R.: Petrology and Geochemistry of Silica Cementation in Some Pennsylvanian Sandstones. Soc. Econ. Palaeontol. Mineral., Spec. Publ. No. 7 "Silica in Sediments", 55—79 (1959).
Stockdale, P. B.: Stylolites: Their Nature and Origin. Indiana Univ. Studies **9,** 55, 1—97 (1922).
— Intraformational Solution of the Floyds Knob Limestone. Proc. Indiana Acad. Sci. **39,** 213—220 (1930).
Taylor, J. M.: Pore-Space Reduction in Sandstones. Bull. Am. Assoc. Petrol. Geologists **34,** 701—716 (1950).
Trurnit, P.: Morphologie und Entstehung von Druck-Lösungserscheinungen während der Diagenese, 498 p., PhD.-Thesis, Heidelberg, Germany, 1967.
Wagner, G.: Stylolithen und Drucksuturen. Geol. Palaeontol. Abhandl. N. F. **11** (15), 1—30 (1913).
Waldschmidt, W. A.: Cementing Materials in Sandstones and their Probable Influence on Migration and Accumulation of Oil and Gas. Bull. Am. Assoc. Petrol. Geologists **25,** 1839—1879 (1941).
Wepfer, E.: Die Auslaugungs-Diagenese, ihre Wirkung auf Gestein und Fossilinhalt. Neues Jahrb. Mineral. **54,** Beil.-Bd. B, 17—94 (1926).
Young, R. B.: Stylolitic Solution in Witwatersrand Quartzites. Trans. Proc. Geol. Soc. S. Africa **47,** 137—141 (1945).

Review on Electron Microscope Studies of Limestones

Erik Flügel, Helmut E. Franz, and Wilhelm F. Ott*

With 9 Figures

Abstract

Electron microscope studies of limestones are reviewed based on own studies and on the evaluation of published data. Further investigations should draw more attention to the possibility of recognition of diagenetic processes.

Introduction

Electron microscope studies have been used for the investigation of limestones for about 10 years. Most of the authors deal with ancient limestones (see Table 1). Major contributions have been made by Shoji and Folk (1964), Harvey (1966), and E. Flügel (1967). Recent carbonates were mostly described in respect to the origin of aragonite needles in shallow marine environments (e.g. Cloud, 1962; Studer, 1963; Illing et al., 1965). Electron micrographs of artificially consolidated carbonate muds have been published only by Hathaway and Robertson (1961). The submicroscopic features of calcite were observed by Pfefferkorn (1952) and d'Albissin (1963).

Most of the studies of Ancient limestones are concerned with the groundmass especially with micrite. Components, e.g. ooids and pellets, have scarcely been described. The principal characteristics presently used to typify micrites are size, contacts, surfaces of grains, and fossils (see Table 1). Nearly all of the work has been done in a more descriptive way only without significant conclusions as to the environment and origin of sediments.

The aim of the present paper is to clarify whether or not the different textural fabrics can be typified in a more exact way in order to find new tools for the interpretation of depositional environment and diagenesis of limestones. For this purpose a critical review of the published electron micrographs has been made, followed by the interpretation of own material.

The samples studied came from different formations and environments, especially from the Mesozoic of the Northern Alps (see Table 2).

Technique

The preparation of limestone samples for electron microscope studies requires two main steps: (1) pre-shadowing phase: The samples are either fractured only [Shoji and Folk, 1964; E. Flügel and Franz, 1967 (2)] or cut, ground, and etched (Gregoire and Monty, 1963; Honjo and Fischer, 1965; Harvey, 1966). (2) Replica and shadowing (Two-Stage-method): Fractured samples are shadowed directly [E. Flügel and Franz, 1967 (2)] or after inserting a replica (Shoji and Folk, 1964).

* Geologisch-Paläontologisches Institut, Technische Hochschule Darmstadt, Germany

From all cut samples a replica was made also before shadowing (e.g., Garrison, 1967; Frost, 1967).

Most authors use replicas made of plastics, while Kahle and Turner (1964) draw attention to faxfilm material. Materials used for shadowing are predominately C, Pt, Pd, Cr, Au, or Al.

Table 1. *Main characteristics used for typifying limestones in electron micrographs*

Authors	Grain-size	Grain-contacts	Grain-surfaces	Roundness	Grain-deformation	Pores	Fossils	Ooids Pellets
Kabelac, 1955	×	×	—	—	—	—	—	—
Seeliger, 1956	×	×	—	—	—	×	—	—
Grunau and Studer, 1956	—	—	—	—	—	—	×	—
d'Albissin, 1959	—	×	×	—	×	—	—	—
Grunau, 1959	—	—	—	—	—	—	×	—
d'Albissin, 1961, 1962	×	×	×	—	—	—	—	—
Grégoire and Monty, 1963	×	×	×	—	—	×	×	—
Kahle, 1964	×	—	—	—	—	—	—	×
Kahle and Turner, 1964	×	—	—	—	—	—	—	×
Shoji, 1964 } Shoji and Folk, 1964 }	×	×	×	×	—	—	×	×
Temmler, 1964	×	—	—	—	—	—	—	—
Honjo and Fischer, 1964, 1965	—	—	—	—	—	—	×	—
Folk, 1965	×	×	—	×	—	—	—	—
Teichert, 1965	×	×	×	—	—	—	—	—
Garrison, 1966 } Garrison and Bailey, 1967 }	×	—	—	—	—	×	×	—
E. Flügel, 1967	×	×	×	—	—	—	×	—
H. Flügel and Fenninger, 1966	×	×	—	—	—	—	×	—
Harvey, 1966	×	×	×	×	—	—	×	—
Garrison, 1967	×	×	—	—	—	×	×	—
Fischer and Garrison, 1967	×	—	—	—	—	×	×	—
E. Flügel and Franz, 1967 (1, 2)	×	×	×	—	—	—	×	—
Minato et al., 1967	×	×	—	—	—	—	×	×
Wolf et al., 1967	×	—	—	—	—	—	×	—

Different pre-shadowing phases lead to various results in the recognition of grain shape, grain surfaces, and fossil contents. Fracturing of samples seems to be preferable for investigations of texture and possible porosity, whereas cut samples show the apparent grain size more clearly.

Texture

Texture as seen in electron micrographs can be determined by grain size, grain shape, grain contacts, and roundness.

Grain size: Grain size of micrites has been measured by counting the longest diameter of about 100 to 200 well defined grains. These data show a range in grain size mostly between 0.3 and 10 microns with predominating frequency maxima below 3.5 microns (Figs. 1 and 2). Maxima were defined as grain diameters with a frequency of more than 10% of the grain size distribution. Both figures show a general agreement in size range and position of maxima. Most of these maxima lie below the value of FOLK's "micrite curtain" (FOLK, 1965), independent of the environmental origin of the limestones. No sample except one shows a maximum in the microspar range as postulated by FOLK (1965, p. 32).

Plotting of the grain size distribution as representative curves on probability paper results in the recognition of a gradual transition between micrite and microspar fields

Table 2. *Origin of samples*

Formation	Locality	Age
Upper Pseudoschwagerina Ls.	Karnische Alpen, Austria	Lower Permian
Opponitz Ls.	Kirchdorf, Upper Austria	Upper Triassic
Hallstatt Ls.	Salzkammergut, Upper Austria	Upper Triassic
Crenularis beds	Olten, Aargau, Switzerland	Oxfordian
Crenularis beds	Mellikon, Aargau, Switzerland	Oxfordian
Bank Ls.	Urach, Schwäbische Alb, Germany	Malm
Bank Ls.	Tieringen, Schwäbische Alb	Malm
Nusplingen Schieferkalk	Nusplingen, Schwäbische Alb	Malm
Solnhofen Ls.	Solnhofen, Fränkische Alb, Germany	Malm
Sulzfluh Ls.	Mt. Sulzfluh, Rätikon (Vorarlberg/ Graubünden)	Tithonian
Ernstbrunn Ls.	Ernstbrunn near Vienna, Austria	Tithonian
Calpionellid Ls.	Baronnies, Southern France	Tithonian
Calpionellid Ls.	Chartreuse, Southern France	Tithonian
Turriliten beds	Mt. Säntis, Switzerland	Upper Cretaceous
Seewen Ls.	Mt. Säntis, Switzerland	Upper Cretaceous

(Fig. 3). About 80% of all grains in micritic limestones have diameters below 3.5 microns.

The data of curve 1 have been taken from the micrographs of a Devonian algal limestone as given by WOLF et al. (1967, pl. 1 A). Similar divergent particle sizes were observed in Lower Permian algal limestones of the Karnische Alpen (Southern Alps) within a sparite matrix. These aberrant data may be due to a primary difference in the size of grains of algal origin.

Curve 9 was drawn after plate 8 of GARRISON and BAILEY (1967) who described pelagic Cretaceous limestones extremely rich in coccoliths together with minute anhedral calcite grains. Only those grains thought to be deposited chemically were measured. The sediment seems to be a typical example of particle-by-particle deposition.

All other curves measured are located within the limits of curves 2 and 8. Anyway, the curves seem to show no significant difference between shallow- and deep marine carbonates.

According to FOLK (1965, p. 33) micrite grains show a very good "sorting". In contrast to this the calculated sorting index (standard deviation σ_I FOLK and WARD,

1957) lies between 0.375 and 0.935, i.e., the samples are well to moderately well sorted following Friedman (1962). Sorting in micrites does not mean mechanical separation of grains, but is only the result of competition between space and grain growth. Therefore, the degree of sorting (moderately well sorted) can be used as a measure of textural maturity of micrites.

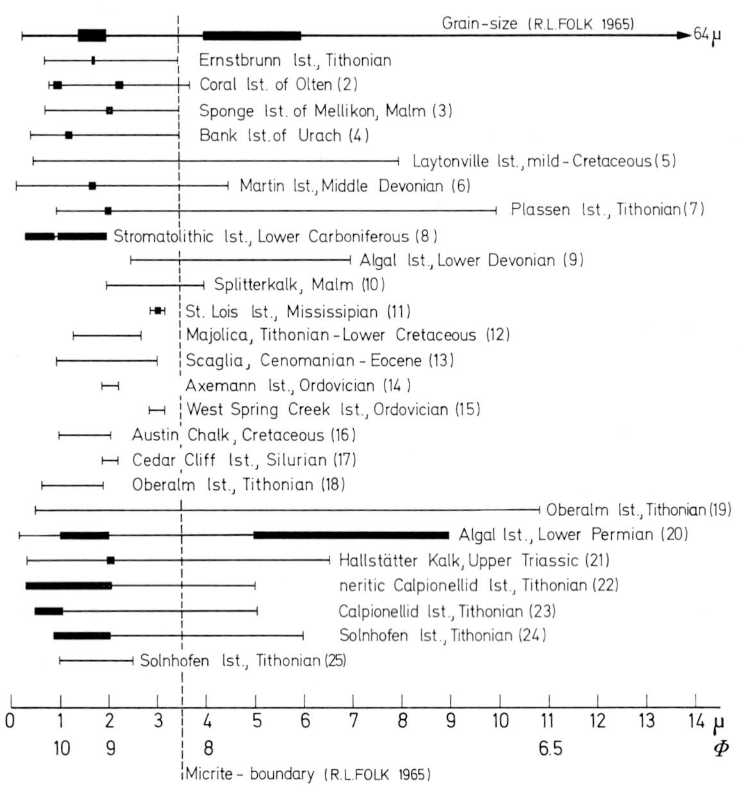

Fig. 1. Size-range of micrite grains in electron micrographs. Frequency maxima (more than 10% of the grain size distribution) are shown as thickened lines. Grain size of micrites as obtained by Folk (1965) from thin-section measurements are indicated by the uppermost line. Date were taken from the following references: 1—4 E. Flügel and H. E. Franz, 1967 (2); 5 Garrison and Bailey, 1967; 6 Teichert, 1965; 7 H. Flügel and A. Fenninger, 1966; 8 Grégoire and Monty, 1963; 9 Wolf et al., 1967; 10 Kabelac, 1955; 11 Harvey, 1966; 12—13 Farinacci, 1964; 14—17 Shoji and Folk, 1964; 18 Honjo and Fischer, 1964; 19 Garrison, 1967; 20—24 E. Flügel and H. E. Franz, 1967 (1); 25 Hathaway and Robertson, 1961

Grain shape: Most authors describe grain shape of micrites as rounded to polyhedral blocks with curved or straight contacts. Following Friedman (1965) most of the micrite grains are anhedral to subhedral forming a hypidiotopic fabric. True euhedral grains seem to be rare perhaps due to a special fracture mechanism which is controlled by the distribution of clay minerals not as envelope but in pores between the grains. This can be assumed for some samples of the Hallstatt Limestone (see pl. 1, Fig. 1, E. Flügel, 1967) and of Sulzfluh Limestone (H. E. Franz, unpublished).

Roundness: Roundness seems to be of greater value for the characterization of grain shapes than the crystallization patterns mentioned above. As shown in Fig. 5, most of the micrite grains are angular to subangular following the comparison chart by KRUMBEIN and SLOSS (1963). Roundness in micrites is caused mainly by intergranular solution during diagenesis, indicating BATHURST's "overgrowth" mechanism. Similar degrees of roundness have been observed by HATHAWAY and ROBERTSON (1961) during experimental compaction of aragonite needles.

Grain contacts: HARVEY (1966, p. 13) draws special attention to three different contact types: curvilinear, straight, and serrate. Curvilinear boundaries were observed

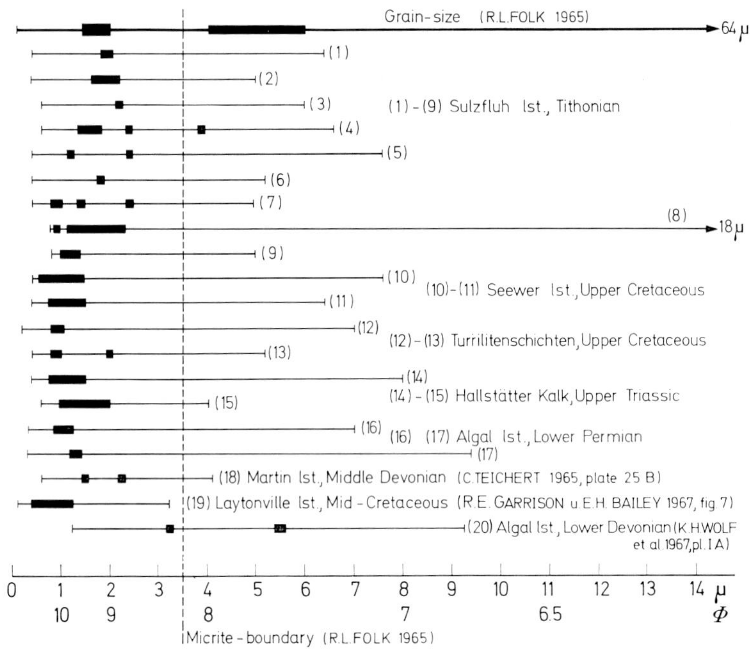

Fig. 2. Size-range of micrites according to measurements mostly from own electron micrographs. Almost all maxima (black lines) are located below FOLK's micrite boundary

mainly in fine-grained micrites, whereas straight contacts predominate in coarse-grained and sparry limestones. Serrate contacts are thought to be caused by solution and reprecipitation due to ground water (HARVEY, 1966). Since this contact type can only be seen in clastic fossil particles such as brachiopods, it is more probable that serration depends on a specific fracture of these fossil tests. Similar conclusions as to the occurrence of straight contacts in sparites and concavo-convex boundaries in micrites have been drawn by SHOJI and FOLK (1964).

Examination of own electron micrographs from different sources show that in micrites there exists almost an equilibrium between straight and curved contacts. No clear differences in contact types can be seen in samples from shallow and deep marine environments though neritic and bathyal limestones show slightly predominating curvilinear contacts (Hallstatt Limestone; Seewen Limestone). The method used for counting contact types was the one suggested by J. M. TAYLOR (1950).

Differences in width of contacts as postulated by HARVEY (1966, p. 13) seem to be due to effects of shadowing during preparation rather than to true distances between grains.

Grain surfaces: The most obvious characteristics of grain surfaces of limestones as seen in the electron microscope are fracture surfaces and solution figures.

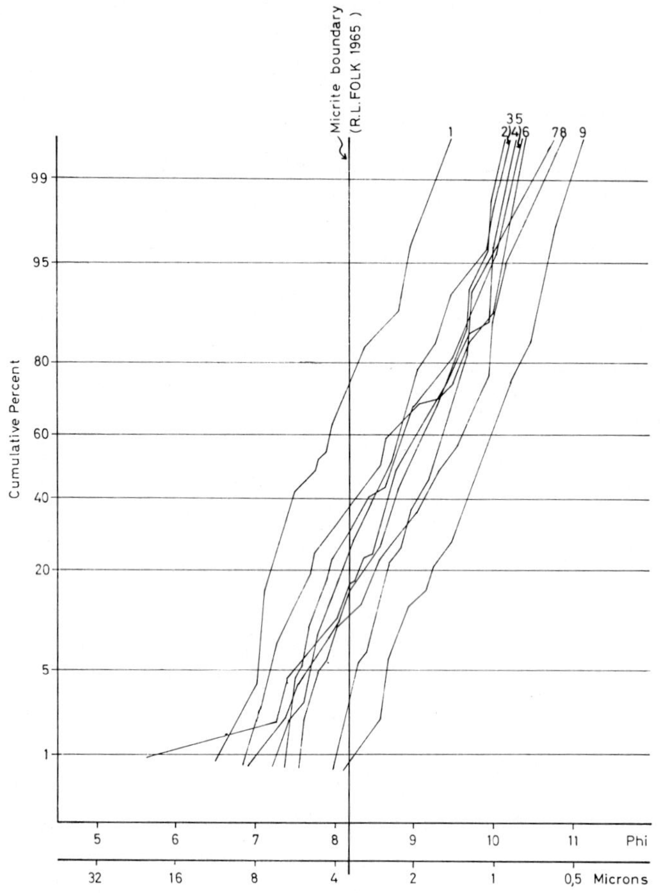

Fig. 3. Grain-size distribution of representative micrite samples from different environments 1 Algal Ls. (Lower Devonian), data from WOLF et al., 1967, pl. I A; 2—4, 6, 8 Sulzfluh Ls. (Tithonian, Rätikon, shallow marine biomicrite); 5 Algal Ls. (Lower Permian, Karnische Alpen); 7 Ernstbrunn Ls. (Tithonian, Lower Austria, reef-talus micrite); 9 Coccolith-bearing Laytonville Ls. (mid-Cretaceous, California), data from GARRISON and BAILEY, 1967, pl. 8

Fracture surfaces: Most of the micrite samples observed show tensile fractures following grain boundaries or more rarely across grains. Shear fractures as observed by HARVEY (1966) seem to be restricted mainly to sparry limestones. There is no general agreement now on whether or not the distances of cleavage steps can be used for typifying different mechanical properties of limestones as assumed, for example, by TEICHERT (1965) and HARVEY (1966).

Solution marks: Several authors (e.g. HARVEY, 1966) have published micrographs of surfaces structures which can be attributed to etching during preparation by com-

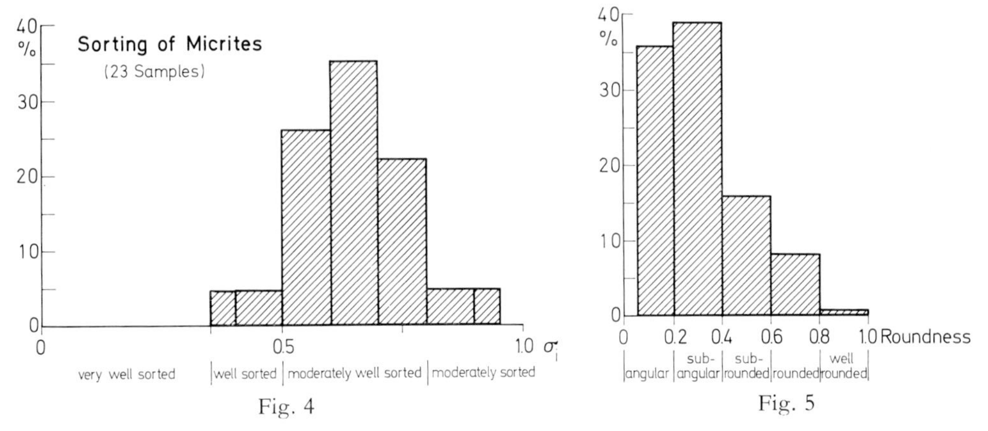

Fig. 4. Sorting (standard deviation) of micrites. Classification after FRIEDMAN (1962)

Fig. 5. Roundness of micrites, measured with the comparison chart of KRUMBEIN and SLOSS (1963)

Plate 1 Fig. 1 Fig. 2

Fig. 1. Electron micrograph of fractured micrite. Sulzfluh Ls. (Tithonian), Sulzfluh, Rätikon

Fig. 2. Detail view of micrite. Sulzfluh Ls. (Tithonian), Sulzfluh, Rätikon. The grains are in part moderatly well rounded, the contacts are curvilinear and straight, fractures are developed as tensile fractures. Etching marks on the grains are indicated by arrows

Plate 2 Fig. 1 Fig. 2

Fig. 1. Coccolith with etching marks, Turriliten beds (Upper Cretaceous), Mt. Säntis, Switzerland. The etching figures can be seen not only on the coccolith plates but also on anhedral grains surrounding the coccolith (arrows)

Fig. 2. Pellet in micritic matrix, Hallstatt Ls. (Upper Triassic), Gosausee, Upper Austria. The well defined pellet consists of minute anhedral grains without regular arrangement

parison of electron micrographs of solution figures of calcite as described by Pfefferkorn (1952) and d'Albissin (1963).

It is of great interest that similar shaped etching figures have been observed in samples, which have not been etched during preparation (plates 1/2, 2/2). The symmetry of these wedge-shaped figures corresponds clearly with the rhombohedral crystal shape of the grains. These etching marks were found not only in the Upper Jurassic, shallow marine Sulzfluh Limestone but also on the plates of coccoliths in pelagic Cretaceous Limestones. Etching by pore-filling mechanisms seems to be unlikely, since according to Usdowski (1967) the existence of pore fluids would result in dolomitization. The samples studied have been taken from non-dolomitized sequences. Solution by biological weathering affects the uppermost millimeters of carbonate rocks only (see Jones, 1965). Therefore it is obvious that these figures must have been formed during diagenesis, (see Chave, 1966).

Other features observed on grain surfaces, mainly in sparites, have been described by Shoji and Folk (1964) as so-called "Spanish moss inclusions" which were attributed to primary water-filled cavities in crystals.

Porosity

Porosity of some micrites studied submicroscopically has been determined by courtesy of Prof. Dr. Füchtbauer (Bochum). Measurements were made by CCl_4 saturation. The data are given below (Table 3). Most of the data lie within the range of 0.5 and 2.9% similar to the results of Füchtbauer (1964) who studied micritic Zechstein limestones. The data compiled in Table 3 indicate a small but obvious porosity which probably would increase up to duplication if measurements were

Table 3. *Porosity of Micrites*

Formation	Microfacies	Locality	Porosity (%)
Kohlenkalk (Lower Carboniferous)	micrite	Aachen, Germany	1,1
Upper Pseudoschwagerina Ls. (Lower Permian)	micrite with fusulinids and algae	Karnische Alpen, Austria	1,2
Opponitz Ls. (Karnian)	sparite	Kirchdorf, Upper Austria	2,0; 2,2; 2,8
Hallstatt Ls. (Norian)	cephalopod micrite	Steinbergkogel near Hallstatt, Austria	1,1; 2,9
Hallstatt Ls. (Norian)	cephalopod micrite	Sommeraukogel near Hallstatt, Austria	3,2
Hallstatt Ls. (Norian)	cephalopod micrite	Siriuskogel near Bad Ischl, Austria	0,9
Hallstatt Ls. in Dachstein reef Ls. (Norian)	intercalation of basin sediment within reefs	Gosausee, Upper Austria	1,2
Crenularis beds (Oxfordian)	micrite with sponge bioherms	Mellikon, Aargau, Switzerland	1,1
Plassen Ls. (Tithonian)	biomicrite	Jaintzen near Bad Ischl, Upper Austria	0,9
Ernstbrunn Ls. (Tithonian)	biomicrite of reef talus	Ernstbrunn near Vienna, Austria	0,5
Calpionellid Ls. (Tithonian)	biomicrite with coccoliths	Baronnies, Southern France	2,3
Solnhofen Ls. (Tithonian)	lagoon-micrite	Solnhofen, Germany	2,9
Bank Ls. (Malm)	biomicrite with cephalopods	Urach, Schwäbische Alb, Germany	7,1
Sulzfluh Ls. (Tithonian)	biomicrite	Mt. Sulzfluh, Rätikon (Vorarlberg/Graubünden)	0,7; 0,8; 0,9; 1,0; 1,3

made with a gasporosimeter (Füchtbauer, 1964, p. 518). Contrary to this, micrographs show no clear features which could be interpreted as pores. Additional measurements of effective porosity by an Erba Sorptomat (courtesy of Dr. D. Heling, Heidelberg) of a sample of a Solnhofen micrite revealed that the pore-size goes down to 0.070 microns, and that the bulk of pore-spaces has a diameter of 0.140 microns. Of course these pore-spaces are too small as they could be detected on the electron micrographs used in our investigations. It seems likely that these pores are located between grain-contacts rather than within grains because only effective porosity was measured.

Fossils

One of the main results of electron microscope studies of Mesozoic limestones is the recognition of the importance of nannofossils, i.e. coccoliths and nannoconids, as rock-builders in neritic and bathyal limestones (FARINACCI, 1964; GARRISON, 1967; FISCHER and GARRISON, 1967) as well as in shallow marine limestones of littoral environments [E. FLÜGEL and H. E. FRANZ, 1967 (1), (2)]. Moreover, shell fragments of Foraminifera, Lamellibrancha, and others can be identified. In this way a distinction between biodetrital and more or less chemically deposited micrites becomes possible (E. FLÜGEL, 1967, Fig. 1).

Algal structures have been studied scarcely under the electron microscope (stromatolithic limestone, GREGOIRE and MONTY, 1963; Renalcis Limestone, WOLF et al., 1967; Girvanella Limestone, E. FLÜGEL, unpublished). Structures attributed to algal origin show distinct patterns of equigranular and euhedral grains lying within a coarser matrix. Rod-like elements, interpreted as detrital algal needles, have been described from Lower Permian algal limestones (E. FLÜGEL, 1967, plate 2).

Further studies require investigation of the various kinds of pellets which may be algal remains or fecal pellets. Electron micrographs of Hallstatt Limestone show well defined globular bodies, 10—15 microns in diameter, consisting of anhedral, minute grains within a size-range between 0.60—1.00 microns, see plate 2/1. The pellets show no deformation, indicating an almost complete lack of compaction in this environment (see FRUTH et al., 1966).

Conclusions

(1) Frequency maxima of grain sizes of micrites lie mostly below FOLK's "micrite curtain" (3.5 microns) independent from the environment of deposition. A second maximum as suggested by FOLK between 5 and 6 micron seems to be absent.

(2) Sorting of micrites is the result of competition between space and growth and can be used as an indicator of textural maturity of micrites.

(3) Angular to subangular roundness of micrite grains indicates overgrowth mechanisms during diagenesis.

(4) Grain contacts are straight and curvilinear in equal amounts giving no significant hints for the recognition of micrite origin or environment.

(5) Following FOLK's ideas of neomorphism of carbonate muds to micrite, uniformity of grain sizes, roundness, and low porosity indicate porefilling overgrowth as the main mechanism of lithification independent of primary mineralogy.

(6) Special attention should be given to characteristic surface features such as primary etching marks which have been found in littoral as well as in bathyal deposits. As these solution marks may be seen both on coccoliths and on micrite grains precipitated after particle-by-particle sedimentation in interspaces between nannofossils the etching figures must have been formed during the final stage of lithification.

(7) Samples studied show small porosities according to saturation measurements. The measured pore-size of about 0.140 microns is too small as that pores could be detected on the electron micrographs used in the published investigations.

(8) Nannofossils and minute skeletal detritus observed in electron micrographs of limestones can be used as indicators for the bioclastic origin of micrites but not as bathymetrical marks.

(9) Electron microscope studies of limestones seem to be of greater value for the recognition of new data dealing with lithification and diagenesis than for typifying texture and depositional environment of micrites.

Acknowledgements

The studies have been supported by the *Deutsche Forschungsgemeinschaft* in connection with its Sediment Research Program. Financial aid was given to one of the authors (W. F. Ott) by "Stiftung Volkswagenwerk". Electron micrographs were taken at the Elektronenmikroskopisches Zentrallaboratorium of the University of Zürich (Director: Dr. A. Vogel). Special thanks are due to Prof. Dr. A. G. Fischer of Princeton University, and Dr. R. E. Garrison of the University of British Columbia, Vancouver, for sending unpublished manuscripts, and to Dr. H. Füchtbauer of the Ruhr-Universität Bochum, and Dr. D. Heling of the Universität Heidelberg for providing porosity measurements.

References

Albissin, M. d': Une application du microscope électronique à l'étude de la déformation des calcaires. Rev. Géogr. phys. Géol. dyn. (2), **2**, 35—38 (1959).

— Les traces de la déformation dans les roches calcaires. Rev. Géogr. phys. Géol. dyn. (1961—62), fasc. suppl. **1963**, 3—155 (1963).

—, et C. de Rango: Etude de la microstructure des roches calcaires par l'observation au microscope électronique de l'orientation des figures de corrosion. Bull. Soc. franç. Min. Cristall. **85**, 170—176 (1962).

Alonso, J., et B. Agulleiro: Etude des roches carbonatées espagnoles. Sixth Internat. Congress Electron Microscopy, Kyoto, p. 647—648 (1966).

Chave, K. E.: Early diagenesis of carbonate particles in calstic sediments. Bull. Am. Ass. Petrol. Geol. **50**/3, 607 (1966).

Cloud, P. E.: Environment of calcium carbonate deposition West of Andros Island, Bahamas. U.S. Geol. Survey Profess. Papers **350**, 138 p. (1962).

Farinacci, A.: Microorganismi dei Calcari "Maiolica" e "Scaglia" osservati al microscopio elettronico (Nannoconi e Coccolithophoridi). Boll. Soc. Pal. Ital. **3**, 172—181 (1964).

Fischer, A. G., and R. E. Garrison: Carbonate lithification on the sea floor. J. Geol. (1967) (In press).

Flügel, E.: Elektronenmikroskopische Untersuchungen an mikritischen Kalken. Geol. Rundschau **56**, 341—358 (1967).

—, u. H. E. Franz: (1) Elektronenmikroskopischer Nachweis von Coccolithen im Solnhofener Plattenkalk (Ober-Jura). Neues Jahrb. Geol. u. Paläontol. Abhandl. **127**, 245—263 (1967).

— — (2) Über die lithogenetische Bedeutung von Coccolithen in Malmkalken des Flachwasserbereiches. Eclogae geol. Helv. **60**, 1—17 (1967).

Flügel, H., u. A. Fenninger: Die Lithogenese der Oberalmer Schichten und der mikritischen Plassen-Kalke (Tithonium), Nördliche Kalkalpen. Neues Jahrb. Geol. u. Paläontol. Abhandl. **123**, 249—280 (1966).

Folk, R. L.: Some aspects of recrystallization in ancient limestones. Soc. Econ. Paleont. Min., Spec. Publ. **13**, 14—38 (1965).

—, and Ward: Brazos River bar, a study in the significance of grain-size parameters. J. Sediment. Petrol. **27**, 3—27 (1957).

Friedman, G. M.: On sorting, sorting coefficients, and the lognormality of the grain-size distribution of sandstones. J. Geol. **70**, 737—753 (1962).

— Terminology of crystallization textures and fabrics in sedimentary rocks. J. Sediment. Petrol. **35**, 643—655 (1965).

Frost, J. G.: A negative replication for electron microscopy study of carbonates. J. Sediment. Petrol. **37**, 692—695 (1967).

Füchtbauer, H.: Fazies, Porosität und Gasinhalt der Karbonatgesteine des norddeutschen Zechsteins. Z. deut. geol. Ges. **114**, 484—531 (1964).

Garrison, R. E.: Electron microscopy of Franciscan Pelagic Limestones, California. Geol. Soc. Am., Ann. Meetings Program 1966, San Francisco, 75—76 (1966).

— Pelagic limestones of the Oberalm beds (Upper Jurassic Lower Cretaceous), Austrian Alps. Bull. Canadian Petrol. Geol. **15**, 21—49 (1967).

—, and E. H. Bailey: Electron microscopy of limestones in the Franciscan formation of California. U.S. Geol. Survey Profess. Papers **575-B**, B 94—B 100 (1967).

GRÉGOIRE, CH., et CL. MONTY: Observation au microscope électronique sur le calcaire à pâte fine entrant dans la constitution des structures stromatolithiques du Viséen Moyen de la Belgique. Ann. soc. géol. Belg. **10**, 389—397 (1963).

GRUNAU, H. R.: Mikrofazies und Schichtung ausgewählter jungmesozoischer, Radiolarit-führender Sedimentserien der Zentralalpen mit Berücksichtigung elektronenmikroskopischer und chemischer Untersuchungsmethoden. Intern. Sediment. Petrogr. Ser. **4**, 179 p. Leiden: E. J. Brill 1959.

—, u. H. STUDER: Elektronenmikroskopische Untersuchungen an Bianconekalken des Südtessins. Experientia **12**, 141—150 (1956).

GYGI, R.: Zur Stratigraphie der Oxford-Stufe (untere Malm-Serie) der Nordschweiz und des süddeutschen Grenzgebietes. Diss., Univ. Zürich 1967.

HANCOCK, J. M., and W. J. KENNEDY: Demonstration: Photographs of hard and soft chalks taken with a scanning electron microscope. Geol. Soc. London **136**, 2. (1967).

HATHAWAY, J. C., and E. C. ROBERTSON: Microtexture of artificially consolidated aragonite mud. U.S. Geol. Survey Profess. Papers **424-C**, 301—304 (1961).

HARVEY, R. D.: Electron microscope study of microtexture and grain surfaces in limestones. Illinois State Geol. Survey **404**, 1—18 (1966).

HONJO, S., and A. G. FISCHER: Fossil coccoliths in limestones examined by electron microscopy. Science, **144**, No. 1620, 837—839 (1964).

— — Paleontological investigation of limestones by electron microscope. In: KUMMEL, B., and D. RAUP: Handbook of Paleontological Techniques, p. 326—334, San Francisco—London: Freeman 1965.

ILLING, L. V., A. J. WELLS, and C. M. TAYLOR: Penecontemporary dolomite in the Persian Gulf. Soc. Econ. Paleont. Min., Spec. Publ. **13**, 89—111 (1965).

JONES, R. J.: Aspects of the biological weathering of limestone pavement. Proc. Geologists Ass. (Engl.) **76**, 421—434 (1965).

KABELAC, F.: Beiträge zur Kenntnis und Entstehung des unteren Weißjuras am Ostrand des südlichen Oberrheingrabens. Ber. naturforsch. Ges. Freiburg Breisgau **45**, 5—57 (1955).

KAHLE, CH.: Aspects of diagenesis in oolithic limestones. Program 1964. Ann. Meet., Miami Beach, Geol. Soc. America, p. 106.

KAHLE, CH. F., and M. D. TURNER: A rapid method of making replicas of rock and minerals surfaces for use in electron microscopy. J. Sediment. Petrol. **34**, 604—609 (1964).

KRUMBEIN, W. C., and L. L. SLOSS: Stratigraphy and Sedimentation, 2nd Ed., 660 p., San Francisco and London: W. H. Freeman and Company 1963.

MINATO, M., M. ISHII, and S. HONJO: Electron microscopic study of oil bearing calcilutite, Khafji Oil Field, Neutral Zone. Proc. VIIth World Petrol. Congress, **2**, No. 3, 447—456 (1967).

PFEFFERKORN, G.: Untersuchungen zum Realbau von Kalkspat. Neues Jahrb. Mineral. Geol. Abhandl. **84**, 281—326 (1952).

PITTMAN, J. S., JR.: Silica in Edwards Limestone, Travis County, Texas. Spec. Publ. Soc. Econ. Paleont. Min. **7**, 121—134 (1959).

SEELIGER, R.: Übermikroskopische Darstellung dichter Gesteine mit Hilfe von Oberflächenabdrücken. Geol. Rundschau **45**, 332—336 (1956).

SHOJI, R.: Electron-Microscopic Study of Limestones. J. Geol. Soc. Japan **70**, 154—169 (1964) (Japan. with English summary).

—, and R. L. FOLK: Surface morphology of some limestone types as revealed by electron microscopy. J. Sediment. Petrol. **34**, 144—155 (1964).

STUDER, H. P.: Electron microscope study of aragonite crystals in marine sediments. Amer. Ass. Petrol. Geol., Meeting Program, **1963**, 54 (1963).

TAYLOR, J. M.: Pore-space reduction in sandstones. Bull. Am. Ass. Petrol. Geologists **34**, 701—716 (1950).

TEICHERT, C.: Devonian rocks and paleogeography of Central Arizona. U.S. Geol. Survey Profess. Papers **464**, 181 p. (1965).

TEMMLER, H.: Über die Schiefer- und Plattenkalke des Weißen Jura der Schwäbischen Alb (Württemberg). Arb. Geol. Paläont. Inst. TH Stuttgart, N.F. **43**, 106 p. (1964).

USDOWSKI, H.-E.: Die Genese von Dolomit in Sedimenten. Mineralogie und Petrographie in Einzeldarstellungen **4**, 95 p. (1967).

WOLF, K. H., A. J. EASTON, and S. WARNE: Examination and analysis of sedimentary carbonates. In: CHILLINGAR, G. V., J. J. BISSELL, and R. W. FAIRBRIDGE, Carbonate Rocks, Dev. Sedimentology, **9b**, 253—341. Amsterdam: Elsevier 1967.

Appendix

Some important papers were published after the manuscript has been sent to the editor (July, 1967). These papers are cited in the following list with brief critical remarks. Since during the last time besides the electron microscope the scanning electron microscope is increasingly used for the study of carbonates some papers dealing with this new method are included.

COCOZZA, T., C. MAXIA e V. PALMERINI: Il "calcare ceroide" del Cambrico sardo osservato al microscopio elettronico. Boll. Soc. Geol. Ital. **86**, 725—731 (1967).

The authors describe fractured samples of a Cambrian "micrite" which seems to be strongly recrystallized (probably sparite, grain size 10 to 25 microns).

FARINACCI, A.: La tessitura della micrite nel valcare "Corniola" del Lias medio. Accad. Naz. Lincei, Rendiconti Cl. Sci. Fis., Mat. Nat., Fasc. 2, Ser. 8, **44**, 284—289 (1968).

Together with a description of a shallow-marine micrite the author proposes the term "nannomicrite" for micrites with abundant nannofossils, and tries to distinguish four diagenetic stages during the lithification of lime-mud.

FISCHER, A. G., S. HONJO, and R. E. GARRISON: Electron micrographs of limestones and their nannofossils. Monographs Geol. Paleont. Univ. Princeton **1**, 141 p. (1967).

This valuable contribution contains many important details concerning the technique of preparation, the different fossils as seen in micrographs ("algal" filaments, coccoliths, protozoans, molluscs, and Tunicata), and especially the submicroscopic diagenetic fabrics. Two fundamental fabrics are recognized: the amoeboid mosaic (formed by irregularely interlocked grains), and the pavement mosaic (formed by block-like grains). The samples come from Cambrian to Miocene (mostly bathyal) environments.

FLÜGEL, H. W.: Die Lithogenese der Steinmühl-Kalke des Arracher Steinbruches (Jura, Österreich). Sedimentology **9**, 23—53 (1967).

Some micrographs are used as additional data for typifying the environment of an Upper Jurassic limestone group.

HANCOCK, J. M., and W. J. KENNEDY: Photographs of hard and soft chalks taken with a scanning electron microscope. Proc. Geol. Soc. London. **1967**, No. 1643, 249—252.

The authors use scanning electron micrographs for differentiating the texture of hard chalks (entirely composed of interlocking crystals), and soft chalks (built up by loosely packed coccoliths).

LAFFITTE, R., et D. NOEL: Sur la formation des calcaires lithographiques. C.R. séances Acad. Sci. **264**, 1379—1382 (1967).

Scanning electron micrographs of Upper Jurassic micrites show the fundamental role of coccoliths as rock-building elements.

NOEL, D.: Etude de roches carbonatées par répliques de surfaces examinées au microscope électronique. C.R. séances Acad. Sci. **264**, 544—547 (1967).

The author discusses different methods of prepartion and draws attention to the importance of nannofossils in lithology. The term "nannofacies" is suggested for features as seen in electron microscope.

NOEL, D.: Nature et genèse des alternances de marnes et de calcaires du Barrémien supérieur d'Angles (Fosse vocontienne, Basses-Alpes). C.R. séances Acad. Sci. **266**, 1223—1225 (1968).

Scanning electron micrographs of samples from a limestone-marl-sequence show abundant nannoconids in micrite limestones and abundant coccoliths in marls. Nannoconids therefore must not be indicators of bathyal deposits as suggested by COLOM.

ROBERTSON, E. C.: Laboratory consolidation of carbonate sediment. Marine Geotechnique **1967**, 118—127.

Together with laboratory experiments electron micrographs are used for controlling the effects of artificial consolidation.

Microporosity of Carbonate Rocks

DIETRICH HELING*

With 5 Figures

Abstract

Consolidated carbonate rocks have relative low porosities but high specific surface areas thus showing a fine dispersion of the pore volume. The pore radii distribution derived from Hg-capillary pressure curves show maxima which are in good accordance with the grain size distribution. This and the rather large portion of not communicating porosity which is deduced from the difference between total porosity and pore volumes measured by Hg-capillary pressure and benzole-vapor adsorption leads to the conclusion that the porosity of the investigated carbonate rocks is the rest of the primary intergranular porosity of the sediment before compaction.

The petrographic consequences of pore geometry on material transport and reaction velocities are discussed.

A. Pore Space and Inner Surface

I. Arenites

Porosity in arenites is defined as the free space between grains — the pore space. This intergranular pore space is independent of grain size at constant packing density (coordination number), constant sorting, and uniform grain shape. The diameters of the capillaries forming the pore space diminish with decreasing grain diameter; the free space remains constant per unit volume, however. With increasing "dispersion" of the pore space, the specific inner surface of the system also increases, that is, the sum of grain surfaces per unit of volume or weight

$$S_g = w' \frac{\pi}{d} \text{ cm}^2 \cdot \text{cm}^{-3}$$

S_g = specific surface
d = grain diameter
w' = module for deviations from a sphere

This is valid for packing of spheres assuming constant sorting and packing density (Fig. 1). In an analogous way, the quotient specific surface/porosity increases with decreasing grain diameter.

The specific surface of natural sands can be calculated from detailed grain size analyses with sufficient accuracy after

$$S_g = q \cdot s,$$

where $s = (P_1/r_1 + P_2/r_2 + \ldots + P_e/r_e)$,
$q = 3.0$ for spheres,
$q > 3.0$ for deviations from a sphere,
P_i = fraction with radius r_i in weight-%.

Specific surfaces calculated with $q = 3.5 \ldots 4.0$ in fairly well rounded sand and poorly cemented sandstones with less than 3 weight per cent silt correlate well

* Laboratorium für Sedimentsforschung, Universität Heidelberg, Germany.

with measured values of specific surfaces. Clayrich sands have larger-than-calculated inner surfaces because of the much higher surface of phyllosilicates.

Specific surface is best measured by using the "adsorption isotherm" with the BET-method after Brunauer, Emmet and Teller (1938): the adsorption isotherm between $p/p_0 = 0...0.25$ is approximated by a straight line which is drawn through 0 and a measured point within the BET-range. Thus, the recording of the entire isotherm by measuring at different p/p_0-values is superfluous. The monolayer capacity of the adsorbens is determined by extrapolation, and from the space occupied by each gas molecule the surface area is calculated.

Table 1. *Spec. Surface Area and Petrographic Data of Weakly Consolidated Carbonate Sediments*

	Grain Size Median ⌀	Carbonate Contents		Spec. Surface Area S_g		Porosity
		Total	Quotient	calculated	measured	
	mm	%	Calc./Dol.	$m^2 \cdot g^{-1}$	$m^2 \cdot g^{-1}$	%
Carbonate-Arenite	0,031	77		0,52	4,4	28
Tertiary, Canary Islds.	0,044	74		0,56	5,3	32
	0,041	83	Dol.	0,31	5,4	39
	0,062	72		0,55	27,1	37
	0,077	76		0,41	13,8	36
	0,016	53		0,68	42,5	54
Biocalcilutite	0,027	94	$CaCO_3$ 18	1,40	8,3	87
Recent, Florida Bay	0,220	97	$MgCO_3$ 31	0,13	3,5	
Biocalcilutite	0,012	76			8,5	71
Recent, Lake Constance	0,014	84			5,6	62
	0,011	75	> 50	1,3	8,0	73
	0,012	65			11,6	70
	0,011	79			4,4	61
Loess	0,030	34	~ 50	0,4	7,1	47
Pleistocene, Rhine Rift						
Beersheba, Israel	0,045	20	~ 50	2,2	24,1	28
(Desert Loess)						

A monolayer on the surface of the material being examined is accomplished using nitrogen or other suitable gases at the temperature of liquid nitrogen (77.6°K), because under this condition the gas molecules form a uniform, tightly packed layer. The amount of adsorbed nitrogen is calculated from the pressure decrease with input pressure equal to atmospheric pressure. Thus the relative pressure ($p/p_0 = 0.25$) is just within the BET-range and the systematic error due to inaccuracies is less than 8%; the measured values scatter less than 3%. This method enables measuring surfaces between 0.3 and 1000 $m^2 \cdot g^{-1}$ with sufficient accuracy.

II. Calcarenites

The good correlation between specific surfaces calculated from grain size distribution and measured with the BET-method for quartz sands and powdered quartz is no longer valid for calcareous sands and silts. The relationship between calculated and measured specific surfaces of some calcarenites is shown in Table 1 together with mechanical analysis parameters. Measured values are 10 to 100 times higher than the calculated ones.

Calcarenites from the Canary Islands (samples from ROTHE) contain abundant organic detritus such as foraminifera, siliceous skeletons, sponge needles etc. These components strongly increase the specific surface area of the sediment because they deviate considerably from a sphere. However, the factor 10 to 60 is too high to be due solely to the influence of organic remains.

The same is true for the clayey-silty to fine sandy carbonates of Florida Bay (Cross Bank) which consists to more than 94% of, mostly aragonitic, biogenic detritus (MÜLLER, G., and J. MÜLLER, 1967). Their measured specific surface area is 6 times higher than the calculated value. GINSBURG (1957) reported measured porosities of

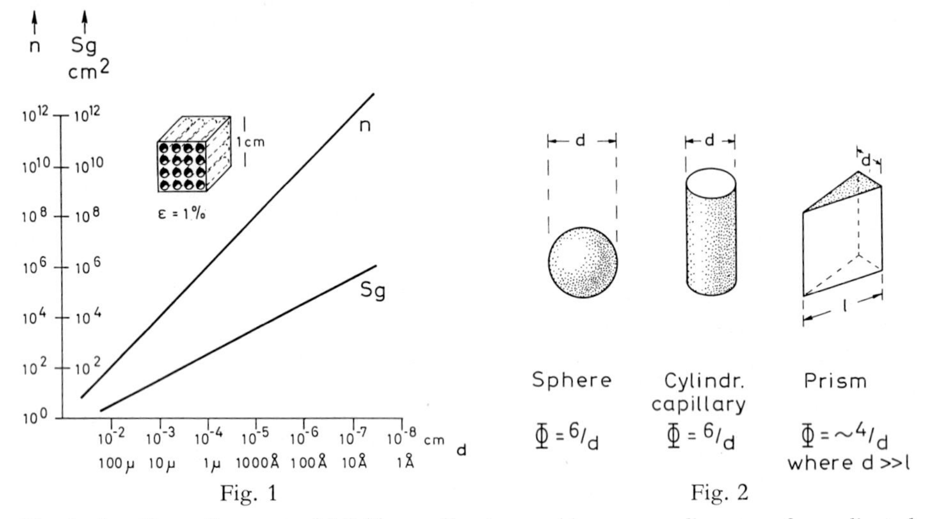

Fig. 1. Specific surface area of 6-fold coordination packing versus diameter of coordinated spheres

Fig. 2. Quotient (Φ) specific surface area (S_g)/Volume (V) of different stereometric forms

80 to 90% in fine grained calcareous muds in Florida Bay. The sediment was not altered after deposition.

The same relationship between measured and calculated specific surface was found for Recent sediments from the Gnadensee (Lake Constance) (SCHÖTTLE, 1967) consisting of about 75% of carbonatic particles. The discrepancy does not depend on the clay portion, so that the higher surfaces are not only caused by the clay minerals.

Loess from the east-slopes of the Oberrheingraben have 31% carbonate particles, about 15% illite and kaoline, the remainder consisting of quartz, feldspar, and micas. Carbonate occurs predominantly as crypto-crystalline calcitic to dolomitic grains, and some as cement. Rounded oolite fragments with radial and concentric textures were also found. They are probably derived from the Hauptrogenstein outcropping in the east flank of the southern Rhine Rift and in the Swiss Jura and were transported by rivers and later possibly by air (ALDINGER, personal communication). Primary oolite grain sizes recalculated from the fragments fit well the grain size distribution curve of the Hauptrogenstein.

The Loess carbonate grains are weakly cemented by calcite. This is the reason for the steep slopes in Loess areas. SCHUMANN (1949) calculated from the Cozeny-Carman-equation from porosity and permeability measured in carbonate-cemented sands higher specific surface areas than in uncemented ones. He concluded that cemented rocks have higher specific surfaces than those without cement. This, however, is only true if the cement in form of euhedral crystals fringes pore walls as reported by FRIEDMAN (1964). Pore spaces completely filled with cement have no inner surface area. Therefore, the larger the completely cemented portion of the pore space, the smaller the surface of a clastic sediment will be.

The investigations reveal that a number of carbonate sediments show a pronounced discrepancy between the specific surface to be expected from the grain size distribution, and the measured surface. These discrepancies are too large to be caused only by deviation of particles from a sphere, by surface roughness, or by partial cementation. It is probable that microscopic cracks, crevices, and capillaries considerably add to the specific surface. This *intra*-granular porosity adds little to the total porosity of the sediment although it increases the specific surface considerably.

The relationship ϕ between volume and surface is illustrated for three stereometric forms (Fig. 2). The surface/volume ratio increases with decreasing dimension r for spherical spaces $1^1/_2$ times that of cylindrical or planar capillaries. There are probably combinations of all three basic forms in carbonate grains, although the wedgeshaped prism may be preferred at the interface between euhedral crystals.

III. Consolidated Carbonate Rocks

Swabian Upper Jurassic carbonates were investigated as examples for consolidated rocks (Table 2). Specific surface areas of these "dense" rocks with 87 to 100% carbonate fraction are between 0.3 and 6 $m^2 \cdot g^{-1}$ with porosities between 0.3 and 8%. If 1% porosity in a cube (1 cm^3) is distributed to cylindrical capillaries with a

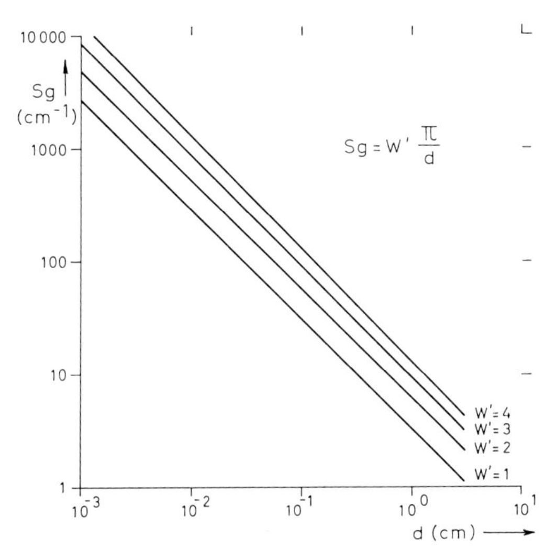

Fig. 3. Specific surface area (S_g) and number of capillaries (n) in 1 cm^3 with 1% porosity versus diameter of capillaries (d). Capillaries straight, parallel and cylindrical

diameter of 100 Å parallel to its edges, then the cube is penetrated by 10^{10} capillaries with an inner surface area of 3 m^2 ($S_g = 1.15$ $m^2 \cdot g^{-1}$) (Fig. 3).

No relationship was found between the measured porosities and specific surface area. This may indicate either a wide range of pore radii distribution due to poor sorting of the carbonate grains or a differing portion of the non-communicating pore space, which does not contribute to the specific surface area. Electronmicrographs (Fig. 4) show a rather well-sorted grain community with a strong frequency maximum at the 1 μ diameter range. (The sample was cut rectangular to bedding and was briefly etched with HCl in order to show the relief.) The tightly intergrown crystals are

Table 2. *Spec. Surface Area of Upper Jurassic Carbonate Rocks of Swabia (SW-Germany)*

Stratigr. Pos.	Locality	Facies	HCl-insoluble Fract. %	Porosity %	spec. Surface Area $m^2 \cdot g^{-1}$
δ_4	Seeburg	Massenkalk	1	0,8	0,26
δ_4	Hülben	Bankfac.	13	3,0	0,43
ξ_3	Hengener Steige	Bank/Schwammfac.	6	1,8	1,00
$\delta_{2/4}$	Neuffen	Bankfac.	6	4,2	1,00
$\delta_{2/4}$	Hengener Steige	Bankfac.	5	1,4	1,27
ξ_1	Nusplingen	Bankfac.	11	2,2	1,35
δ_3	Hülben	Bankfac.	6	3,0	1,53
ξ_3	Wittlinger Steige	Bankfac.	11	4,4	1,86
β	Hochwanger Steige	Bankfac.	4	2,2	1,94
ε_2	Seeburg	Bankfac.	2	2,2	2,26
δ	Hochwanger Steige	Schwammfac.	3	0,3	2,30
ξ_1	Rietheim	Bankfac.	8	2,2	2,57
β	Urach	Bankfac.	10	0,8	2,66
γ	Hochwanger Steige	Bankfac.	0	2,6	2,94
ξ_3	Hengener Steige	Bankfac.	8	4,8	2,96
δ_1	Hülben	Bankfac.	9	4,5	3,54
$\delta_{2/4}$	Neuffen	Bankfac.	13	4,5	4,03
γ	Hochwanger Steige	Schwammfac.	4	4,7	5,40
ξ_1	Solnhofen	Bankfac.	8	8,0	5,54
δ	Hochwanger Steige	Schwammfac.	12	2,9	5,89
α	Urach	Bankfac.	13	3,7	6,02

elongated parallel to bedding. They were probably oriented by overburden pressure while they were still in plastic state, combined with later pressure solution.

The geometry of the micro pore spaces in consolidated carbonate rocks may be considered as a system of well connected cavities of highly variable dimensions — much alike Karst-cavity-systems. Determination of pore sizes is allowed by Mercury penetration. A liquid cannot enter small pores because of the surface tension as the liquid has a contact angle of more than 90°. This tension can be overcome by applying a certain external pressure. The relation between this pressure and pore size is

$$r = -\frac{2\,\sigma \cos \vartheta}{p}$$

where r = pore radius (Å),
σ = liquid surface tension (dynes/cm),
ϑ = contact angle of the liquid used,
p = external pressure applied (kg/cm²).

Using Mercury as penetrating liquid (surface tension = 480 dynes/cm, contact angle = 140°) the following relationship between the pore radius (Å) and pressure (kg/cm²) is pratically obtained

$$r = 75.000/p.$$

Fig. 5 shows a typical pore size distribution of an Upper-Jurassic-sample from Swabia. There are distinctive maxima in the ranges of 10^3 Å and $5 \cdot 10^4 - 10^5$ Å of pore radii. The first mode can be attributed to grain size distribution, which was

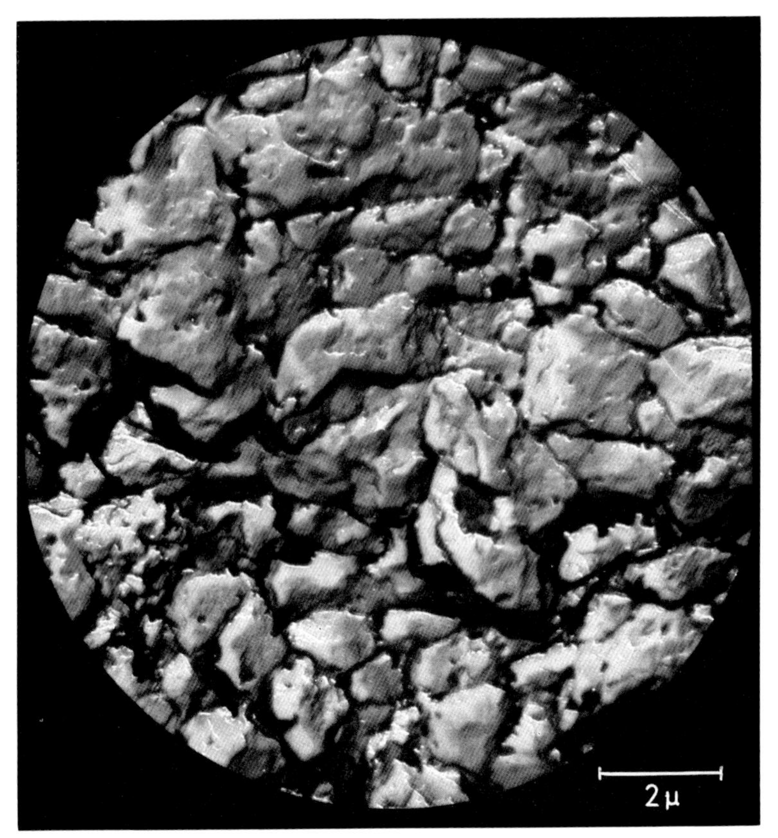

Fig. 4. Electronmicrograph of fine grained carbonate rock (Upper Jurassic, SW-Germany)

derived from electronmicrographs. Here the median grain size proved to be about 2 microns. In sufficient tight packings capillary diameters can be supposed to be one tenth of the grain size. This coincides with the observed pore radii maximum at 10^3 Å.

The pore volume determined by Hg-porosimetry is 0.5%, i.e. 0.5% of the rocks bulk volume is pore space of communicating capillaries with radii larger than 100 Å. Porosity measurements by gasadsorption methods give evidence of the pore space with very small radii (approx. 10 Å) by the effect of capillary condensation. In the example of Fig. 2 the porosity measured by benzole-vapor adsorption is 0.9%, whereas the total porosity measured by sensitive fluid pycnometry is 4,0%.

Adding the porosities gained by Hg-capillary pressure and benzole-vapor adsorption we get not more than about one third of the total porosity. Thereupon we have to conclude that the other two third belong to not communicating pore space. This "dead" pore space might be considered as the rest of the intergranular pore volume which the sediment possessed before compaction started. This is in accordance with the observed maxima of pore radii distribution in correlation to grain size.

Finally the problem has to be considered if tectonic deformation joints are evident in micro fabrics, e.g. the Zechstein-dolomites of NW-Germany known as gas-reser-

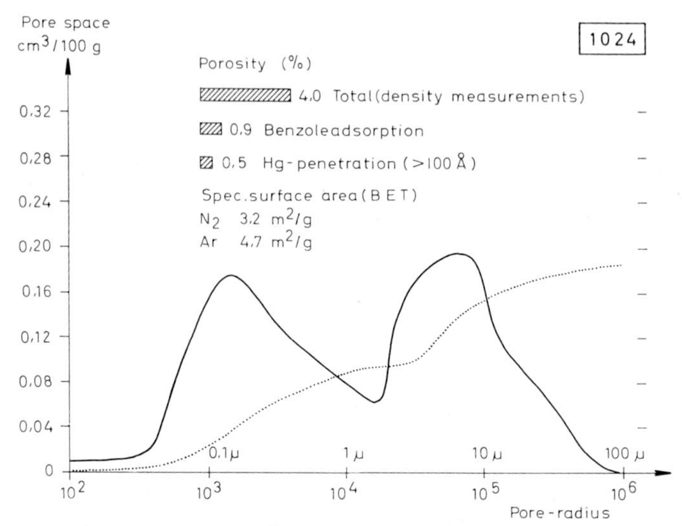

Fig. 5. Pore radii distribution > 100 Å in the communicating pore space of a typical sample of Swabian Upper Jurassic

voir rocks have 0.5% porosity due exclusively to fine tectonic joints (FÜCHTBAUER, 1959). If these also occur on a micro-scale the rock should more easily desintegrate upon pressure application than the investigated Upper-Jurassic rocks.

B. Petrogenetic relations

Which are the petrographic consequences of pore geometry, in particular for the diagenesis of carbonate sediments?

Material transport is accomplished by flow of pore solutions caused by a pressure gradient, provided the pore channels are of sufficient widths. If pore channel width is smaller than the boundary layer thickness (where flow velocity is zero at laminar conditions) flow will cease even at extreme differential pressures. Therefore, in these small capillary pores material transport is accomplished solely by diffusion. The coefficient D which determines diffusion according to

$$J = -D \frac{dc}{dx}$$

depends in porous media on the pore geometry, which is expressed by the tortuosity factor. The solid phase delays diffusion in porous rocks compared with that in free space. Besides, the diffusion coefficient depends on the diameter of the diffundating molecules or ions. Thus a separation effect by diffusion can be imagined.

Dispersion of the pore volume not only influences pore radii but the specific surface area too. The velocity of any kind of reaction — it might be solution or ion exchange — is directly related to specific surface area. The greater the surface area of the communicating pore volume, the greater the reaction velocity, for instance of dolomitisation or solution by pore waters rich in carbon dioxide.

On this more theoretical background we are going now to classify carbonate rocks by their micropore geometry in correlation with their diagenetic properties.

Acknowledgements

The author acknowledges the help of Dr. H. Jüntgen, Bergbau-Forschung GmbH., Essen, in performing gas adsorption measurements.

References

Brunauer, S., P. H. Emmet, and R. Teller: The adsorption of gases in multi-molecular layers. J. Am. Chem. Soc. **60**, 309—316 (1938).

Friedman, G. M.: Early diagenesis and lithification in carbonate sediments. J. Sediment. Petrol. **34**, 777—813 (1964).

Füchtbauer, H.: Zur Petrographie der Zechsteindolomite westlich der Ems. In: Andres, J., E. Brand, W. v. Engelhardt und H. Füchtbauer, Die Erdgaslagerstätten im Zechstein von Nordwestdeutschland. I Giacimenti Gassiferi dell'Europa Occidentale. Vol. I. Academia Nazionale dei Lincei. Roma 1959.

Ginsburg, R. N.: Early diagenesis and lithification of shallow-water carbonate sediments in South Florida. In: Leblanc, and Breeding (Ed.), Regional aspects of carbonate deposition. Soc. Econ. Palaeont. and Mineralogists, Spec. Publ. No. 5, 80—100 (1957).

Müller, G., and J. Müller: Mineralogisch-sedimentpetrographische und chemische Untersuchungen an einem Bank-Sediment (Cross-Bank) der Florida Bay, USA. Neues Jahrbuch Mineral. Abhandl. **106**, 257—286 (1967).

Schöttle, M.: Die Sedimente des Gnadensees. Diss., Heidelberg 1967.

Schumann, H.: Die Raumgestaltung von Gesteinsporen. In: Erdöl und Tektonik in Nordwestdeutschland (Amt für Bodenforschung Hannover-Celle), p. 321 (1949).

Outlines of Distribution of Strontium in Marine Limestones

W. M. Bausch*

With 6 Figures

Abstract

The strontium content of marine limestones is governed by four factors:

Age. The aragonite-calcite transition is the reason for the strong diminution of the strontium content during diagenesis. This is efficient for 50 to 100 mill. years.

Clay mineral content. Diagenetically altered limestones show a relationship between the Sr-content and the insoluble residue. This is explained by the adsorption of Sr by clay minerals during the aragonite-calcite transition.

Salinity. Limestones of evaporitic series show a relative high Sr-content.

Environment. In alternating limestone-marl-series no relationship between the Sr-content and the insoluble residue seems to exist.

Contrary to the relationships found in recent carbonates, in fossil limestones the reef complexes have a minimal Sr-content, for they have small clay mineral content.

A. Introduction

Most investigations concerning trace elements in limestones are concerned with strontium. The several hundred analyses cited in the literature agree in average on the value of about 400 to 500 ppm Sr (Vinogradov and Ronov, 1956; Kulp, Turekian and Boyd, 1952; Fornaseri and Grandi, 1963). But single measurements yield values between 100 ppm and more than 1200 ppm. Thus it seems necessary to discuss the reasons for the differences in distribution of strontium in limestones.

Recent carbonate sediments are also well examined. Several workers have tried to use variations in the strontium content for facies analyses based on recent statements.

B. Comparison of Fossil Limestones with Recent Carbonate Sediments

Recent carbonates differ distinctly from fossil limestones in their absolute Sr-content as well as in the relative strontium contents representing each of the different facies areas. Absolute strontium contents in recent carbonates average about several 1000 ppm, and thus they are about ten times higher than those of fossil limestones (Stehli and Hower, 1961; Siegel, 1961). The relative strontium content of recent carbonates is as follows: reefs have maximum values, also in lagoonal limestones the values are high; but in the correlated basin sediments they are low. This behaviour corresponds to the distribution of aragonite and calcite in these sediments.

In fossil limestones, consisting only of calcite, the distribution is completely reversed: in reef complexes the strontium content is low, while in correlated basin

* Mineralogisches Institut, Universität Erlangen-Nürnberg, Germany

sediments it is high (see Table 1). This was found by FLÜGEL and FLÜGEL-KAHLER (1963), BAUSCH (1965), CHESTER (1965), and FLÜGEL and WEDEPOHL (1967). Contrary to the opinion of FLÜGEL and WEDEPOHL (1967), this seems to be a rule of general validity; contradictory examples are not known.

Evidently, the high Sr-content of recent sediments is reduced diagenetically, and simultaneously the distribution in the facies areas is changed.

Recent carbonate sediments consist of the three phases aragonite, high-Mg-calcite and low-Mg-calcite. Only low-Mg-calcite, that is normal calcite, is stable for longer periods of time. According to FRIEDMAN (1964), no recent sediments can be found, which consist only of the stable calcite phase. Therefore, all carbonate sediments must undergo the diagenetic transition into low-Mg-calcite. In addition, there are processes of solution and pore space filling. Thus the conservation of primary strontium contents is very improbable.

Table 1. *Comparison of strontium contents (in ppm) of reef- and non-reef facies in ancient limestones*

Authors	reef complex lst.	basin limestone
FLÜGEL and FLÜGEL-KAHLER (1963)	300—400	1800 ("fore-reef")
BAUSCH (1965)	200	600
CHESTER (1965)	67	533
FLÜGEL and WEDEPOHL (1967)	100—200	500—3000

C. Crystal Chemistry

The modification of aragonite permits the incorporation of considerable amounts of strontium into the crystal lattice, because in the sequence of the earth-alkali carbonates the aragonite structure is realized with large cations ($r_{Kat} > r_{Ca}$). As is known, strontianite also crystallizes with the aragonite structure. On the other hand, the little cations prefer the calcite structure, and therefore the modification of calcite is less favorable for the incorporation of strontium.

OXBURGH et al. (1959) and HOLLAND et al. (1963, 1964) found distribution coefficients of aragonite/solution and calcite/solution the respective values of which differ about ten times in their magnitude. Therefore, aragonite is able to pick up ten times the amount of strontium that calcite can incorporate under equal conditions.

Recent marine aragonite and calcite differ in the same manner with regard to their Sr-content (as listed by FLÜGEL and WEDEPOHL, 1967). This is the reason for the different strontium contents of different facies areas: in reef- and lagoon-areas the amount of aragonite and the strontium content also are high; on the other hand, in basin sediments calcite prevails and the strontium content is low.

For calcite precipitated in sea water the Sr values should be about 950 ppm according to the distribution coefficients of OXBURGH et al. (1959) and of HOLLAND et al. (1963, 1964) (the statement by FLÜGEL and WEDEPOHL, 1967, of 2300 ppm is erroneous)[1]. — Since Sr-values found in nature are often higher, possibly other ions in sea water are of importance (not NaCl, which has no influence according to HOLLAND et al.). Moreover, organisms are obviously able to influence the strontium content in carbonate phases, because the values for skeletal elements of aragonite- or

[1] Meanwhile corrected: Contr. Min. Petr. **16**, 114 (1967)

calcite building organisms differ from the values expected from the distribution coefficients (as listed by FLÜGEL and WEDEPOHL, 1967). Also, for magnesium such a relationship to the kind of organisms is well known (TUREKIAN and ARMSTRONG, 1964).

The range of the discrepancy between recent and fossil limestones may be explained by a difference in the capacity of aragonite and of calcite to pick up strontium. SIEGEL (1960) found a linear relation in slightly altered carbonates: the strontium content decreases simultaneously with a decrease in the aragonite content.

D. Influencing Factors

I. Age

VINOGRADOV and RONOV (1956) plotted the strontium content against the geologic age of carbonate rocks of the Russian platform (see Fig. 5, p. 542 of the cited paper). It is shown that the strontium curve runs conformably with the sulfate curve, but the curves diverge nearing the younger formations (Cretaceous and Tertiary). This is caused by two effects: salinity (see later) and age.

Disregarding the peaks corresponding with the sulfate curve (for in time of evaporite sedimentation the composition of sea water is not normal) only the increase in Sr-content with a decrease in the geologic age of the formations remains. This may be extrapolated to the still higher values for recent carbonates. The Sr-content of all limestones older than Cretaceous varies within the known average values. There is no doubt that this significant loss of strontium corresponds to the aragonite-calcite transformation (KAHLE, 1965). As is evident from the high values, this transformation is not yet completed in Tertiary limestones, and coupled its consequences, continues in limestones of Cretaceous age. Indeed, this is the range in which aragonite can be met in sediments (aragonite is also found in Mesozoic, even Paleozoic limestones, but these are special cases of fossil shells enveloped by bituminous liquids (FÜCHTBAUER, 1964).

The relatively high values reported by HARDER (1964) from Cretaceous limestones of Münsterland may range within this evolution.

II. Clay Content

As shown by BAUSCH (1965), the strontium content of limestones can be related to the clay content. This was explained by a strontium enrichment in the pore space during the aragonite-calcite transition. This mechanism was already discussed by MÜLLER (1962).

The transition of all aragonite extends over geologic time. Aragonite is stable as long as the sediment is in contact with sea water (FRIEDMAN, 1964). As known, limestones are readily cemented, and the pore space is also readily diminished. Remainders of aragonite will experience transformation in lithified and covered sediments. This is a system neither completely open nor completely closed. This system might be described as a closed system with regard to short-time process (such as transition of a grain of aragonite into calcite) but may be described as an open system with regard to processes of geologic duration (e.g., the exchange of pore solutions).

When part of the present aragonite is dissolved in this nearly closed system while the pore volume is small, then the strontium concentration will increase in the pore space. The clay mineral particles present can adsorb more strontium from this increased availability than they could have adsorbed directly from the sea water. According to the results of WAHLBERG et al. (1965) the distribution coefficient of strontium between solution and the clay minerals (sodium or calcium illites) is constant up to concentrations of 10^{-3} strontium in solution. Because sea water is a 10^{-4} molar strontium solution, clay minerals are able to further increase their strontium content. If new calcite crystallizes in this pore solution, it will also pick up more strontium than when in equilibrium with sea water, due to the increase of strontium in the pore solution (USDOWSKI, personal communication).

Due to the different distribution coefficients of aragonite/solution and calcite/solution, but also to the adsorbing clay minerals, distinctly less strontium will be incorporated by calcite than was contained in the aragonite. In spite of this, one also has to assume calcite-calcite transitions in order to obtain the small amounts of strontium in fossil limestones. This concept of strontium enrichment in pore solutions (and its subsequent adsorption by clay minerals) is substantiated by the result of CHAVE (1960), who in 187 formation waters generally found an enrichment of strontium as compared to sea water, which ranges from 7 to 24 fold. The entrichment of calcium in these formation waters ranges up to 7 fold only. Therefore the Sr/Ca ratio is increased, too. The statement by FLÜGEL and WEDEPOHL (1967), that there is no diagenetic pore solution with increase in the content of strontium as compared to sea water, deduced from a theoretical scheme, cannot be supported. — This concentration of strontium is very high and seems to be higher than can be explained through the compaction effect only (V. ENGELHARDT and GAIDA, 1963).

BAUSCH (1965) postulated that the relationship between strontium content and the amount of clay minerals, which was found for Upper Jurassic limestones in southern Germany, should also be found in other formations. If the dates of FORNASERI and GRANDI (1963) are plotted against the calculated insoluble residues (without regarding the Cretaceous and Tertiary formations), than the same relationship is obvious. Their values and those of BAUSCH (1965) and FLÜGEL and WEDEPOHL (1967) all lie in the same field (Fig. 1). But the lines which deliniate this field should diverge more than those drawn by BAUSCH (1965) for the Upper Jurassic limestones of southern Germany. Although the dots are scattered considerably, an increase in the strontium content coupled with an increase in the clay mineral content seems to be a rule of general validity.

Younger limestones show a still more rapid increase (Fig. 2). This is thought to be due to the influence of geologic age. On the other hand, older diagenetically affected limestones should no longer show any more increase, because after repeated changes in the composition of pore solutions (or through the influence of fresh water) a levelling-off should occur. Fig. 3 may examplify this statement.

This conception is summarized in Fig. 4.

III. Salinity

The evaporite formations of the Russian platform show distinctly higher strontium contents (VINOGRADOV and RONOV, 1956). According to SMYKATZ-KLOSS (1966) the values for northern Germany Zechstein limestones are also higher than "normal"

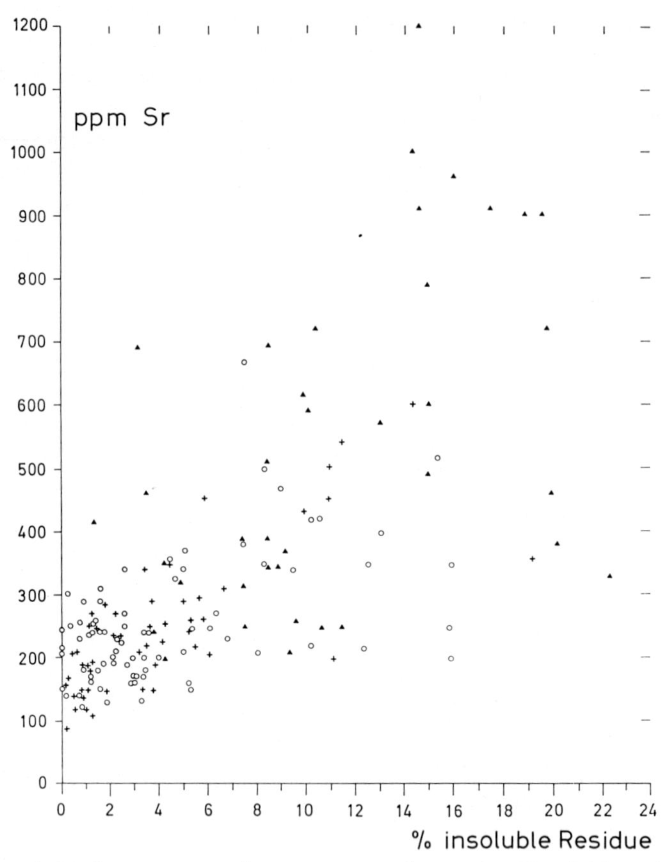

Fig. 1. The relation between strontium contents and amount of insoluble residues in limestones. There are plotted the results of FORNASERI and GRANDI, (1963) (circles; values of different formations, but all older than Cretaceous), BAUSCH, (1965) (crosses; upper Jurassic) and FLÜGEL and WEDEPOHL, (1967) (triangles; upper Jurassic)

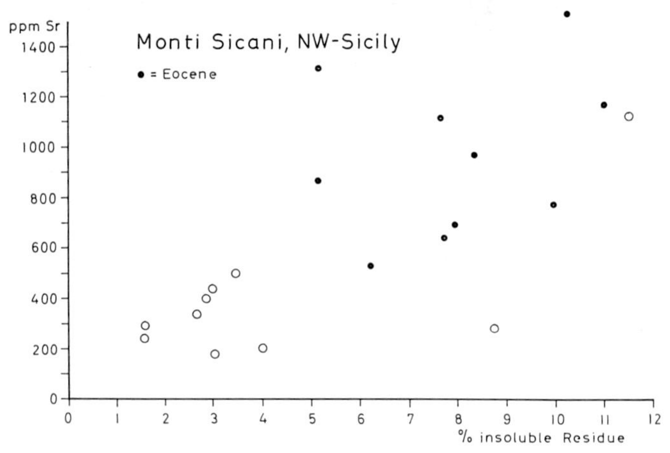

Fig. 2. Relation between strontium contents and amount of insoluble residues in diagenetical young limestones, according to values of FORNASERI and GRANDI (1963), Table V. For orientation are drawn lines between which the dots of Fig. 1. would lie. The increase of strontium contents is higher than in Fig. 1

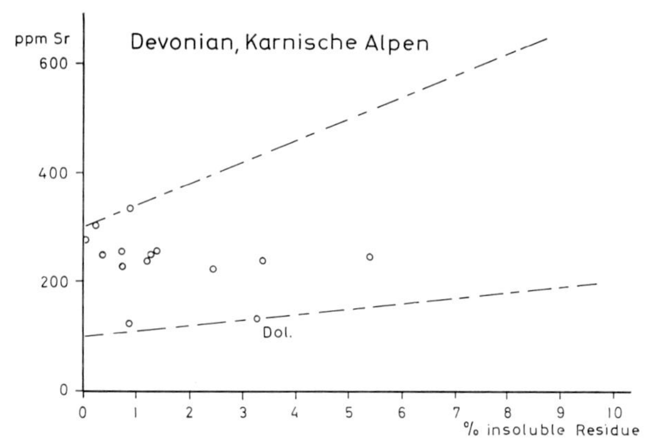

Fig. 3. Strontium contents and amount of insoluble residues in presumably diagenetical old limestones, according to values of FORNASERI and GRANDI (1963), Table IX. No increase seems to occur

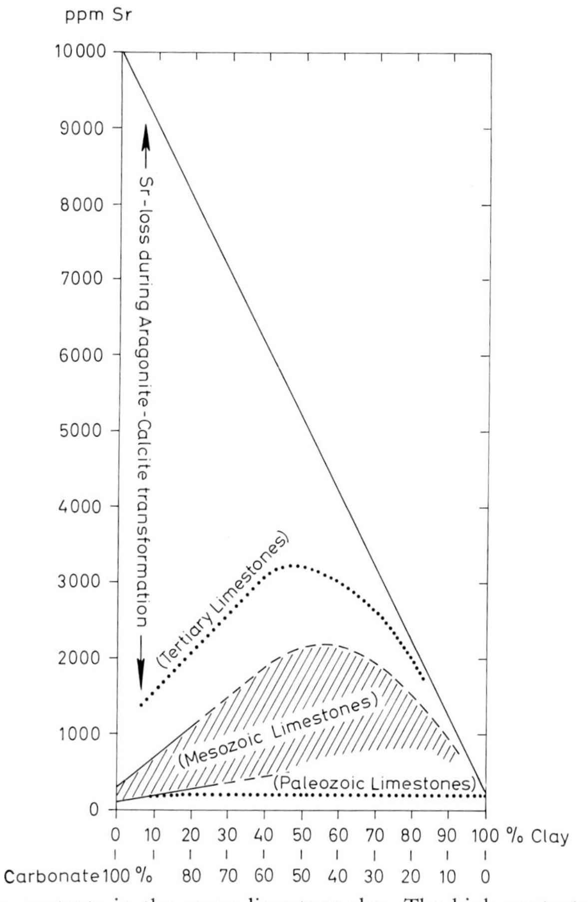

Fig. 4. Strontium contents in the range limestone-clay. The high contents of fresh sediments are lost mostly during aragonite-calcite transformation. Adsorption on clay minerals occurs, but disappears with further diagenetic alteration (repeated exchange of pore solutions)

(Fig. 5). This corresponds to the altered concentrations of sea water in evaporite formations and to the high strontium content in anhydrite (MÜLLER, 1962).

Perhaps, diagrams showing strontium content versus the insoluble residue content could be used to recognize limestones influenced by increased salinity.

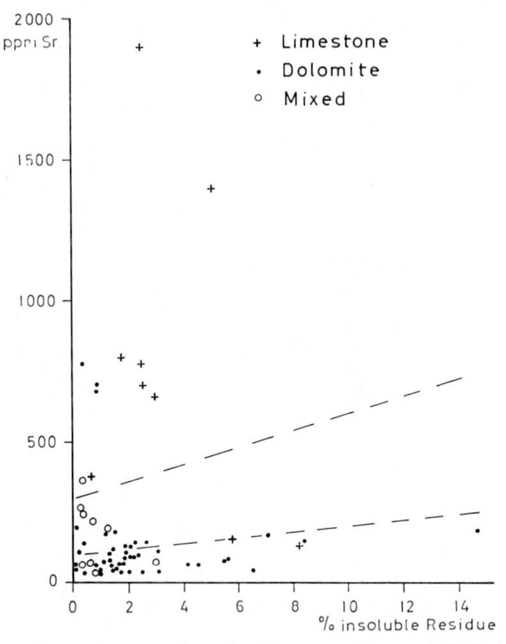

Fig. 5. Values of strontium contents given by SMYKATZ-KLOSS (1966) for northern German Zechstein limestones and dolomites. The limestones of this salinar formation have higher strontium contents, the dolomites lower strontium contents than the "normal" pattern (lines)

IV. Environment

KNOBLAUCH (1963) examined a sequence of alternating limestones and marls. Although the lime and clay contents vary considerably, no relationship between strontium content and the proportion of clay was found. The sequence examined is one of Upper Jurassic limestones from the Swabian Jura. The average values for the single stages of this formation show such a relationship, but the immediate neighbouring layers do not show this relationship.

This may be understood, but not exactly explained by keeping in mind that the pore solution is the medium responsible for removal of the strontium during diagenesis. In close proximity there may be a possible levelling-off. In the diagram where the strontium content has been plotted versus the clay mineral content, this caused a scattering of the points in the direction of the abszissa.

E. Dolomites

According to the results of WEBER (1964), and SMYKATZ-KLOSS (1966) and to our own measurements (not published) dolomites show very low strontium contents, which average about 100 ppm and lower and thus are distinctly lower than those of pure limestones. This can be explained by the complete metasomatic replacement of this carbonate, which causes a further decrease in the trace element content. Dolomites or recalcitized carbonate rocks should be plotted, therefore, separately when the strontium content of carbonate rocks are compared.

F. Discussion

Flügel and Wedepohl (1967) stated that there exists no general relationship between the strontium content and the clay content, and cite the results of Vinogradov and Ronov (1956) as counterevidence. An occasionally observed relationship

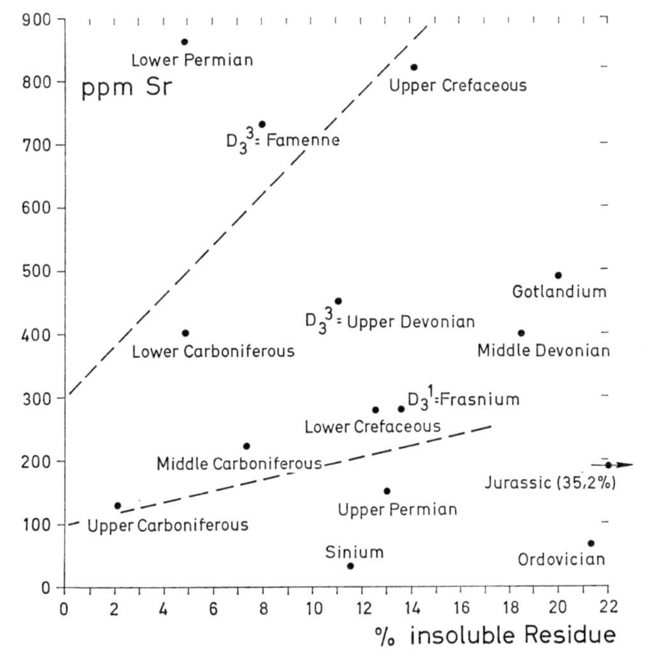

Fig. 6. Values of strontium contents given by Vinogradov and Ronov (1956) for carbonate rocks of the Russian platform. Discussion see in the text

could be explained by a restricted diagenetic evolution through a decrease in the pore space with increasing clay content.

The values of Vinogradov and Ronov (1956), plotted in a diagram, show no relationship between the strontium content and the insoluble residue content at the first glance (Fig. 6). But it should be realized that Vinogradov and Ronov collected and indiscriminantly put marbles and carbonate rocks from evaporate series together with the limestones, sandy marls and dolomites to obtain their average values (198 analyses of 8847 samples). Dolomites and carbonate rocks from evaporite series yield extreme values, as was shown before. Although clay minerals seem to be responsible for the increase in strontium values, quartz sand is inert to strontium-bearing solutions; but quartz sand is comprised with the insoluble residue. Samples with more than 20% insoluble residue will contain quartz sand and therefore will be shifted to the right side of the diagram (example: Jurassic). All points representing carbonate rocks which are dolomitized or metamorphous will be shifted downwards (example: Sinium). Vice versa, the values for evaporite formations will be shifted upwards (example: Upper Devonian and Permian). All other points lie within the encircled field. Fig. 6 should not be used as a proof for the existenc of a relationship between strontium/clay contents, because it is not suitable for this purpose due to the method of sampling utilized.

On the other hand, it can never be applied as counterevidence against the existence of such a relationship.

The pore space does not decrease in the order limestone-marl-clay, as was postulated by FLÜGEL and WEDEPOHL. According to v. ENGELHARDT (1960), the pore volume of clays is 10 to 20%, under considerable pressure, while pure limestones have about 2% pore space. According to recent results obtained by HELING (this volume) the pore space values for pure and clayey upper Jurassic limestones vary from 2 to 10%, and no relationship of the pore volume versus the clay content can be found.

Restricted diagenesis with diminishing pore space rather would cause the opposite effect of what was observed. Therefore, the explanation by FLÜGEL and WEDEPOHL (1967) is not satisfying.

Conclusions

The differences in the strontium content between recent and fossil limestones are essentially caused by the differing capacity of aragonite and calcite to incorporate strontium. The differences in strontium content of recent reef and basin facies correspond to the distribution of the strontium-rich phase, aragonite.

The differences in strontium content of fossil reef and basin limestones can be explained by differences in the clay mineral content of the sediments of these facies areas. Very pure limestones (reef limestones) contain 100 to 300 ppm, limestones with a 5 to 15% clay content contain 400 to 700 ppm, and limestones with more than 20% clay content contain 1000 ppm strontium and more. Diagenetically young limestones show a similar relationship, but higher values; diagenetically old limestones show no relationship of the Sr-content with the clay content. Direct facies analyses, based on comparisons with recent material, are not possible using strontium as a trace element.

References

BAUSCH, W. M.: Strontiumgehalte in süddeutschen Malmkalken. Geol. Rundschau **55**, 86 bis 96 (1965).

CHAVE, K. E.: Evidence on history of sea water from chemistry of deeper subsurface waters of ancient basins. Bull. Am. Assoc. Petrol. Geologists **44**, 357—370 (1960).

CHESTER, A.: Geochemical criteria for differentiating reef from non-reef facies in carbonate rocks. Bull. Am. Assoc. Petrol. Geologists **49**, 258—276 (1965).

ENGELHARDT, W. v.: Der Porenraum der Sedimente. Min. u. Petr. in Einzeldarst. **2**, S. 207. Berlin-Göttingen-Heidelberg: Springer 1960.

—, and K. H. GAIDA: Concentration changes of pore solutions during the compaction of clay sediments. J. Sediment. Petrol. **33**, 919—930 (1963).

FLÜGEL, E., u. E. FLÜGEL-KAHLER: Mikrofazielle und geochemische Gliederung eines obertriadischen Riffes der nördlichen Kalkalpen. Mitt. Mus. Bergb., Geol. u. Techn. (Graz) **24**, 129 (1963).

FLÜGEL, H. W., u. K. H. WEDEPOHL: Die Verteilung des Strontiums in oberjurassischen Karbonatgesteinen der Nördlichen Kalkalpen. Contr. Mineral. Petrol. **14**, 229—249 (1967).

FORNASERI, M., et L. GRANDI: Contenuto in stronzio di serie calcaree Italiane. Giorn. geol., Ann. museo geol. Bologna 2 **31**, 171—198 (1963).

FRIEDMAN, G. M.: Early diagenesis and lithification in carbonate sediments. J. Sediment. Petrol. **34**, 777—813 (1964).

FÜCHTBAUER, H., u. H. GOLDSCHMIDT: Aragonitische Lumachellen im bituminösen Wealden des Emslandes. Beitr. Mineral. Petrol. **10**, 184—197 (1964).

HARDER, H.: Geochemische Untersuchungen zur Genese der Strontianitlagerstätten des Münsterlandes. Beitr. Mineral. Petrol. **10**, 198—215 (1964).

HOLLAND, H. D., M. BORCSIK, J. MUNOZ, and U. M. OXBURGH: The coprecipitation of Sr^{+2} with aragonite and of Ca^{+2} with strontianite between 90° and 100°C. Geochim. Cosmochim. Acta **27**, 957—977 (1963).

—, H. J. HOLLAND, and J. L. MUNOZ: The coprecipitation of Sr^{+2} with calcite between 90° and 100°C. Geochim. Cosmochim. Acta **28**, 1287—1301 (1964).

KAHLE, C. F.: Strontium in oölitic limestones. J. Sediment. Petrol. **35**, 846—856 (1965).

KNOBLAUCH, G.: Sedimentpetrographische und geochemische Untersuchungen an Weißjurakalken der geschichteten Fazies im Gebiet von Urach und Neuffen, S. 106., Diss., Tübingen 1963.

KULP, J. L., K. K. TUREKIAN, and D. W. BOYD: Strontium content of limestones and fossils. Bull. Geol. Soc. Am. **63**, 710—716 (1952).

MÜLLER, G.: Zur Geochemie des Strontiums in ozeanen Evaporiten unter besonderer Berücksichtigung der sedimentären Coelestinlagerstätte von Hemmelte-West (Süd-Oldenburg). Beih. Geol. Jahrb. **35**, 90 (1962).

OXBURGH, U. M., R. E. SEGNIT, and H. D. HOLLAND: Coprecipitation of strontium with calcium carbonate from aqueous solutions. Program GSA-Meeting Pittsburgh, 95 A—96 A (1959).

SIEGEL, F. R.: The effect of strontium on the aragonite-calcite ratios of pleistocene corals. J. Sediment. Petrol. **30**, 297—304 (1960).

— Variations od Sr/Ca ratios and Mg contents in recent carbonate sediments of the northern Florida Keys area. J. Sediment. Petrol. **31**, 336—342 (1961).

SMYKATZ-KLOSS, W.: Sedimentpetrographische und geochemische Untersuchungen an Karbonatgesteinen des Zechsteins. Teil II. Contr. Mineral. Petrol. **13**, 232—268 (1966).

STEHLI, F. G., and J. HOWER: Mineralogy and early diagenesis of carbonate sediments. J. Sediment. Petrol. **31**, 358—371 (1961).

TUREKIAN, K. K., and L. ARMSTRONG: Magnesium, strontium and barium concentrations and calcite-aragonite ratios of some recent molluscan shells. J. Marine Research (Sears Foundation) **18**, 198—215 (1964).

VINOGRADOV, A. P., and A. B. RONOV: Composition of the sedimentary rocks of the Russian platform in relation to the history of its tectonic movements. Geochemistry **6**, 533—559 (1956).

WAHLBERG, J. S., J. H. BAKER, R. W. VERNON, and R. S. DEWAR: Exchange adsorption of strontium on clay minerals. U.S. Geol. Survey Bull. **1140-C**, 26 p. (1965).

—, and R. S. DEWAR: Comparison of distribution coefficients for strontium exchange from solutions containing one and two competing cations. U.S. Geol. Survey Bull. **1140-D**, 10 p (1965).

WEBER, J. N.: Trace element composition of dolostones and dolomites and its bearing on the dolomite problem. Geochim. et Cosmochim. Acta **28**, 1817—1868 (1964).

Exceptionally High Strontium Concentrations in Fresh Water Onkolites and Mollusk Shells of Lake Constance

German Müller*

With 5 Figures

Abstract

Calcitic onkolites and aragonitic shells of Lake Constance show unusually high Sr/Ca-ratios ranging from 0.89 to 3.38; mean values for algal calcite are 1.06, for Gastropods 1.52, and for Pelecypods 2.19. For mollusks, most of these ratios fall into the range of marine environment.

The high Sr/Ca-ratios in the carbonate material can be explained by the high Sr/Ca-ratio of Lake Constance water derived mainly from the Alpenrhein. In the Alpenrhein, ratios even higher than in sea water were observed at low water levels.

The distribution coefficient for Strontium between water and organic calcium carbonate ranges from 0.177 in calcite algal material (with very low amounts of aragonite) to 0.25 in Gastropods and 0.36 in Pelecypods. Similar coefficients for mollusks are known from marine shells. Our observations clearly demonstrate that Sr-contents and Sr/Ca-ratios of organic carbonates are *not* necessarily indicators of salinity but of the Sr/Ca-ratio in the depositional environment.

A. Introduction

In the littoral and sublittoral zones of Lake Constance, great amounts of carbonates are continuously produced by organisms and after deposition may become an essential part of the bottom sediment (Müller, 1966, 1967; Schöttle, 1967; Schöttle and Müller, 1968).

Blue-green algae, secreting calcite during the assimilation process, are to be found mainly in the Untersee. They form the biscuit-shaped "onkolites", also known from marine environments. Mollusk shells, mainly consisting of aragonite, are to be found anywhere in the littoral of the lake.

Geochemical investigations of Lake Constance carbonate sediments of the Untersee revealed Strontium contents being too high for a normal fresh water environment. These findings led to the following investigations on the Sr/Ca-ratio of skeletal material of carbonate secreting organisms and to a geochemical study of the waters causing this high Sr-content; they also raised the question whether the Sr/Ca-ratio in carbonates can be used as salinity indicator for a depositional environment (Landergren and Manheim, 1963; Krejci-Graf, 1966).

B. Selection of Samples, Analytical Procedures

Through courtesy of Dr. Meier-Brook of the Tropenmedizinisches Institut of Tübingen University, we received an extensive collection of well-determined Gastro-

* Laboratorium für Sedimentforschung, Universität Heidelberg

pods and Pelecypods from Lake Constance. In addition, 21 onkolites from the Untersee (Insel Langenrain, Gnadensee) were analyzed which had been collected by us during the course of other investigations.

During the period of one year, water samples of the most important influent of Lake Constance, the Alpenrhein, were taken at a two weeks interval. Besides, a great number of analyses of the water of the Obersee and Untersee were carried out. Fig. 1 shows the locations where carbonate- and water samples were taken.

Concentrations of Strontium both in carbonates and in waters were measured by atomic absorption spectroscopy (Perkin Elmer 303), the content of calcium and magnesium was either determined by atomic absorption spectroscopy and/or by

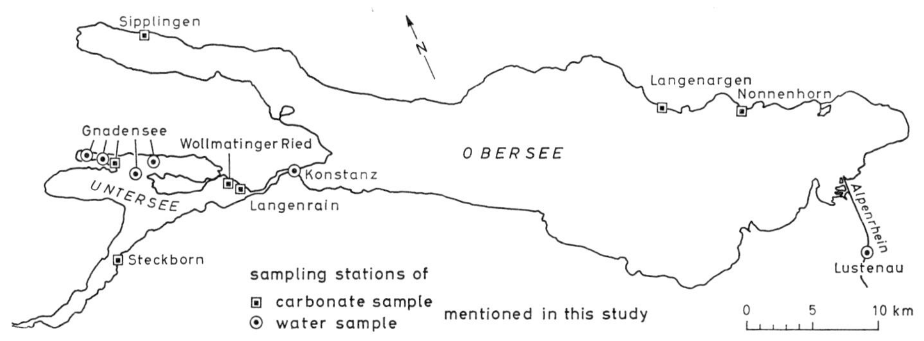

Fig. 1. Sampling stations of carbonate and water samples in Lake Constance and at Lustenau

EDTA titration. The accuracy and precision of the Strontium determinations (duplicate) is about $\pm$ 2%.

From all carbonate samples, X-ray diffractometer analyses were made in order to estimate the proportions of calcite and aragonite.

C. Concentration of Strontium and Calcium in Samples of Onkolites and Mollusk Shells of Lake Constance

The concentration of Strontium and calcium in altogether 57 samples is presented in table 1.

The lowest Strontium concentration found was 790 ppm, the highest 2776 ppm, the calcium concentration ranging between 29.20 and 38.75%. In order to obtain values independent from the fluctuating calcium contents, in the following only the Sr $\times$ 10^3/Ca atomic ratio (abbreviated: Sr/Ca-ratio) will be employed. As is to be seen from table 1, this Sr/Ca-ratio varies between 0.89 and 3.38.

The values are by no means arranged arbitrarily, they rather indicate that certain Sr/Ca-ratios are characteristic for certain species and genera. This is expressed very clearly in Fig. 2 and 3: the lowest Sr/Ca-ratios are to be found in calcitic onkolites, the highest Sr/Ca-ratios appear in pelecypods consisting mainly of aragonite.

Generally it can be said that higher aragonite contents usually imply higher Sr/Ca-ratios. With the exception of Pisidium (many different species!) and Planorbis (differences in the aragonite content!), the scattering area within a definite species is relatively limited. Thus, the Sr/Ca-ratio of a certain fresh water environment may cause a characteristic Sr/Ca-ratio in a certain species.

Table 1. *Chemical composition of onkolites, onkolitic sands and mollusk shells in Lake Constance*

No.	Species, material	Location	Year	% Ca	% Mg	ppm Sr	$\frac{Sr}{Ca}$ 1000	Aragonite : Calcite	
			A. Algae						
1				36.9	0.30	717	0.89		+++++
2				37.2	0.25	813	1.01		+++++
3				27.8	0.30	538	0.89		+++++
4				37.2	0.28	921	1.13		+++++
5				36.3	0.26	1033	1.28	(+)	(+)++++
6				37.2	0.23	813	1.01		+++++
7				24.1	0.37	541	1.03		+++++
8				37.2	0.27	1006	1.25		+++++
9				36.3	0.27	734	0.91		+++++
10		Insel Langen-rain (Untersee)	1966	37.8	0.29	914	1.11		+++++
11	Onkolites produced by blue-green algae (Schizophycea, Chlorophycea)			39.1	0.31	606	0.72		+++++
12				35.3	0.29	632	0.82		+++++
13				35.3	0.28	912	1.17	(+)	(+)++++
14				37.8	0.25	817	0.99		+++++
15				38.1	0.30	1007	1.21		+++++
16				36.2	0.25	915	1.16		+++++
17				39.4	0.27	765	0.91		+++++
18				37.2	0.23	929	1.14		+++++
19				37.5	0.27	1021	1.24		+++++
20		Gnadensee (Untersee)	1967	36.11	n. det.	946	1.20		+++++
21			1967	34.20	n. det.	901	1.21	(+)	(+)++++
1—21	Average			36	0.28	832	1.06		
		B. Sediments mainly derived from algal calcite							
22	I 3e		1965	34.83	0.15	917	1.24	+	++++
23	V 2g	Gnadensee (Untersee)	1965	37.88	0.24	1156	1.41	+	++++
24	B 7 Calcareous onkolitic sands		1965	31.70	0.49	851	1.23	+	++++
25	E 9		1965	26.32	0.63	756	1.31	(+)+	(+)+++
22—25	Average			32.68	0.38	920	1.30		

Table 1 (continued)

		C. Gastropods							
26	Bathyomphalus contortus	Sipplingen	1961	23.00	0.25	948	1.35	++++	+
27	Valvata piscinalis ssp. antigua	Nonnenhorn	1961	29.50	0.24	790	1.22	(+)+++	(+)+
28	Valvata piscinalis ssp. antigua	Nonnenhorn	1961	29.20	0.34	1013	1.58	(+)+++	(+)+
26—28	Average			29.30	0.28	901	1.40		
29	Lymnaea stagnalis	Sipplingen	1961	33.23	0.23	980	1.35	(+)+++	(+)+
30	Lymnaea (Radix) sp.	Steckborn	1962	37.65	0.14	1131	1.37	(+)+++	(+)+
31	Lymnaea (Radix) sp.	Sipplingen	1961	36.16	0.16	1102	1.39	++++	+
32	Lymnaea (Stagnicola) palustris	Sipplingen	1961	33.08	0.21	1068	1.47	++++	+
33	Lymnaea (Stagnicola) palustris	Sipplingen	1959	37.30	0.18	1278	1.56	(+)+++	(+)
29—33	Average			35.48	0.18	1112	1.43		
34	Planorbis carinalus	Sipplingen	1959	38.75	0.20	936	1.11	(+)+++	(+)+
35	Planorbis carinalus	Sipplingen	1961	30.05	0.29	976	1.48	(+)+++	(+)+
36	Planorbis planorbis	Sipplingen	1961	35.48	0.15	1184	1.52	++++	(+)
37	Planorbis planorbis	Wollmatinger Ried	1962	38.02	0.12	1750	2.10	+++++	
34—37	Average			35.58	0.19	1212	1.55		
38	Bithynia tentaculata	Steckborn	1962	37.10	0.19	1278	1.57	++++	+
39	Bithynia tentaculata	Sipplingen	1959	35.68	0.22	1290	1.64	++++	+
40	Bithynia tentaculata	Nonnenhorn	1961	35.00	0.18	1340	1.75	++++	+
41	Bithynia tentaculata	Sipplingen	1961	33.43	0.28	1322	1.82	++++	+
38—41	Average			35.30	0.22	1308	1.69		
42	Gyraulus acronicus	Sipplingen	1961	31.28	0.35	1025	1.49	(+)+++	(+)+
26—42	Average Gastropods			33.76	0.22	1142	1.52		

Table 1 (continued)

No.	Species, material	Location	Year	% Ca	% Mg	ppm Sr	$\frac{Sr}{Ca}$ 1000	Aragonite : Calcite	
	D. Pelecypods								
43	Anodonta cygnea	Sipplingen	1961	36.05	0.09	1486	1.85	++++	+
44	Anodonta cygnea	Langenargen	1966	36.60	0.08	1588	1.98	(+)++++	(+)
45	Anodonta cygnea	Langenargen	1966	36.35	0.11	1694	2.12	++++	+
43—45	Average			36.33	0.09	1589	1.98		
46	Sphaerium corneum	Nonnenhorn	1961	37.33	0.10	1628	1.99	(+)++++	(+)
47	Sphaerium corneum	Sipplingen	1961	37.30	0.13	1783	2.18	(+)++++	(+)
48	Sphaerium corneum	Sipplingen	1961	38.21	0.12	1945	2.32	(+)++++	(+)
46—48	Average			37.61	0.12	1785	2.16		
49	Pisidium amnicum	Nonnenhorn	1961	38.04	0.11	1144	1.37	(+)++++	(+)
50	Pisidium subtruncatum	Nonnenhorn	1961	37.80	0.15	1390	1.68	+++++	
51	Pisidium amnicum	Sipplingen	1961	36.11	0.25	1412	1.78	+++++	
52	Pisidium subtruncatum	Sipplingen	1961	33.08	0.36	1425	1.96	(+)++++	(+)
53	Pisidium nitidum	Nonnenhorn	1961	37.30	0.14	1880	2.30	+++++	
54	Pisidium casertanum ssp. ponderosum	Nonnenhorn	1961	30.76	0.10	1560	2.32	+++++	
55	Pisidium nitidum	Sipplingen	1961	30.76	0.41	1818	2.69	+++++	
56	Pisidium henslowanum	Nonnenhorn	1961	37.83	0.17	2430	2.93	+++++	
57	Pisidium henslowanum	Sipplingen	1961	37.41	0.28	2776	3.38	(+)++++	
49—57	Average			36.45	0.22	1759	2.27		
43—57	Average Pelecypods			36.06	0.17	1731	2.19		

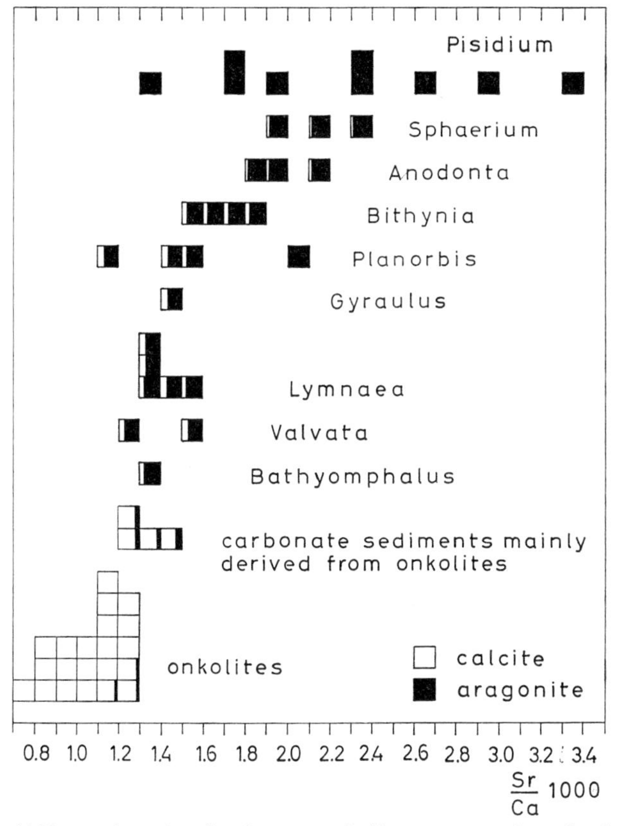

Fig. 2. Sr/Ca × 1000 atomic ratio of carbonate onkolites, gastropods and pelecypods as well as of sediments mainly derived from onkolites in Lake Constance

D. Comparison of Sr/Ca-Ratios with other Environments (Table 2)

The Sr/Ca-ratios of Lake Constance carbonates are considerably higher when compared with the mean values for fresh water carbonates (Odum, 1957). In comparison with the mean values for sea water mollusk shells, however, (Thompson and Chow, 1955; Odum, 1957; Waskowiak, 1962) the average Sr/Ca-ratios are lower. It is impossible, however, to obtain a clear differentiation in an individual case: if the "lowest marine values" indicated by Thompson and Chow (1955) and Odum (1957) are plotted in Fig. 3, the result will be that almost all individual values of Lake Constance mollusks could still fall within the marine range indicated by Thompson and Chow (1955). Relying on the values of Odum (1957), this applies to still more than half of the values, above all Pelecypods. Only algal calcite is clearly differenciated.

E. Composition of Lake Constance and Alpenrhein Waters

The composition of Lake Constance and Alpenrhein waters as well as that of the Seerhein (connexion between Obersee and Untersee) is well known as a result of investigations made by Müller (1965) and Müller and Röper (1967). The total concentration of Lake Constance water shows seasonal changes and is related to the varying composition of the Alpenrhein water (table 3).

Table 2. *Mean Sr/Ca-ratios for fresh water and marine environment calcite algae and aragonite mollusks as compared with Lake Constance material*

	Fresh waters of humid regions (ODUM, 1957)		Lake Constance this study		Marine environment (ODUM, 1957)			Marine environment (THOMPSON and CHOW, 1955)		
	Number of cases	Mean Sr/Ca	Number of cases	Mean Sr/Ca	Number of cases	Mean Sr/Ca	Lowest marine value	Number of cases	Mean Sr/Ca	Lowest marine value
Calcite algae	3	0.58	21	1.06	9	3.96	3.60	9	3.20	2.93
Aragonite gastropods	33	0.75	17	1.52	14	2.34	1.70	53	1.68	1.25
Aragonite pelecypods	38	0.87	15	2.19	8	2.51	1.70	64	1.85	1.01

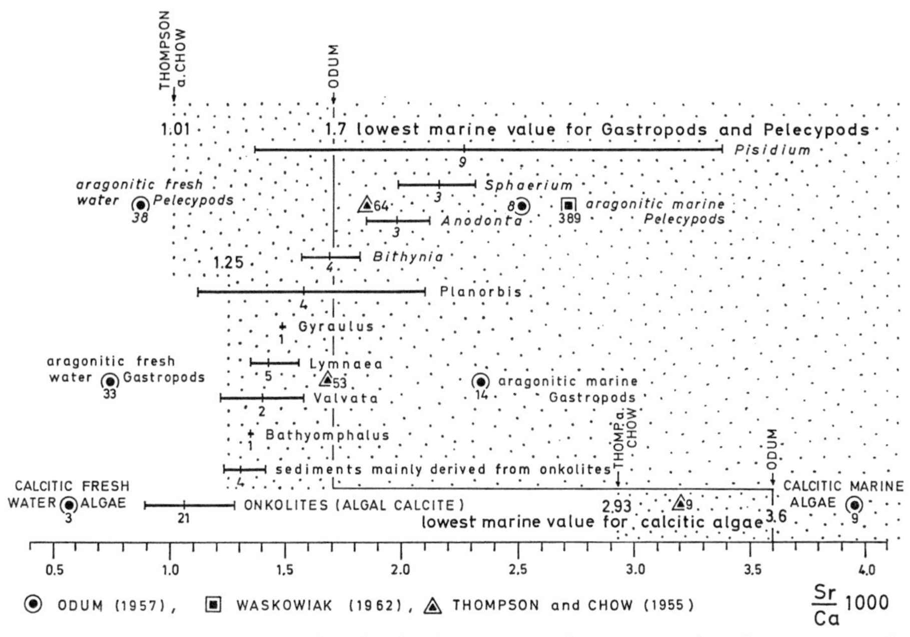

Fig. 3. Sr/Ca × 1000 atomic ratio of Lake Constance carbonate secreting algae, gastropods and pelecypods as compared to corresponding marine values

Table 3. *Major ions in Lake Constance and Alpenrhein waters*

	Lake Constance (Obersee) mean values from 28 stations (MÜLLER and RÖPER, 1967)		Alpenrhein (Lustenau) mean values (MÜLLER, 1965)	
	2. 7. 1965	18. 3. 1966	June 1963	October 1962
HCO_3^-	146.4	165.3	117.0	112.1
SO_4^{--}	36.8	40.6	39.2	74.4
Cl^-	3.1	4.2	1.1	2.1
Ca^{++}	47.4	47.0	38.2	48.2
Mg^{++}	8.4	6.7	10.4	12.4
Na^+	2.9	3.4	1.3	2.9

Table 4. *Sr/Ca-ratios in the water of the Untersee (Lake Constance) as measured on July 7, 1967*

Location	Ca mg/l	Sr mg/l	$\frac{Sr}{Ca}$ 1000
Seerhein, Konstanz	32.2	0.42	5.9
Mettnau, Wiesenrain, depth 1.5 m	36.0	0.43	5.6
Markelfinger Winkel, depth 5 m	37.0	0.42	5.3
Gnadensee between Mettnau-Reichenau, depth 21 m	34.4	0.42	5.6
Reichenau-Mettnau, depth 4.5 m	34.5	0.43	5.7

Table 4 comprises a number of typical analyses revealing that the waters have low- to medium electrolyte concentrations and are characterized by relatively high bicarbonate- and sulfate content. During a low water level, the sulfate content of the Alpenrhein may rise to more than 100 mg/l (MÜLLER, 1965).

After having found the high Strontium contents in the carbonates, the investigations were extended to Strontium contents of Lake Constance and Alpenrhein waters; it was to be expected, that a high Sr/Ca-ratio in the water might have caused the high Sr/Ca-ratios in the carbonates.

Fig. 4 shows the high water level mark of the Alpenrhein during the period of investigation from June 1966 to July 1967 as well as the observed Sr-values and the Sr/Ca-ratios. The Sr-contents vary between 0.35 and 0.77 mg/l, the Sr/Ca-ratios between 4.7 and 10.7. It may be clearly seen from Fig. 4 that a relationship exists between the Sr-contents and the Sr/Ca-ratio on one hand and the high water level mark of the Alpenrhein on the other hand: with decreasing water discharge, Sr-contents and Sr/Ca-ratio increase.

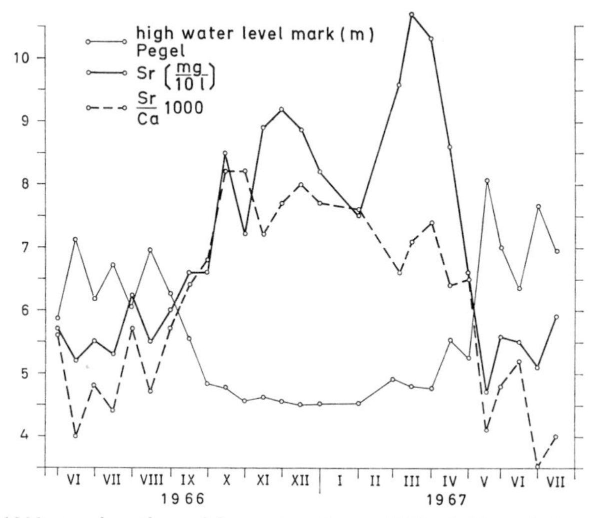

Fig. 4. Sr/Ca × 1000 atomic ratio and Sr-content (in mg/10 l) of Alpenrhein-water at Lustenau

As the curve for the sulfate content of the Alpenrhein water shows a similar relationship to the water discharge, the curve for the Sr-content runs also ± parallel to that of the sulfate content.

The Sr/Ca-ratios found in the Alpenrhein are much higher than those observed in rivers and lakes of humid climates (Odum, 1957 [2]; Livingstone, 1963; Skougstad and Horr, 1963). The question of the origin of these high Sr/Ca-ratios, which even sometimes exceed those of sea water, will be discussed in detail elsewhere.

The highest Sr/Ca-ratios appear only at low water level of the Alpenrhein. At high water level much more water (with a lower Sr/Ca-ratio) is led into the lake. As a result, the water of the lake is homogenized and the Sr/Ca-ratios are relatively uniform, fluctuating according to the season and the location in the lake between 4.5 and 7 (Müller, in preparation). Table 4 contains — as an example — the Sr/Ca-ratios for the Gnadensee of Lake Constance on July 7, 1967. It must be considered, however, that these values were observed during a high water period of the Alpenrhein; higher values might thus be expected in winter. According to our investigations, a Sr/Ca-ratio of 6 might be assumed as average for Lake Constance water.

F. The Distribution Coefficient for Strontium between Water and Calcium Carbonate

From Sr/Ca-ratios for onkolites and mollusk shells and from the Sr/Ca-ratio of the water the distribution coefficient of Strontium between water and the biogenically precipitated carbonates was calculated according to the formula

$$D\frac{\text{solid}}{\text{liquid}} = \frac{(\text{Sr/Ca})\ \text{solid}}{(\text{Sr/Ca})\ \text{liquid}}$$

The results are given in table 1 and in the graph of Fig. 5.

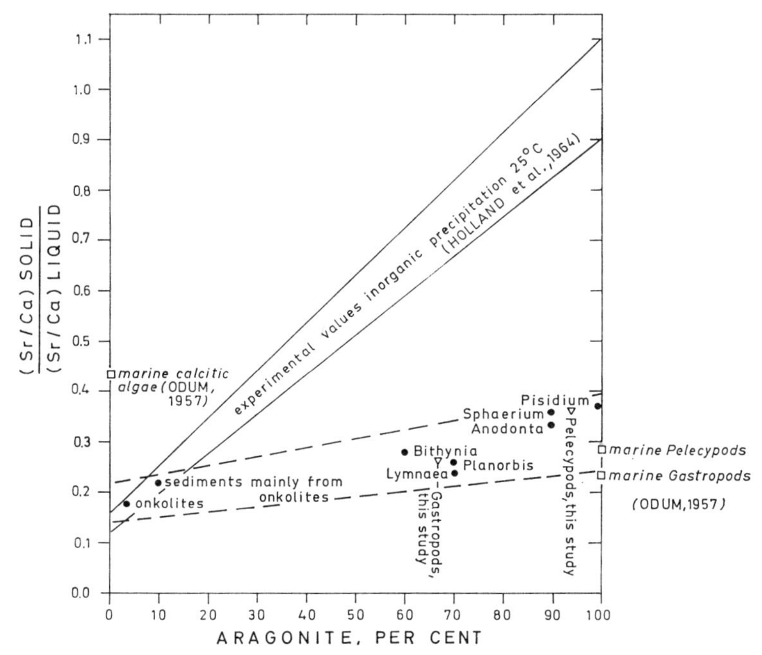

Fig. 5. Distribution coefficient for strontium between water and calcium carbonate in Lake Constance shell material

The lowest values were found for algal calcite, the highest for Pelecypods consisting mainly of aragonite.

During laboratory experiments by Holland et al. (1964), distribution coefficients of inorganically precipitated calcium carbonate were determined. For calcite, precipitated at 25 °C, $D = 0.14 \pm 0.02$ was found; for aragonite at 25 °C, $D = 1.0 \pm 0.1$ was determined.

For the calcitic onkolites $D = 0.177$ is in close accordance with the experimental value if one considers, that small amounts of aragonite are often present (aragonitic shell debris as nuclei of the onkolites) and also some Sr is revealed from non-carbonate minerals during the dissolution of carbonates.

The distribution coefficients for the shell material consisting of aragonite or mixtures of aragonite and calcite are significantly lower than those observed in inorganically precipitated material, they range from $D = 0.24$ in Lymnaea to $D = 0.37$ in Pisidium.

The distribution coefficients in organic marine shells (Fig. 5) are in the same magnitude (Odum, 1957), thus demonstrating that in both cases not the external Sr/Ca-ratio but that of the body fluid has to be considered as the liquid phase of the system.

The values obtained in mollusk shells are therefore "a measure of the *effective* partitioning of Sr^{2+} by the organism" (Faure et al., 1967).

G. The Magnesium Content of the Carbonates

The magnesium concentrations show great uniformity within the examined material. The concentrations range between 0.09 and 0.41%. In general, the Mg-concentrations seem to be slightly higher in Gastropods (0.22%) than in Pelecypods (0.17%). The onkolites exhibit the highest values (0.28%). These high values, however, might be due to inclusions of detrital dolomite grains from the normal sediment on which they grow.

H. Discussion of Results

The examination of Lake Constance algal and mollusk shell carbonates proved Sr/Ca-ratios unusually high for fresh water environments. These high Sr/Ca-ratios, however, can be contributed to a high Sr/Ca-ratio in Lake Constance water which is mainly supplied from the Alpenrhein. In most cases, the observed Sr/Ca-ratios for mollusk shells fall into the range of marine mollusks.

These findings reveal that Sr-contents and Sr/Ca-ratios of carbonates are *not* indicators of salinity but of the Sr/Ca-ratio in the depositional environment.

Acknowledgements

We are very indebted to Dr. Meier-Brook for putting his excellent material at our disposal. Thanks are also due to Manfred Gastner for carrying out the chemical analyses.

References

Faure, G., J. H. Crocket, and P. M. Hurley: Some aspects of the geochemistry of strontium and calcium in the Hudson Bay and the Great Lakes. Geochim. et Cosmochim. Acta **31**, 451—461 (1967).

Holland, H. D., T. V. Kirsipu, J. S. Huebner, and U. M. Oxburgh: On some aspects of the chemical evolution of cave waters. J. Geol. **72**, 36—67 (1964).

Krejci-Graf, K.: Geochemische Faziesdiagnostik. Freiberger Forschungsh. C 224 Geologie, 80 p. (1966).

Landergren, S., u. F. T. Manheim: Über die Abhängigkeit der Verteilung von Schwermetallen von der Fazies. Fortschr. Geol. Rheinld. u. Westf. **10**, 173—192 (1963).

Livingstone, D. A.: Chemical composition of rivers and lakes. In: Data of Geochemistry, 6th ed., Chapter G. Geol. Survey Profess. Papers **440—G**, 64 p. (1963).

Logan, B. W., R. Rezak, and R. N. Ginsburg: Classification and environmental significance of algal stromatolites. J. Geol. **72**, 68—83 (1964).

Müller, G., u. H.-P. Röper: Erste Ergebnisse hydrochemischer Untersuchungen am Bodensee. Gas- u. Wasserfach **107**, 826—830 (1967).

— — Ergebnisse einjähriger systematischer Untersuchungen über die Hydrochemie von Alpen- und Seerhein (mit Einzeluntersuchungen an weiteren Bodenseezuflüssen). Fortschr. Wasserchem. u. Grenzgeb., H. 2, 33—99 (1965).

— — Die Sedimentbildung im Bodensee. Naturwissenschaften **53**, 237—247 (1966).

— — Beziehungen zwischen Wasserkörper, Bodensediment und Organismen im Bodensee. Naturwissenschaften **54**, 454—466 (1967).

ODUM, H. T.: Biogeochemical deposition of strontium. Inst. Marine Sci. IV, 38—114 (1957).
— Strontium in natural waters. Inst. Marine Sci. IV, 22—37 (1957).
SCHÖTTLE, M.: Die Sedimente des Gnadensees. — Ein Beitrag zur Sedimentbildung im Bodensee. Unpubl. Diss., 104 p., University of Heidelberg 1967.
— and G. MÜLLER: Recent carbonate sedimentation in the Gnadenssee (Lake Constance) In: MÜLLER, G., and G. M. FRIEDMAN (Edts.): Recent Developments in Carbonate Sedimentology in Central Europe, p. 148—156. Berlin-Heidelberg-New York: Springer 1968.
SKOUGSTAD, W., and C. A. HORR: Occurrence and distribution of strontium in natural waters. Geol. Survey Water-supply paper 1496-D, 55—97 (1963).
THOMPSON, T. G., and T. J. CHOW: The Strontium-calcium ratio in carbonate-secreting marine organisms. In: Papers in Marine Biology and Oceanography, Supplement to Deep-Sea Research, **3**, 30—39 (1955).
WASKOWIAK, R.: Geochemische Untersuchungen an rezenten Molluskenschalen mariner Herkunft. Freiberger Forschungsh., C 136, 155 p. (1962).

Ca-Mg-Distribution in Carbonates from the Lower Keuper in NW-Germany

Hannelore Marschner *

With 4 Figures

Abstract

55 samples of carbonate sediments from a vertical stratigraphic section through the Lower Keuper Formation (Triassic) in the Weser area (North Western Germany) were evaluated as to their Mg/Ca ratio, taking their genetic aspects under consideration. Subsequently, the dolomite/calcite ratio, the Ca/Mg ratio of the dolomites, and the Mg-content of the calcite are discussed.

Results:

1. The Ca/Mg ratio of the early diagenetically formed solid solution members between calcite and dolomite depends on the Ca/Mg ratio of the depositional environment, i.e. on its salinity.
2. In the fossiliferous carbonates investigated, this relationship is only reflected by the dolomite, the $CaCO_3$-excess of wich decreases a) within a locally developed carbonate bank, b) within the entire section with increasing salinity until it disappears completely.
3. Accordingly, stoichiometric dolomites are formed during early diagenesis only in a hypersaline environment.
4. The unstable $CaCO_3$-excess in the dolomite appears to be unaffected by secondary processes; consequently, it can be considered as metastable for the relationship under consideration.
5. Because of its instability, the Mg-content of the early diagenetically formed calcite does not show a dependence on the stratigraphy and facies.
6. Much of the calcite is secondary. Its Mg-content is influenced by the Mg-content of the pore-solutions and is an indicator for the degree of dedolomitization.
7. Dolomite and calcite often belong to different phases of formation, so that their quantitative ratio cannot be used to determine facies. The Ca/Mg-ratio in the dolomites may be used instead.

A. Introduction

X-ray investigations of Chave (1952), Graf and Goldsmith and others (1955 and later) revealed that: a) calcite can incorporate Mg^{2+} distributed irregularly in its crystal lattice up to a quantity equivalent to that in dolomite, b) in dolomite, up to approximately 10 mole-% Mg^{2+} can be substituted by excess Ca^{2+} while an ordered Ca-Mg-distribution is retained. Thus a continuous solid solution series calcite-dolomite was established. Natural and experimentally produced intermediate members were described since then in numerous publications. The series is unstable under normal PT-conditions. The Mg-content of the calcite appears to be considerably less stable than the Ca-excess of the dolomite with respect to time and changes in the surrounding solution.

---

* Laboratorium für Sedimentforschung, Universität Heidelberg, Germany.

The Ca/Mg-ratio of these carbonates, i.e., their position within the continuous solid solution series, corresponds to definite stability conditions during their formation and is thus of particular interest.

Synthesis (SIEGEL, 1961; GLOVER and SIPPEL, 1967) and study of Recently-formed natural associations of Mg-calcites and Ca-dolomites (V. D. BORCH, 1965) indicate that a) the total Mg/Ca-ratio of the carbonates, in which the dolomite/calcite-ratio is included, increases with the Mg/Ca-ratio of the precipitating solution, b) the dolomite/calcite-ratio is proportional to the cation-ratio Mg/Ca in these carbonates. E.g., SKINNER (1962) found in Recent material, that the Mg-content of calcite in a particular bed increases proportionally to the amount of associated dolomite. FÜCHTBAUER and GOLDSCHMIDT (1965) made use of the fact that the Ca-excess in fossiliferous, early diagenetic dolomites increases with the amount of the co-existing calcite.

Under natural conditions, the Mg/Ca-ratio increases during evaporation of sea water, e.g., in saline environments. The dolomite/calcite-ratio was used up to now for deducing such environments in sedimentary rocks. This ratio commonly undergoes changes in older carbonate rocks; it remains to be proved, however, to what extent it can be replaced or modified by using the relationships outlined above.

In the present study, carbonate sediments of a vertical section through the Lower Keuper were studied under these aspects. Because the Lower Keuper is transitional between the Upper Muschelkalk (limestone facies) and the Middle Keuper (anhydrite facies) with respect to its stratigraphy and facies, these carbonate sediments are of particular interest.

B. Materials and Methods

I. Vertical Section Studied

The beds of the Lower Keuper include the entire spectrum of shallow water sediments — sand-silt-clay-marl-carbonate — in an irregular, alternating sequence. They are stratigraphically sub-divided by GRUPE (1907):

km_1	
ku_2	Grenzdolomitregion
ku_1,	Hauptlettenkohlensandstein
$ku_1(_3)$	Anoplophora-Schichten
$ku_1(_2)$	Hauptdolomit
$ku_1(_1)$	Unterer Lettenkohlensandstein
mo_2	

The stratigraphic symbols serve to schematically represent the section in the following parts of this study.

55 surface samples were examined from the carbonate rich beds of a composite vertical section through the Lower Keuper, encompassing 4 partial sections from a limited area west of the Weser, approximately 15 km south of Bad Pyrmont[1]. The total thickness of the section is about 40 m. In the lower part of the section ($ku_{1(1)}$—$ku_{1(3)}$), relatively coarse-grained carbonate banks with skeletal fragments

[1] Sections and their geographical coordinates: Bödexen (R 35_{237}, H 57_{448}), Bönekenberg (R 35_{197}, H 57_{472}), Fürstenau (R 35_{222}, H 57_{448}), Löwendorf (R 35_{194}, H 57_{455}).

predominate. Their carbonate content fluctuates between 85 and 97%. In intertratified, fine-grained marl layers, the carbonate content can decrease to below 50%. In the upper portion of the section (upper ku_1,—ku_2), the carbonate fraction of the carbonate beds amounts to 75% or less. The material is fine-grained, with little variation in grain-size and without skeletal fragments.

Apparently the depositional environment changed within the section toward saline conditions: A decrease in the supply of fresh sea water during deposition of the stratigraphically higher rocks is indicated by: 1. the absence of intercalated sand layers, 2. a pronounced decrease in the quartz-fraction in favor of clay minerals in the non-carbonate fraction of the carbonate layers, and 3. a decrease in the content of metallic trace elements. „Residual structures" which become more common upwards and which occur exclusively in ku_2, consist generally of small calcite-filled druses containing authigenic quartz with inclusions of anhydrite. They indicate the transition to the $CaSO_4$-facies of the Middle Keuper.

II. Method

The Ca/Mg-distribution in the carbonates was determined by the X-ray diffractometer method. Technical data: Müller Mikro 111, Philips diffractometer, PW 1051, Co-tube.

The dolomite/calcite-ratio was obtained from the intensity-relationship of both the principal reflections $d_{(104)\ \text{dolomite}} = 2{,}886$ Å/$d_{(104)\ \text{calcite}} = 3{,}035$ Å by comparison with standard mixtures. From the peak shifts in the position of the reflections measured against quartz d (101) = 3,343 Å the deviations from the theoretical cation content of these minerals were determined following GRAF and GOLDSMITH (1958, p. 97). The possible incorporation of Fe^{2+} and Mn^{2+} in the crystal lattice is not taken into consideration. The standard deviation is $\pm$ 0.4 mole-% $CaCO_3$.

C. Results and Discussion

I. Dolomite/Calcite-Ratio of the Sediments

The dolomite/calcite ratio varies from 100/0 to 30/70 independent of the position of the sample within the section; Dolomite percentages of 80 to 100 predominate (Fig. 1, points measured).

Fabric investigations show, however, that both minerals can often be assigned to various phases of formation. Thus their bulk ratio alone is of no interpretative value. Dolomite and some of the calcite formed during early diagenesis. Another portion of the calcite — particularly in the calcite-rich samples — largely crystallized during late diagenesis or is secondary (i.e., by weathering). The dashed line in Fig. 1 only represents the dolomite/calcite-ratio for the early diagenetic carbonate minerals. In the carbonate sediments of the lower portion of the section, calcite always occurs (about 10%), although it is missing in the upper part indicating a decrease of calcite with time.

Moreover, the amount of calcite is generally larger in the lower part of a carbonate bank than in the upper. The same is true for the content of non-carbonate minerals. Thus the sequence: marl-calcite-dolomite is only weakly indicated within each carbonate bank, because the first two members are almost totally suppressed. One must be careful, however, using this interpretation, because weathering processes can lead to the same results: After dissolution of carbonate (calcite + dolomite), reprecipitation took place only in the form of calcite and chiefly in the lowermost zones of the banks.

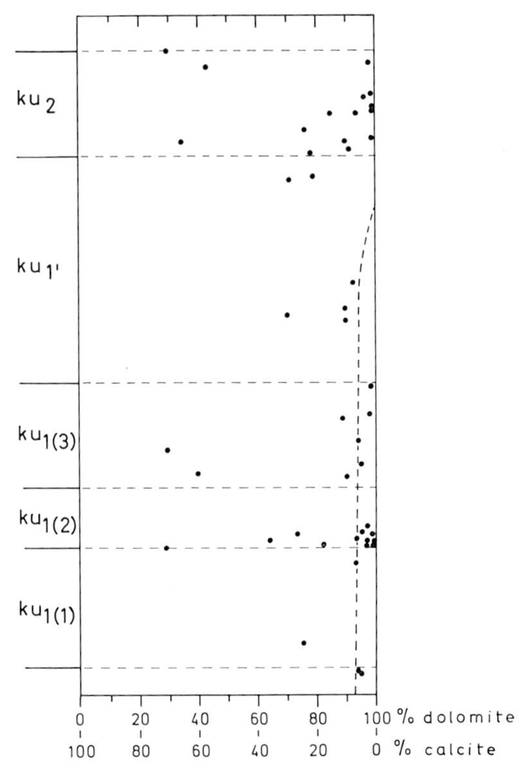

Fig. 1. Dolomite/calcite-ratio of the sediments depending on their stratigraphic position. The filled symbols shows the dolomite/calcite ratios of the total carbonate fraction regardless of time of formation. The dashed line indicates the approximate ratio of the early diagenetically formed carbonates

II. Ca/Mg-Ratio of the Dolomites

The $CaCO_3$-values in the dolomite vary between 8.6 mole-% excess and 1.3 mole-% deficiency. They correspond remarkably with the detailed stratigraphic position of the samples (Fig. 2). Generally, the dolomite from the lower portions of the section ($ku_{1(1)}$—$ku_{1(3)}$) has a more or less constant $CaCO_3$-excess of 6.5 to 5.5 mole-%, which decreases only slightly upwards. The values diminish in the middle ku_1, to approximately 2 mole-% and finally disappear completely in the ku_2-horizon. In some cases they reach negative values. The values deviating from the lower part of $ku_{1(2)}$ are shown in Fig. 3.

Thus, the $CaCO_3$-excess in the dolomite generally decreases from the base to the top of the section showing the same trend as the calcite that is syngenetic with dolomite, a trend, also found by Füchtbauer and Goldschmidt. This indicates, as does the fabric, that these carbonates were formed during early diagenesis and that their Ca/Mg-ratio was influenced by the depositional environment. Therefore, the degree of salinity in the depositional environment seems to have increased with time, i.e. upwards in the stratigraphic section.

Füchtbauer and Goldschmidt described dolomites with considerable Ca-excess from rocks deposited in deeper water environment in the Zechstein ("Hangfazies");

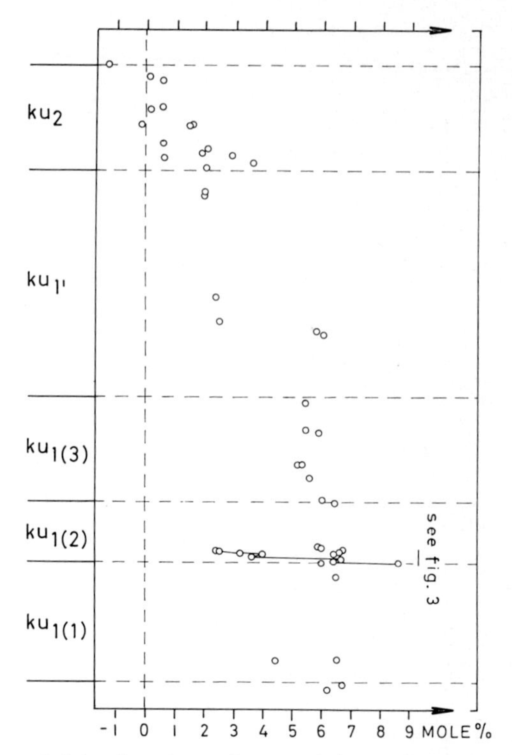

Fig. 2. $CaCO_3$-excess of dolomites depending on their stratigraphic position. The linearly connected values are enlarged in Fig. 3

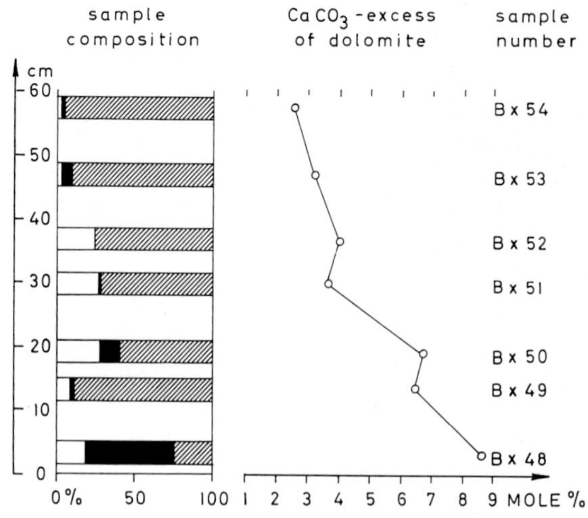

Fig. 3. Change in the composition of the samples and of the $CaCO_3$-excess in the dolomite within a carbonate bank (enlarged portion of Fig. 2, base $ku_{1(2)}$). ▭ non-carbonates, ■ calcite, ▨ dolomite. To the left, the composition of the samples depending on their exact stratigraphic position (schematically drawn in cm); to the right the $CaCO_3$-excess of their dolomites

such dolomites also occur in the lower part of the section. Even on the basis of organic remains and clastic intercalations, a normal marine or slightly restricted environment may be assumed. In the upper part of the section, the fine-grained, marly dolostones which partly contain residues of gypsum are probably deposited in a hypersaline environment and are thus of purely inorganic origin.

Therefore, a stochiometric composition of early diagenetic dolomite appears to indicate hypersaline conditions.

A decrease of the Ca/Mg ratio in dolomite with increasing salinity can also be observed within some particularly thick carbonate banks, which are regionally restricted. In their upper parts they show Ca/Mg-ratios in dolomite which are much lower than those which one would expect from their stratigraphic position. Such locally developed Ca/Mg-ratio are in the total depositional area first attained in the upper parts of the section.

A 60 cm thick carbonate bank from the base of $ku_{1(2)}$ is shown schematically in Fig. 3. The individual layers of this bank differ petrographically. Notably, calcite, $CaCO_3$ content in dolomite, and non-carbonate fraction exhibit similar changes[2]. On the whole, the Ca/Mg ratio decreases from 8.6 mole-% at the base to 2.5 mole-% at the top of the bed, thus reflecting a gradually more restricted environment. However, the Ca/Mg-ratio of dolomite increases upwards in some layers, paralleled by an increase in the amount of calcite and non-carbonate matter. This indicates higher supply of fresh sea water and therefore a decrease in the concentration of the solution. The repeated influx of fresh sea water (regressive tendency) was unimportant, however, compared with the overall increase in salinity (progressive tendency).

III. Mg-Content of the Calcites

Mg-contents in calcite vary from 1.2 to 5.6 mole-%, but are generally around 2 mole-%. In contrast to the Mg-contents of the dolomite they do not depend on the stratigraphic position (Fig. 4), even when only the early diagenetic calcite is considered. A relationship to the dolomite content such as exists in Recent calcite is also lacking. This was expected considering the age of these sediments. It is surprising, however, that the calcite formed by weathering likewise contains Mg in quantities of approximately 2 mole-%. This seems to depend on the Mg-content of the pore-solution, since the values are highest where a dissolution of dolomite by weathering can be substantiated.

IV. Facies Development within the Section

The lower part of the section reflects constant and nearly normal marine conditions. During deposition of the upper ku_1, a significant chemical restriction of the environment took place, leading to the hypersaline environment of deposition for ku_2. The interruption of the carbonate sedimentation by clastic intercalations within the section had no appreciable effects on the composition of the carbonate fraction. With

[2] As these parameters are mutually dependent, the absence of calcite in sample Bx 52 (Fig. 3), therefore, is probably due to secondary dissolution. The original calcite content might have been larger than that in sample Bx 51 judging from the higher $CaCO_3$-excess in the dolomite. The high calcite content of sample Bx 48 was evidently enlarged through later formation of the mineral.

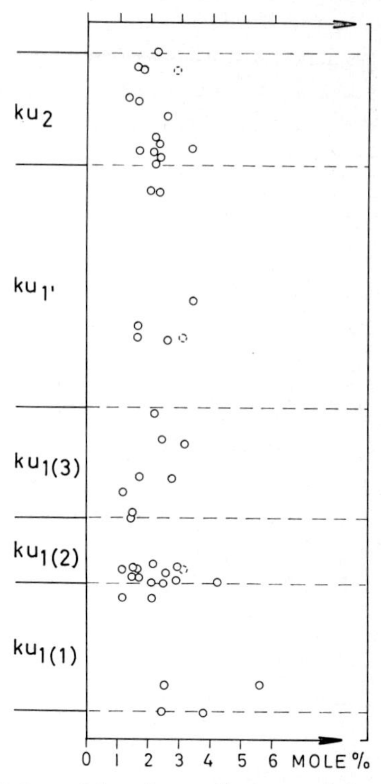

Fig. 4. $MgCO_3$-content of the calcites from all times of formation depending on their stratigraphic position

the exception of local peculiarities, a uniform environment of precipitation can be deduced.

Within the progressive facies development, a rhythmic development could not be ascertained.

Acknowledgements

The present report is part of a Ph. D. thesis. The work was done in 1964—1965 in the Laboratory for Sedimentary Petrography of the Geologisches Staatsinstitut, Hamburg. I am indepted to Prof. I. Valeton who suggested the study for her encouragement and advice. I am also grateful to Prof. H. Saalfeld for his permission to make use of space and laboratory equipment of the Mineralogisch-Petrographisches Institut, University of Hamburg.

References

v. d. Borch, C.: The distribution and preliminary geochemistry of modern carbonate sediments of the Coorong area, South Australia. Geochim. et Cosmochim. Acta **29**, 781—799 (1965).

Brown, G.: The x-ray identification and crystal structures of clay minerals. 544 p. London: Mineralogical Society 1961.

Chave, K. E.: A solid solution between calcite and dolomite. J. Geol. **60**, 190—192 (1952).

Füchtbauer, H., u. H. Goldschmidt: Beziehungen zwischen Calciumgehalt und Bildungsbedingungen der Dolomite. Geol. Rundschau **55**, 29—40 (1965).

Glover, E. D., and R. F. Sippel: Synthesis of magnesium calcites. Geochim. et Cosmochim. Acta **31**, 603—613 (1967).

GOLDSMITH, J. R., and D. L. GRAF: Relation between lattice constants and composition of Ca-Mg-carbonates. Am. Mineralogist **43**, 84—101 (1958).

— —, and O. I. JOENSUU: The occurence of magnesian calcites in nature. Geochim. et Cosmochim. Acta **7**, 212—230 (1955).

GRAF, D. L.: Some hydrothermal synthesis of dolomite and protodolomite. J. Geol. **64**, 173—186 (1956).

GRUPE, O.: Der Untere Keuper im südlichen Hannover. Adolph v. Kuenen-Festschrift, Stuttgart: E. Schweizerbartsche Verlagsbuchhandlung, S. 64—134, 1907.

INGERSON, E.: Problems of the geochemistry of sedimentary carbonate rocks. Geochim. et Cosmochim. Acta **26**, 815—848 (1962).

KÜBLER, B.: Etude pétrographique de l'Oehningen (Tortonien) du Locle (Suisse occidentale). Beitr. Mineral. Petrog. **8**, 267—314 (1962).

MARSCHNER, H.: Mineralogisch-petrographische Untersuchungen an karbonatreichen Gesteinen aus dem Unteren Keuper des Weserberglandes, 137 p. Diss., Univ. Hamburg, priv. circ. 1966.

SIEGEL, F. R.: Factors influencing the precipitation of dolomitic carbonates. State Geol. Survey Kansas, Bull. **152**, 127—158 (1961).

SKINNER, H. C. W.: Precipitation of calcian dolomites and magnesian calcites in the Southeast of South Australia. Am. J. Sci. **261**, 449—472 (1963).

WURSTER, P.: Krustenbewegungen, Meeresspiegelschwankungen und Klimaänderungen der deutschen Trias. Geol. Rundschau **54**, 224—240 (1965).

Proof and Significance of Amino Acids in Upper-Jurassic Algal-Sponges-Reefs of the Swabian Alb (SW-Germany)

Karl Hiller*

Calcareous crusts and/or stromatolites take an essential, often even predominant part in the composition of the upper-jurassic algal-sponges-reefs (Hiller, 1964; see Aldinger, 1968) of the Swabian Alb. According to Hiller's (1964) sedimentological studies, these calcareous crusts and stromatolites are very probably calcareous precipitations of primitive sea algae (presumably blue and green algae). To obtain another clue to the organic nature of these forms it was attempted to prove the presence of amino acids by paper chromatography, for the most according to the methods of Degens and Bajor (1960) and Porter, Margolis, and Sharp (1957) (see Hiller and Kull, 1967). The most important data of the analysis are listed in Table 1.

Results

In reef limestones, there are appreciable differences both of total amino acid content of some particular amino acids in the $\pm$ micritic matrix on the one hand and the organic components (siliceous sponges, calcareous crusts, stromatolites) on the other, provided that these samples come from the immediate vicinity.

The relatively high content of amino acids in reef limestones as compared to bank limestones and the higher concentration of amino acids in reef-building (calcareous crusts, stromatolites) or probably only reef-triggering (siliceous sponges) components are indicative of a relation between higher amino acids content and the presence of reef-building components, as well as of the fact that the amino acids were not evenly distributed in the sediment by diagenesis.

These results provide another important indication to the organic nature of the calcareous crusts and stromatolites whose origin is seen in connection with primitive marine algae and gave rise to the designation algal-sponges-reefs of the Swabian Upper-Jurassic.

References

Aldinger, H.: Ecology of algal-sponge-reefs in the Upper Jurassic of the Schwäbische Alb, Germany. Müller, G., and G. M. Friedman (eds.): Recent developments in carbonate sedimentology in Central Europe, p. 250—253. Berlin-Heidelberg-New York: Springer 1968.

Degens, E. T., u. M. Bajor: Die Verteilung von Aminosäuren in bituminösen Sedimenten und ihre Bedeutung für die Kohlen- und Erdölgeologie. Glückauf **96**, 1525—1534 (1960).

Hiller, K.: Über die Bank- und Schwammfazies des Weißen Jura der Schwäbischen Alb (Württemberg). Arb. geol. paläont. Inst. TH Stuttgart, N.F. **40** (1964) (cum lit.).

—, u. U. Kull: Über den Nachweis von Aminosäuren in Kalksteinen des Weißen Jura der Schwäbischen Alb. Neues Jahrb. Geol. u. Paläontol. Monatsh. **3**, 150—158 (1967) (cum lit.).

Porter, A. C., D. Margolis, and P. Sharp: Quantitative determination of amino acids by paper chromatography. Contrib. Boyce-Thompson Inst. **18**, 465—476 (1957).

* Gewerkschaft Brigitta, Hannover, Germany

Table 1. *Contents of amino acids in ppm; traces = less than 0,05 ppm. I-V = reef limestones, VI = bank limestone.*

		asparagine-acid	glutamic-acid	serine	glycine + ? threonine	alanine	lysine + arginine	valine	leucine + isoleucine	γ-aminobutyric-acid	total amino acids
I	micrit. matrix	0,55	0,29		trace	0,15	0,23	trace	trace	?	1,22
	siliceous sponge	0,50	0,37		trace	0,19	0,31	0,20	trace	0,19	1,76
II	micrit. matrix	0,43	0,20	0,13	trace	0,12	0,30	trace	trace	?	1,18
	calcareous crusts	0,52	0,19	0,13	0,20	0,11	0,45	0,10	0,15	0,12	1,97
III	micrit. matrix	0,48	0,30		0,17	0,14	0,14	0,14	0,17	0,11	1,65
	calcareous crusts	0,49	0,25		0,29	0,12	0,29	0,15	0,23	0,23	2,05
IV	micrit. matrix	0,54	0,33		0,20	0,16	0,28	0,10	0,13	—	1,74
	stromatolites	0,61		0,66		0,22	0,24	0,12	0,19	—	2,04
V	micrit. matrix	0,48		0,28		0,12	0,35	0,08	0,18	—	1,49
	calcareous crusts	0,51		0,42		0,11	0,25	0,12	0,26	—	1,67
	stromatolites	0,59		0,45		0,20	0,26	0,14	0,25	—	1,89
VI	micrit. bank limestone	0,20		0,25		0,08	0,25	0,10	0,05	—	0,93

Geomicrobiology and Geochemistry of the "Nari-Lime-Crust" (Israel)

WOLFGANG E. KRUMBEIN*

With 5 Figures

Abstract

Two samples of Nari-lime-crust and underlying substrate (Turonian limestone) from Israel were investigated microbiologically and geochemically. After determination of quantitative values for certain groups of microorganisms, laboratory experiments on the isolated flora were caried out. We obtained the following results:

(1) The Nari-lime-crust contains a well-developed microflora consisting of large amounts of autotrophic and heterotrophic bacteria, fungi, actinomycetae, green and blue-green algae. The floral association was dominated by the algae.

(2) The numbers of organisms per gram dry weight differed by about 3 exponents from limestone to lime-crust.

(3) Biological cycles (nitrogen, sulfur, carbon) were not as well and homogeneously developed as compared to normal soils. Biochemical reactions may well influence the geochemistry of the Nari-crust.

(4) During culture experiments with the isolated flora, recrystallized grains of calcite in the order of 20 to 110 μ were observed in bacterial aggregates. Within the sterile controls no recrystallisation was shown to occur.

(5) The above mentioned recrystallisation lead to calcitic crusts on limestones and other surfaces.

A. Introduction

Our results on biological alteration [KRUMBEIN and POCHON, 1964; KRUMBEIN, 1966 (2)] and on the influence of the bacterial microflora on the genesis of moonmilk a soft calcareous secretion on cave walls (CHALVIGNAC, KRUMBEIN, and POCHON, 1964), encouraged us to investigate the possible microflora of semi-aried to arid lime-crusts which at first look seem to be sterile products of evaporation.

Prof. Dr. M. EVENARI (Dept. of Botany, Hebrew University Jerusalem) kindly gave us two samples of the Nari-lime crust and the underlying substrate. Our first results are recorded here. They encouraged us to prepare a stage in Israel for further investigations which will be published later on.

The samples we received were obtained under sterile conditions. We made quantitative evaluations of the microflora populating the lime-crusts and the underlying lime-stone. Later, we made geochemical analyses to see if the chemistry of the lime-crusts was influenced biologically. Culture experiments with the isolated flora followed subsequently.

B. The Samples

Calcareous crusts (caliches) are considered as primarily indurated limestones which develop under the influence of freshwater. These may be products of evaporation or of precipitation by inorganic, organic or both influences.

* Geologisches Institut, Universität Würzburg, Germany.

The "Nari-lime-crust" is "a hard calcareous crust, which coats different sediments and rocks (especially limestones and marls) in various parts of Israel. This crust is especially widespread in the drier parts of the Mediterranean climatic zone as well as the semi-arid and arid parts of the country" (DAN, 1962). The Nari-lime-crust usually is a fossil crust which developed in the Pleistocene. RUTTE (1958, 1960) states nevertheless, that caliches may also develop today in climatic zones with less than 500 mm precipitation.

Fig. 1. Nari-lime-crust on Turonian limestones from the Nahal Revivim betwen Ber-Sheba and Dimona (Israel). The limestone is coated by 1 to 4 cm of caliche (Photo Evenari)

In the case of desert caliches, the evaporation theory is the probable mechanism of development. We have some excellent papers on the morphology and genesis of desert lime-crusts by KAISER (1928), LINCK (1930), KREJCI-GRAF (1936, 1960), and KNETSCH (1937).

The caliches usually have a high carbonate content (up to 98%). According to RUTTE (1960) and KREJCI-GRAF (1960) the optimum conditions for development of such caliches occur when there is heavy rain fall, followed by long periods of dryness and evaporation. The rain-soluble elements then will be transported to the surface and precipitated as a surface crust. In other cases, cementation horizons may develop.

The distribution of such crusts nevertheless is not restricted to semi-arid or arid regions, they may well develop under special microclimatic conditions, e. g. on building stones in humid zones, if they are exposed to the sun [KAISER, 1928; KRUMBEIN, 1966(2)]. The underlying substrate usually is limestone or marl. But they may also develop on other substrates. BLANCK (1926) investigated the "Nari-crusts" in Israel and found similar crusts on dolerites.

The material investigated (Fig. 1) was sampled in the Nahal Revivim between Ber-Sheba and Dimona. The substrate is a pure limestone of the Shivta Formation (Turonian). The flat slopes of the Wadi consists of many kilometers of the rudist-limestone banks which are slightly inclined due to the "Yersuchem anticline" and are coated by the Nari-lime-crust. This leads to a quick run-off of eventual rain water. The samples we examined were (1) the Turonian limestone substrate, and (2) the Caliche on the limestone. The surface of the caliche itself was eroded by influence of karst formation. Sometimes the thin (1 to 4 cm) crust desintegrates with the development of terra rossa (DAN, 1962).

C. Methods

The main purpose of our work was to find out if the Nari crust may be influenced biologically. So we examinated the two samples with methods developed by Soil Microbiology.

Various media were employed to get an impression of the total flora and the biological cycles which might take place within the crusts.

All microbial values were obtained by inoculating selective media with various suspension-dilutions of the samples.

10 g of sterile ground material suspended in 100 ml of sterile distilled water served as suspension-dilution 10^{-1}. From this we produced dilutions down to 10^{-10}. We employed either 5 or 10 ml of the media in test tubes or we prepared plates with agarsolidified media. We always inoculated 5 tubes of the media per dilution with 1 ml of the suspension-dilutions from 10^{-1} to 10^{-8} or 10^{-10}. A sixth tube per dilution was inoculated and sterilized afterwards to serve as a blank.

Generally the tubes were incubated at 28 °C. We tried to incubate at 32 °C, but we had no better results.

After 14 days to up to 3 weeks the positive tubes were counted and the most probable number of organisms per gram of dry material was calculated using the tables of MCCRADY. We will not specify most of the well-known media which are listed in POCHON and TARDIEUX (1964). The medium for total microflora was modified according to KRUMBEIN [1966 (1)]. Because the fact that the crusts contained a large amount of algae, we note here the various media used for counting the Algae.

I. (*Sol. of* CHU *cited in* POCHON and TARDIEUX, 1964): Ca $(NO_3)_2$: 40 mg; K_2HPO_4: 10 mg $MgSO_4 \cdot 7H_2O$: 25 mg; $NaCO_3$: 20 mg Na_2SiO_3: 25 mg; iron citrate: 3 mg; citric acid: 3 mg made up to 1000 ml with distilled water.

II. *Sol of* DENNFER: $NaNO_3$: 200 mg; Na_2HPO_4: 40 mg; Na_2SiO_3: 100 mg; soil extract: 200 ml; dist. water 800 mg.

III. *Sol. of* LEFÈVRE *(written communication):* KNO_3: 100 mg; K_2HPO_4: 40 mg; $MgSO_4 \cdot 7H_2O$: 20 mg; $Ca(NO_3)_2$: 100 mg; Fe_2Cl_6 (Codex): 1 drop; Soil extract: 10 ml; dist. water: 1000 ml.

IV. *Sol. of* PRINGSHEIM, *cited in* KOCH (1964): KNO_3:200 mg; $(NH_4)_2HPO_4$:20 mg; $MgSO_4 \cdot 7H_2O$:10 mg; $CaSO_4$:20 ml of saturated sol.; dist. water 1000 ml adding 1 ml of solution of Winogradsky of rare elements.

V. *Sol. of* POCHON and TARDIEUX (1964): Ca $(NO_3)_2$:100 mg; K_2HPO_4:40 mg; $MgSO \cdot 7H_2O$:30 mg; KNO_3:100 mg; $FeCl_3$:traces; soil extract:20 ml; 1000 ml dist. water.

VI. *We made up a special solution where no Ca is introduced when not in the form of $CaCO_3$*: KNO_3:200 mg; K_2HPO_4 40 mg; $NaNO_3$:100 mg; $MgSO_4 \cdot 7H_2O$:20 mg Sol. of Winogradsky: 1 ml; $CaCO_3$:5 g; dest. water 1000 ml. In some cases we added no $CaCO_3$ but put a slice of Muschelkalk-limestone to the plates.

After biological examination the samples were ground to a size below 100 μ. For determination pH we suspended 10 g in 10 ml of boiled distilled water. Than we extracted 2 g of the material with 250 ml bidistilled water. The residue was extracted with 250 ml of 1N HCl in 24 h at 90 °C. The insoluble residue was collected and weighed.

In both extracts we quantitatively determined the following cations: Na^+, K^+, Ca^{2+} (flame-photometry); Ca^{2+}, Mg^{2+}, (complexometry with EDTA, calconcarbon acid, murexid) Fe (colorimetry with orthophenantroline). In the water extracts we looked for the following anions Cl^- (titration) NO^- (destillation) SO_4^{2-} gravimetry and titrimetry according FISKE). CO_3^- was determined by gas-volumetry; Organic C and organic N by the classical methods. All methods are described and references given in KRUMBEIN (1966 [2]). From the first isolated microorganisms we took several suspension fractions and examined them later on medium VI and Difco nutrient broth[1].

D. Results and Discussion of the Quantitative Microbial and Geochemical Analyses

The poor substratum which consisted of 96% $CaCO_3$ and which for long times is dependent on condensation water, developed a rich microflora despite the fact that no organic material is furnished by higher plants. Certainly the metabolism of most of the microorganisms is suspended for a long time but the high number of different organisms proves that the crust is an environment in which microbial life may well develop.

Table 1 shows the results of the microbial counts. We found a high developed heterotrophic microflora with large amounts of algae, fungi and autotrophic bacteria. The values for the Nari crust differ significantly from the underlying limestones. Some biochemical groups which are found in the crust are lacking in the substrate.

Fig. 2 shows that the development of ammonification — usually a good evidence for soil microbial activity —, differs from the crust to the underlying limestone by log 4. The same tendency can be stated for the gelatin liquifying heterotrophs and algae.

The characteristics of the algal countings on the different media were also highly interesting. Soil algae usually reach their maximum within three weeks of culture from dilution-suspensions. The algae of the examinated crusts took up to ten weeks.

[1] The analytical work was done partially in the laboratory of the Geological Institute, partially in the laboratory of the HNO. We are indebted to the DFG for apparatus and material. I am grateful for assistance in analyses of organic C and Ca/Mg to U. GRÜTZE as well as to Dipl.-Chem. I. KEESMANN for X-Ray examinations.

After three weeks only dilution 10^{-1} of the media was positive. But the counts reached dilution 10^{-4} and 10^{-5} within 8 weeks. There are two possibilities to explain this phenomena:

(1) Algae living in such extreme conditions, as the arid desert crusts, develop possibly some mechanism to arrest their metabolism during long times of dryness, so that the renewal of growth periods might take a longer time.

Table 1. *Microbial data obtained on various media*

Physiological groups of organisms classified by selective media	Organisms per gram dry weight	
	crust	underlying limestone
Total microflora on poor soil-extract	2.5×10^7	5×10^4
algae (maximum from different media)	1.3×10^4	95
fungi (medium enriched with antibiotics)	2.1×10^5	3.2×10^3
ammonificating organisms (l-asparagine)	9.5×10^6	2.5×10^3
aerobic fixation of nitrogen (*Azotobacter*)	11	0
anaerobic fixation of nitrogen (*Clostridium*)	0	0
Oxidation of ammonia to nitrite (*Nitrosomonas*)	4.5×10^3	9
Oxidation of nitrite to nitrate (*Nitrobacter*)	2.5×10^3	150
Liquification of gelatin	4.5×10^3	4
Mineralization of organic S (dl-Methionine)	9	0
Reduction of sulphates (e. g. *Desulfovibrio*)	0	0
Oxidation of sulfur flower (*Thiobacillus*)	95	0
Oxidation of hydrogen sulfide (*Thiobacillus*)	15	0
Reduction of ferric iron (heterotrophs unspecif.)	1.3×10^5	2.5×10^2

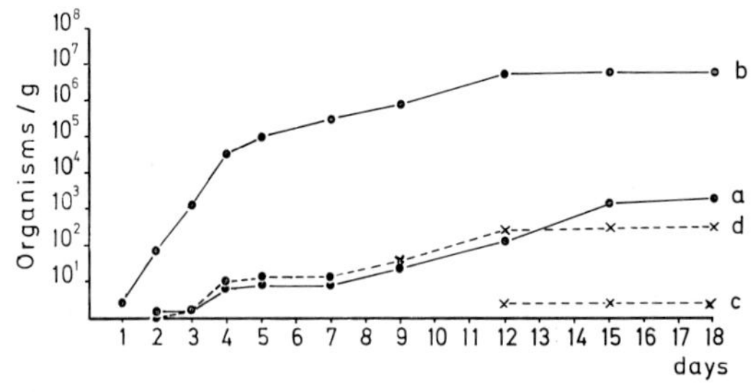

Fig. 2. Ammonification (●——●) and gelatin liquification (●·····●) graphs of the microflora in the caliche (b, d) and the limestone-substrate (a,)c

(2) More probably, we suppose, that large amounts of the algae are living in symbiosis with high numbers of isolated fungi, building up endolithic lichens. It is well known, that lichenisated algae take a very long time to develop in subculture (going up to several months).

The ratio of nitrogen fixating bacteria, like *Azotobacter* and *Clostridium*, was very low. This observation is similar to the counts we did on weathering building-stones [KRUMBEIN, 1966 (2)]. Most of the nitrogen which is needed by the heterotrophic bacteria must then be fixed by green and blue-green algae. Upon their destruction, the ammonifying bacteria take up the amino-acids and oxidize them to ammonia.

The ammonia will partially diffuse and partially be oxidized by autotrophs such as *Nitrobacter* and *Nitrosomonas*. Some of the blue-green algae then will be the next generation of algae, because a large amount of them need nitrates. Most of the blue-green algae isolated by us, grew on the nitrate-rich media II and IV. The same observation was made by FRIEDMANN (1962) with cave algae.

The sulfur cycle was not too well developed. Chemical analyses also showed little evidence of sulfur.

The heterotrophic microflora capable of reducing ferric iron was well developed and is possibly connected with the diminishing iron content from substrate to caliche.

The chemical analyses are combined in Table 2. The values are the sum of water extract and HCl-extract. The organic content of the crust is higher than that of the substrate as expected taking into regard the content of microorganisms. The C/N ratio is lower than in normal soils (8 to 11). Na, K, Ca increase, Fe and Mg diminish relatively. The water-soluble Na, K, and Ca also increased towards the crust. The iron loss may be due to reducing organisms. Summing up the results of microbial and chemical analyses regarding the high numbers of microorganisms, the increase of organic material, the differentiation in ionic composition and, the well established cycle of nitrogen we may conclude, that the caliches are influenced microbiologically. In fact they developed in the Pleistocene, so we can say that at least they are modified by biological agents. This impressions has been amplified by the following culture experiments.

E. Culture Experiments with Isolated Algal and Bacterial Flora

Regarding the literature on experimental carbonate solution and precipitation (BAVENDAMM, 1932; CLOUD, 1962; CHALVIGNAC et al., 1964; LALOU, 1957; OPPENHEIMER, 1961; ROHRER et al., 1959) there is much evidence, that biochemical and biological reactions may influence carbonate solution and precipitation. VOGEL et al. (1960), and MÜNNICH et al. (1959) proved by examination of the light isotopic C-fraction, that freshwater limestones develop biologically. But little is known of the reactions involved in special cases.

We tried to examine the development of carbonate solubility in medium VI by inorganically changing the pH and by culturing the isolated flora on the medium.

Table 2. *Chemical Analysis by HCl-Extraction (in %)*

	insoluble residue	pH	Na^-	K^-	Ca^{--}	Mg^{--}	Fe	Cl^-	SO_4^{--}	NO_3^-	CO_3^{--}	org. C	org. N	C/N
Crust	0.7	9.5	0.13	0.07	37.5	0.31	0.43	0.3	0.3	0.40	55.8	0.4	0.074	5.4
underlying limestone	1.2	9.3	0.08	0.04	35.8	0.47	0.8	0.2	0.1	0.35	53.0	0.2	0.032	6.25

In the carbonate enriched media we observed, after inoculation, a good development of algae connected with a lowering of the pH from 7.0 to 6.8 and an increase of the pH after 2 days up to 9 and 9.5. Meanwhile the Ca^{++} in the solution increased slightly diminishing very quickly afterwards. The highest value was 1.3 g/l, the lowest 400 mg/l.

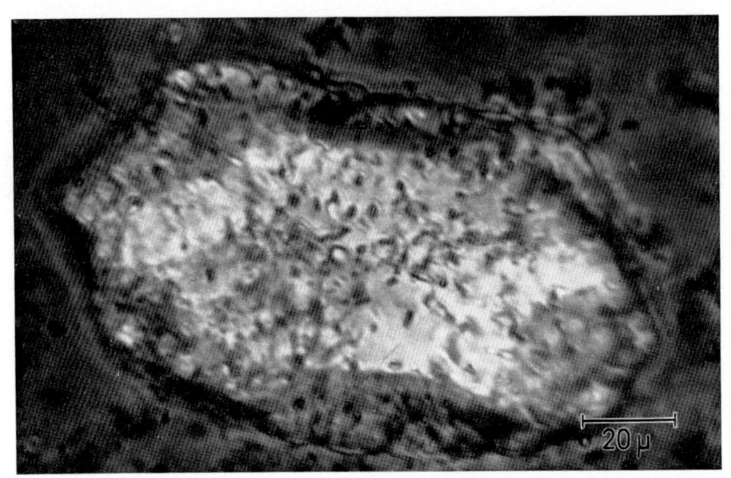

Fig. 3. Calcite aggregate which developed in the gelatinous red aggregates of microorganisms. Circle marks original calcite rhombohedra before recrystallization. The bacterial cell aggregates are visible as a shadow

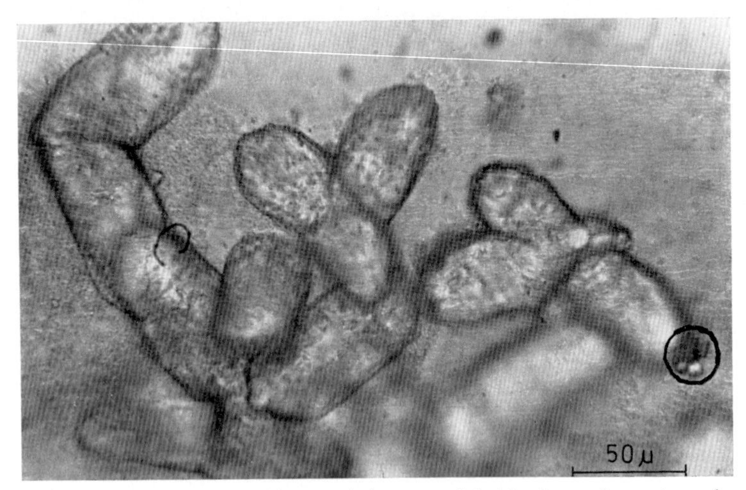

Fig. 4. Single crystal of calcite with pseudohexagonal habit. Long bacteria rods surround the crystal and are attached to its surface

After 10 days the blue-green algae lost color and began to hydrolyze. At the same time bacterial counts reached high numbers and a special red-pigmented flora developed big gelatinous aggregates. Up till now it was impossible to get pure cultures of these organisms which were mixed up with *Pseudomonads*, rests of algae, long rods, spores and unidentified cell aggregates. In every flask we observed at this point and only in the red aggregates, the generation of crystalline aggregates ranged in size between 30 and 110 μ (Fig. 3 and 4). The cristals often were surrounded by the large motile and unmotile rods, which were also attached to the crystal surface (Fig. 4).

The aggregates were isolated and examined by X-Ray analysis. They turned out to be pure Calcite. In Fig. 3 the encircled region shows the difference between original $CaCO_3$ and recristallyzed $CaCO_3$. In some Petri-dishes we cultivated the same flora not by adding $CaCO_3$ but with slices of a sterilized limestone instead. The red gelatinous bacteria developed on the limestone-surface and soon the slices were coated by a film of white calcite crystals. With sterility controlled we lowered the pH from 8 to 6.8 for 6 h and increased it to 9.5 afterwards by adding buffer solutions. We obtained no recrystallizations.

Fig. 5. Thin-section of the Nari-lime-crust. The circle denotes some bacteria-like structures

A thin-section of the caliche (Fig. 5) shows big crystals as well as very little ones. But in some places we found crystals similar to our cultured calcite. In many places we found structures like bacteria (encircled region in Fig. 5).

It is certainly possible that the recrystallisation features we observed in our cultures are only due to changes in pH. But then they ought to take place everywhere in the medium and not only in the cell aggregates. We believe, that the transport and transmission of Ca^{++} is due to chelating substances generated by algae, lichen and bacteria. SCHEFFER et al. (1963) propose the same mechanism for the development of desert varnish influenced by blue-green algae. These algae are very abundant in desert soil (SCHWABE, 1963).

In conclusion we can say, that we found a very well developed microflora on desert caliches, which may influence the chemistry of the lime-crusts even long after deposition. With our culture experiments in the laboratory with mixed cultures we proved that the isolated flora was able to produce large amounts of calcite. It is possible that such a flora may produce or be involved in the production of recent caliche. To understand the mechanism of transport of Ca and its deposition we have to conduct more detailed experiments. But probably biochemical reactions such as the nitrogen cycle and the sulfur cycle (Bavendamm, 1932) or chelating agents (Scheffer et al., 1963) are involved.

References

Bavendamm, W.: Die mikrobiologische Kalkfällung in der tropischen See. Arch. Mikrobiol. **3**, 205—276 (1932).

Blanck, E., S. Passarge, u. A. Rieser: Über Krustenböden und Krustenbildungen wie auch Roterden, insbesondere ein Beitrag zur Kenntnis der Bodenbildungen Palästinas. Chem. Erde **2**, 348—395 (1926).

Chalvignac, M. A., J. Pochon, et W. E. Krumbein: Recherches biologiques sur le Mondmilch. Compt. rend. acad. sci. Paris **258**, 5113 (1964).

Cloud: Environment of calcium carbonate depos. West of Andros Island Bahamas, 46 fig., 10 tab. Geol. Survey Profess. Papers **350**, 138 p. (1962).

Dan, J.: The desintegration of Nari limecrust in relation to relief, soil and vegetation. Photointerpretation, p. 189—194 (1962).

Degens, E. T., u. E. Rutte: Geochemische Untersuchungen eines Kalkkrustenprofils von Altkorinth-Griechenland. Neues Jahrb. Geol. Paläontol. Monatsh. S. 263—276 (1960).

Friedmann, I.: Progress in the biological exploration of caves and subtarrenean waters in Israel. Intern. J. Speleology **1**, 30—33 (1964).

Goldberg, A. A.: Contributions to the study of Nari in Israel and especially in the eastern Esdrealon valley. Diss., Jerusalem 1958.

Kaiser, E.: Über edaphisch bedingte geologische Vorgänge und Erscheinungen. Sitz.-Ber. bayer. Akad. Wiss. Math.-naturw. Kl. (1928).

Knetsch, G.: Beiträge zur Kenntnis von Krustenbildungen. Z. deut. geol. Ges. **89**, 177—192 (1937).

Krejci-Graf, K.: Zur Geologie der Makaronesen. 4. Krustenkalke. Z. deut. geol. Ges. **112**, 37—61 (1960).

—, u. W. Wetzel: Die Gesteine der rumänischen Erdölgebiete, 10 Abb., 9 Taf. Arch. Lagerstättenforsch. **62**, 1—220 (1936).

Koch, W.: Cyanophyceenkulturen, Anreicherungs- und Isolationsverfahren in Anreicherungskultur u. Mutantenauslese. Zentr. Bakteriol. Abt. I. Suppl. **1**, 415—431 (1964).

Krumbein, W. E.: Zum Problem der Bestimmung der Gesamtmikroflora mittels eines Bodenextraktes. Biologie du Sol (Bull. internat. d'information) **5**, p. 19 (1966).

— Zur Frage der Gesteinsverwitterung (über geochemische und mikrobiologische Bereiche der exogenen Dynamik). Diss. unpublish., Würzburg 1966.

—, et J. Pochon: Ecologie bactérienne des pierres altères des monuments. Ann. inst. Pasteur **107**, 724—732 (1964).

Lalou, C.: (2) Studies on bacterial precipitation of carbonates in sea water. J. Sediment. Petrol. **27**, 190—195 (1957).

Linck, G.: Die Schutzkrusten im Handbuch der Bodenlehre, S. 490—505, Berlin: Springer 1930.

Münnich, K. O., u. J. C. Vogel: ^{14}C-Altersbestimmungen von Süßwasserkalkablagerungen. Naturwissenschaften **46**, 168—169 (1959).

Neher, J., u. E. Rohrer: Dolomitbildung unter Mitwirkung von Bakterien. Eclogae Geol. Helv. **51**, 2, 213—215 (1959).

Oppenheimer, C. H.: Note on the formation of spherical aragonitic bodies in the presence of bacteria. Geochim. et Cosmochim. Acta **25**, 295 (1961).

Orni, E., u. E. Efrat: Geographie Israels. Israel Universities Press, 410 pp. (1966).

Pochon, J., et H. Barjac: Traite de microbiologie des Sols, 685 pp. Paris: Dunod 1958.

—, et O. Tardieux: Techniques d'Analyse en microbiologie du Sol, 108 s. Paris: Ed. la Tourelle 1964.

Rutte, E.: Kalkkrusten in Spanien. Neues Jahrb. Geol. u. Paläontol. Abhandl. **106**, 52—138 (1958).

— Kalkkrusten im östlichen Mittelmeergebiet. Z. deut. geol. Ges. **112**, 81—90 (1960).

SCHEFFER, F., B. MEYER u. E. KALK: Biologische Ursachen der Wüstenlackbildung. Zeitschr. Geomorph. **7**, 112—119 (1963).

SCHWABE, A.: Blaualgen aus ariden Böden. Forsch. u. Fortschr. **34**, 194—197 (1963).

SEIBOLD, E.: Untersuchungen zur Kalkfällung und Lösung am Westrand der Great Bahama Bank. Sedimentology **1**, 50—75 (1962).

VOGEL, J. C.: Über den Isotopengehalt des Kohlenstoffs in Süßwasser-Kalkablagerungen. Geochim. et Cosmochim. Acta **16**, 236—242 (1960).

Recent Carbonate Sedimentation in the Gnadensee (Lake Constance), Germany

MANFRED SCHÖTTLE, and GERMAN MÜLLER*

With 5 Figures

Abstract

The sediments of the Gnadensee are subdivided according to grain size into three zones parallel to the shore line: littoral, lake slope, and lake bottom. The mean grain size decreases gradually from the shore towards the middle of the lake. Carbonates are chief constituents of the sediments, except for those samples taken from the near-shore-area. The carbonates consist of calcite, dolomite, and aragonite. The dolomite is detrital-clastic and the aragonite biogenic-clastic in origin. Calcite is

1. detrital-clastic like the dolomite, and
2. calcite crystallized in situ through chemical-biogenic processes.

The precipitation of calcite in the lake is due to numerous carbonate-producing plants, such as macrophytae (schizophyceae), microphytae (blue-green algae). The formation of onkoids, produced by blue-green algae, is discussed in detail. Annual biogenic production of carbonates is ten times larger than detrital-clastic production.

A. Introduction

Lake Constance is the second largest of the lakes surrounding the Alps. It is divided into a) the Obersee including Lake Überlingen (Lake Constance in the strict sense) and b) the Untersee (Fig. 1). The Untersee differs from the Obersee by its morphology, biology, and its bottom sedimentation (KIEFER, 1955; MÜLLER, 1966). The Untersee comprises a number of basins with particular names: the Rheinsee (the Untersee sensu stricto), the Zeller See, and the Gnadensee. Carbonate production is most pronounced in the Gnadensee (MÜLLER, 1966). The origin of these carbonate sediments is the object of the present summary report. For a more detailed discussion see SCHÖTTLE (1967).

B. General Data

I. Topography

The Gnadensee is bordered on the north by the "Bodanrücken", by the Reichenau Island on the south which is connected to the mainland by a dam, and by the Mettnau peninsula (Fig. 1). The Gnadensee is separated from other parts of the Untersee by a shallow strait which has a water depth of only 0.7 m. The maximum depth of the lake is 22 m. The maximum length is 11.3 km and the average width between Reichenau-Mettnau and the mainland is 1.5 km, resp. 0.8 km. The surface area at water is 13.5 km^2. The volume is 0.4 km^3. Two small tributaries flow into the lake: the Allensbach and Markelfingen creeks. Their water and sediment influx is insignificant.

---

* Laboratorium für Sedimentforschung, Universität Heidelberg, Germany.

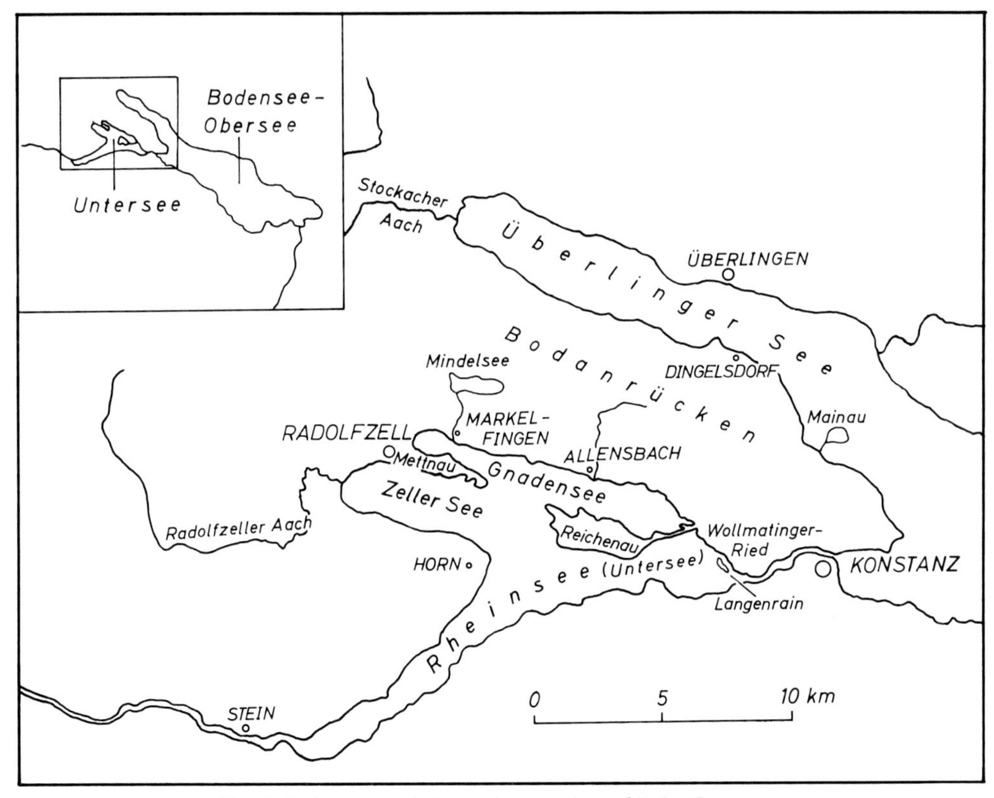

Fig. 1. Index map of the western region of Lake Constance

II. Morphology

The basin of the Gnadensee has a characteristic shape like most of the larger lakes. The rocks on the shore were eroded by wave action and the material transported further into the lake. This explains the origin of the flat shore bank, and of the abrupt slope.

III. Geology

The Untersee is mostly underlain by rocks of the Upper Freshwater Molasse, generally covered by glacial deposits of the Würm glacial period.

C. Sampling

During July and August, 1964 and 1965, 140 undisturbed samples were taken of the upper 10 to 15 cm of the sediment cover using the Auerbach-Wippermann bottom sampler (Auerbach, 1953). Nine traverses were sampled across the lake, supplemented by additional samples between the traverses. The results of the following laboratory analyses are thus representative of the upper 10 to 15 cm of the lake sediment cover.

D. Grain Size

The sediments are classified according to their position in the ternary diagrams whose end members are sand, silt, and clay; resp. sand, silt, and gravel. The sediment

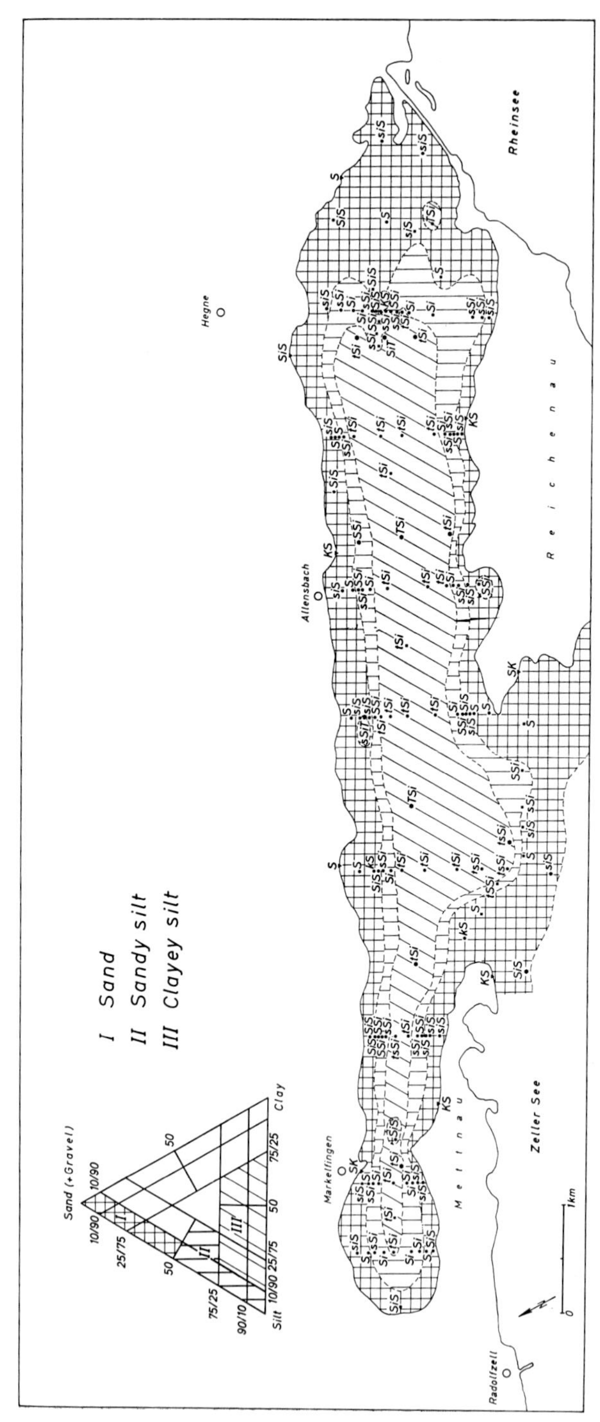

Fig. 2. Mechanical composition of Gnadensee sediments

of the Gnadensee can be subdivided into three zones parallel to the shore-line — littoral, lake slope, and lake bottom (Fig. 2).

Type I: sand (gravelly, silty) and siltsand;

Type II: sandsilt (clayey) and silt (sandy);

Type III: clayey silt (sandy), claysilt, and siltclay.

Grain size gradually decreases from the shore toward the middle of the lake.

E. The Carbonate Constituents of the Sediments

I. Methods

The total carbonate content was determined gasometrically by the Scheibler method (Müller, 1964). The calcite/dolomite-ratio was determined by X-rays.

II. Origin of Carbonate Minerals

The sediments are composed chiefly of carbonates, except for samples from the shore zone. Calcite ist most common, followed by dolomite and aragonite.

The carbonates fall into 3 groups, according to their origin:

1. Detrital-clastic carbonates.

The detrital-clastic carbonates, consisting of calcite and dolomite, were derived from older rocks (Tertiary Molasse and Quaternary glacial deposits) that crop out around and, locally, also within the lake basin. X-ray analyses show this calcite to be a low-Mg calcite. The predominant peak of dolomite lies at 3.04 Å, corresponding to a dolomite with stoichiometric composition (Goldsmith, Graf and Joensuu, 1955).

2. Biogenic carbonates.

The biogenic carbonates of the lake are low Mg-calcite (produced by plants) and aragonite (produced by molluscs).

a) Carbonate production by macrophytic plants.

Many macrophytic plants, such as Najas, Potamogeton, Chara and others (Lang, 1964), that occur in the littoral zones of the Gnadensee are known to precipitate $CaCO_3$ through assimilation, mainly during the summer and fall. Thus, leaves and stems of these plants are gradually covered with crusts of carbonate which easily peels off, crumbles and is deposited in the form of small, microcrystalline grains.

The amounts of carbonate thus produced are considerable because, under extreme conditions, a plant with 30 leaves can precipitate 123 g of carbonates (Wesenberg-Lund in Bärtling, 1922).

Kerner (1922) calculated that every dm^2 of the lake bottom, densely covered with Potamogeton lucens, is covered by approximately 5 g of carbonate deposit. He estimated, that carbonate crusts falling off the leaves would deposit a layer 1 m thick in approximately 5000 years.

The plants are widespread in the Gnadensee (Lang, 1964, 1967) so that a large portion of the carbonates found in the sediments has been derived from these plants.

b) Carbonate production by microphytic plants.

Another type of carbonate sediment consists of onkoids. These are locally known as "Schneggli" sands, because individual onkoids commonly contain snail shells. These onkoids are generated by blue-green algae, generally Schizophyceae (Schizothrix, Rivularia, Calothrix, Plectonema) (Schmidle, 1910). The algae occur as unicellular organisms or thread-like rows of cells which are surrounded by jelly.

They are important sediment producers because they precipitate $CaCO_3$ through assimilation and because the jelly and the fine fabric tissues catch or collect suspended detrial material.

c) Carbonate production by animals.

Skeletal debris of gastropods and pelecypods, consisting of aragonite as shown by X-rays, is nearly restricted to the outer littoral zone where the coarse fraction of the sediment is generally composed exclusively of molluscs.

III. Onkoids

1. Size and form of onkoids.

The size of onkoids in the Gnadensee ranges from 0.05 to 25 mm. The smallest onkoids have ovoid shapes, whereas the larger onkoids are partly disc or muffin-shaped and nearly spherical. The latter always have a plane or concave top layer. The shape of the smaller, ovoid onkoids is probably primarily determined by the generally elongate shape of the nucleus (fragments of molluscs, shells, etc.). The circular cross-section is probably due to constant movement of the smaller onkoids in the water. This prevents the algae from growing in a preferred direction. Nevertheless, if the size of an onkoid reaches a certain limit, its weight makes it practically stationary. Consequently, growth at the sediment-onkoid interface is retarded or interrupted except for a small rim around the margin. The algae on the surface of the onkoids, however, grow unhindered in all directions. This unequal growth yields the characteristic subspherical forms. Occasionally, such onkoids are turned upside down by wave-action or organisms, so that growth can start anew on the flat surface. In this way, symmetric, disc-shaped onkoids form, because first one and then the other side is overgrown. There are all transitions between disc- and muffin-like forms. GINSBURG (1960) made similar observations on Recent onkoids in Florida and the Bahamas.

2. Internal structure of the onkoids.

In cross-section, onkoids show a concentrically layered structure, with alternating porous and less porous micritic carbonate layers. These layers are probably due to seasonal growth, and are controlled by the depth of water. Most onkoids contain a nucleus on which the algae were first deposited; such a nucleus may consist of a snail or mussel shell, detrital grains, or plant particles.

3. Chemical and X-ray analyses.

Onkoids contain on the average 37% Ca, which corresponds to a calcite content of 92%. X-ray analysis of the HCl-insoluble residue indicates quartz, minor amounts of feldspar, mica, and clay minerals. Grain size distribution of the HCl-insoluble residue is similar to that of the surrounding sediment. Thus the onkoids grow on the same substratum during the entire period of their growth and incorporate detrital particles of the surrounding sediment.

4. Porosity and density of onkoids.

The porosities range from 48.9 to 54.1%. Their density varies between 2.61 and 2.66 g/cm^3.

5. Former and present distribution of onkoids.

BAUMANN (1911) has already described the distribution on onkoids in the Gnadensee. He mentions their presence only at the east point of the Mettnau, on the northern shore, and in the eastern part of the lake around the dam leading to Reichenau Island.

This is also confirmed by the present investigation. Moreover, small onkoids mostly a few mm in size were found in almost the entire littoral zone, indicating that the distribution of onkoids was considerably enlarged since 1911.

IV. "Crumbly" Carbonate Sands

The onkoids in the Gnadensee are generally found only on or in "crumbled" sediments, rich in carbonates (Fig. 3). Their origin is polygenetic as already discussed by SCHMIDLE (1910) and BAUMANN (1911).

Fig. 3. Onkoids growing on crumbly carbonate sand

Many individual "crumbs" of these sediments are detrital silicate minerals encrusted with carbonate. However, after treatment with dilute acetic acid, an insoluble jelly remains, apart from the silicates. This jelly is similar to that of the well-shaped, larger onkoids and their origin is thus similar. Crumbs which completely dissolve appear to be fragments of carbonate-encrusted plants. Thirdly there are crumbs that form by fragmentation of larger onkoids during freezing of the water. Laboratory experiments have shown that, under freezing conditions, onkoids will indeed break apart forming fragments similar to some crumbs.

In many samples these onkolitic crumbly carbonate sands overlie fine grained calcareous ooze (Fig. 4). The larger onkoids are restricted to the uppermost layer. Thus facies changes have taken place in recent years, caused by the increasing appearence of microphyta, with their high production of carbonates. The rapid recent growth of the flora probably results from increasing eutrophication of the lake water.

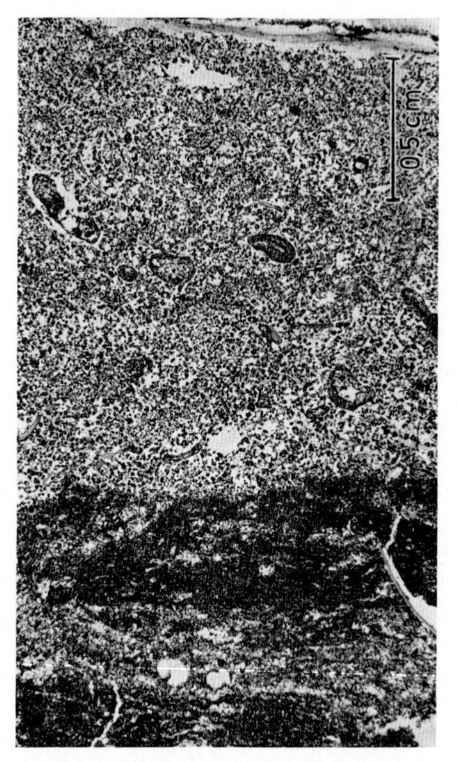

Fig. 4. Onkolitic crumbly carbonate sand overlying fine-grained calcareous ooze

F. Regional Distribution of the Carbonate Components and Origin of Sediments

The regional distribution of the carbonate components is illustrated in Fig. 5. The carbonate content ranges from 21 to 92%. Near-shore samples show values between 18 and 60%. The lake bottom can be divided into three zones based on the carbonate content of the samples.

Zone I: Near-shore-area with values between 20 and 60%.

Zone II: Littoral region and area of slope, with values between 71 and 92%.

Zone III: Lake bottom, with values between 61 and 70%.

Thus the carbonate content greatly increases away from the shore, but decreases again towards the central, deep water portions, where fine-grained sediments occur. By determining the calcite/dolomite ratio, proceeding from the shore line towards the middle of the lake, three zones can be distinguished with values < 10, between 10 to 15, and > 50, corresponding to the zones of carbonate distribution.

The origin of this zonal distribution can be explained as follows:

Zone I: Representing the lowest carbonate content as well as the lowest calcite/dolomite ratio, receives detrital material from Quaternary and Tertiary rocks that

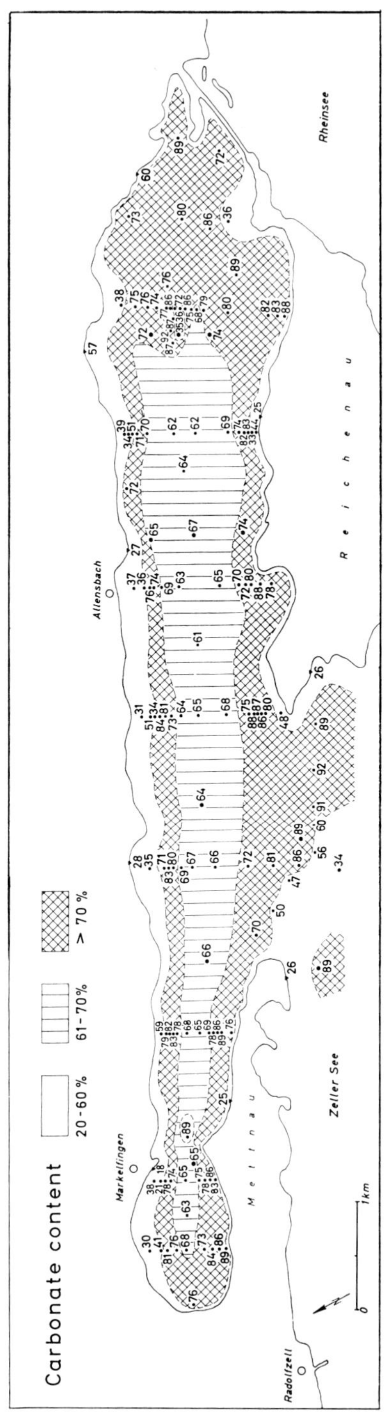

Fig. 5. Carbonate content of Gnadensee sediments

crop out on the mainland. Carbonate content and calcite/dolomite ratio of the sediments correspond in general with those of the glacial and Molasse deposits.

Fine-grained, suspended material, derived both from older rocks as well as from the newly formed carbonates in near-shore areas is transported into the center of the lake (Zone III) which is characterized by fine-grained sediments. However, these carbonates are in part dissolved in the deeper water because of the abundance of free, "aggressive" CO_2 formed by decay of organic matter during the summer. This dissolution explains the low carbonate content of zone III compared to that of zone II. A higher percentage of carbonate content and a higher calcite/dolomite ratio in Zone II are due to carbonate-producing plants and animals (macrophytae, blue green algae, and molluscs).

Acknowledgement

We wish to thank Dr. Lehn, Constance, and Dr. Lang, Karlsruhe, for their assistance during the sampling of the material. The present study was made possible through the financial support of the Deutsche Forschungsgemeinschaft.

References

Auerbach, M.: Ein quantitativer Bodengreifer. Beitr. naturkundl. Forschg. Südwestdeutschland **XII,** 17—22 (1953).

Bärtling, R.: Die Seen des Kreises Herzogtum Lauenburg mit besonderer Berücksichtigung ihrer organogenen Schlammabsätze. Abhandl. preuss. geol. Landesanstalt N. F. **88,** 60 S. (1922).

Baumann, E.: Die Vegetation des Untersees (Bodensee). Eine floristisch-kritische und biologische Studie. Arch. Hydrobiol. Suppl. Bd. **1.** (1911).

Füchtbauer, H.: Sedimentpetrologie, Teil II. Karbonatgesteine. Stuttgart: E. Schweizerbart'sche Verlagsbuchhandlg. 1968. In Vorbereitung.

Ginsburg, R. N.: Ancient analogues of recent stromatolites. Int. Paleontol. Union **XXII,** 26—35 (1960).

Goldsmith, J. R., D. L. Graf, and O. Joensuu: The occurrence of magnesian calcites in nature. Geochim. et Cosmochim. Acta **7,** 212—230 (1955).

Kerner, A. von: Pflanzenleben, 3. Aufl. Leipzig und Wien: 1922.

Kiefer, F.: Naturkunde des Bodensees. 169 p. Lindau und Konstanz: Thorbecke Verlag 1955.

Lang, G.: Vegetationsforschung am Bodensee. Umschau **9,** 270—275 (1955).

— Die Ufervegetation des westlichen Bodensees. Arch. Hydrobiol. Suppl. **XXXII,** 4, 437—574 (1967).

Müller, G.: Das Sand-Silt-Ton-Verhältnis in rezenten marinen Sedimenten. Neues Jahrb. Mineral. Petrol. **7,** 148—163 (1961).

— Sedimentpetrologie Teil I. Methoden der Sedimentuntersuchung, 303 p. Stuttgart: E. Schweizerbart'sche Verlagsbuchhandlung 1964.

— Die Sedimentbildung im Bodensee. Naturwissenschaften **10,** 237—247 (1966).

Schmidle: Postglaziale Ablagerungen im Nordwestlichen Bodenseegebiet. Neues Jahrb. Mineral. **II,** 104—122 (1910).

Schöttle, M.: Die Sedimente des Gnadensees. Ein Beitrag zur Sedimentbildung im Bodensee, 104 p. Diss., Heidelberg 1967.

Mineralogy, Petrology and Chemical Composition of Some Calcareous Tufa from the Schwäbische Alb, Germany

GEORG IRION, and GERMAN MÜLLER*

With 16 Figures

Abstract

Calcareous tufas from the Schwäbische Alb collected and studied botanically by A. STIRN are investigated mineralogically, petrographically, and chemically.

Plant tufas are formed by incrustation of plant fragments (mostly algae and mosses) by calcite; the calcite is deposited around the plants from spring waters, rich in hydrogene carbonate. They commonly exhibit high porosity. Dense calcareous sinter is formed covering calcareous tufas and rubble in water during periods of stagnating plant growth. Mineralogic investigations revealed that more then 99% of crystallized material consists of calcite, small amounts of quartz, kaolinite, montmorillonite and illite are also present. These minerals are detrital and derived from the Upper Jurassic rocks of the drainage area. Some of the tufas still contain small amounts of organic matter.

Mg, Sr, Mn, and Fe only make up a small portion of the HCl-soluble part of tufas. The average is 0.16%, a maximum of 0.6 was found. A clear relationship between Fe and Mn and the amount of HCl-insoluble residue exists. The Sr-content is about ten times lower in comparison with marin carbonate rocks.

Essentially, the fabric of the plant tufas depends on morphology of the incrusted plants. The size of calcite crystals varies between several microns and several millimetres. The larger calcite crystals are mostly euhedrally shaped.

Diagenetic changes (filling of pore spaces, recrystallization) mostly take place in moss tufas and calcareous sinter.

A. Introduction

For many years, the "Institut für Spezielle und Angewandte Botanik" of Tübingen University has preformed systematic investigations of the flora of Recent and Subrecent calcareous tufa (e.g. STIRN, 1964; GRÜNINGER, 1965).

Prof. Dr. MÄGDEFRAU kindly put at our disposal the samples used by STIRN for his dissertation, and we want to express our sincerest thanks to Prof. MÄGDEFRAU.

In the present study, fabric and composition of some important types of tufa are investigated by means of mineralogic, petrographic and chemical methods.

B. Types of Tufa from the Schwäbische Alb (Fig. 1)

The different types of tufa are named according to the predominant types of plants they contain. Algal and moss tufas are predominant in the area. Anorganically formed calcareous sinter also occurs.

* Laboratorium für Sedimentforschung, Universität Heidelberg, Germany.

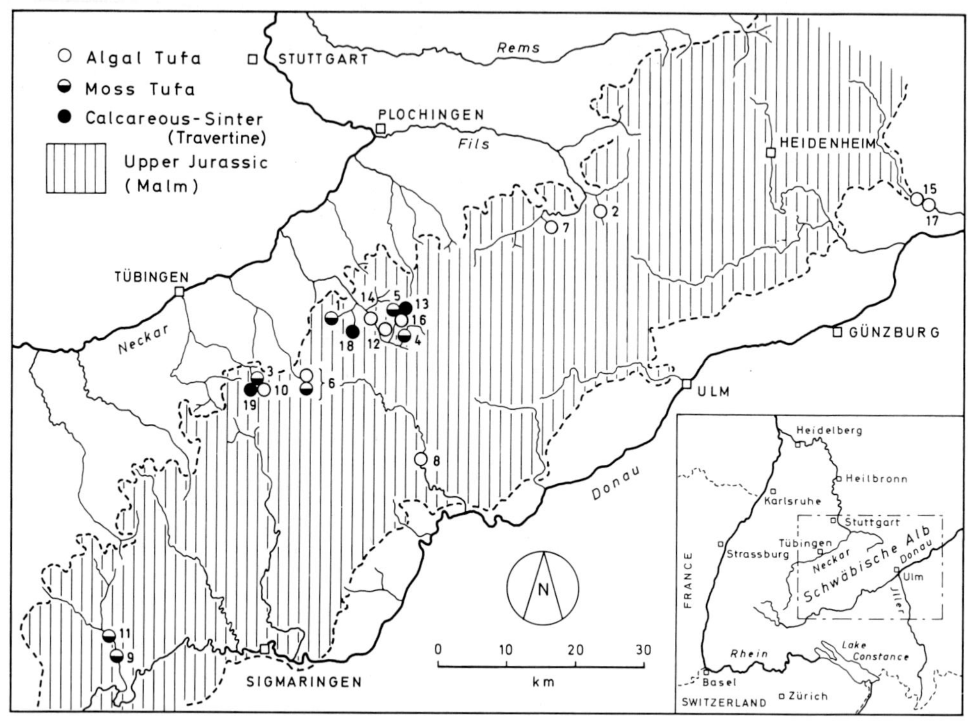

Fig. 1. Sample localities. The samples from the Neckar and Fils area are from creeks and quarries located at the NW-scope of the Schwäbische Alb. The other samples are from the Alb-plateau

I. Algal Tufas

The nomenclature of algal tufa does not always coincide with the botanic nomenclature; it uses the family as well as the genera names.

1. Cyanophyceae

Cyanophyceae (blue algae) make up a particularly large part of the tufa. They are subdivided into Oscillatoriaceae and its genera Lyngbya Phormidium and Schizothrix. As it is difficult to divide Lyngbya and Phormidium, their tufas will be named Oscillatoriaceae tufa. This tufa mantles the beds of small creeks. Continuous layers may be several meters long and 10 to 30 cm thick. In creeks with steep gradient the cascades are covered with Oscillatoriaceae tufa.

In the area, Schizothrix tufa occurs only fossilized. A particularly large occurrence is in the headwaters of the Wiesaz near Gönningen.

2. Xantophyceae

Vaucheria, belonging to the family of Xantophyceae, is another genus of algae frequently found in the area. As a result of its capacity of rapid regeneration, it frequently occurs in springs. In flowing water, the thalli of Vaucheria are aligned parallel to the current direction. Aligned fragile tubes remain after dying off and decay of the algae. The individual Vaucheria layers are commonly imbricated.

3. Chlorophyceae

Chlorophyceae (Green algae) are represented by Cladophora giomerata.

4. Rhodophyceae

In the area, the Rhodophyceae (Red algae) are not rock-forming; Chantransia chalybea however, which belong to the family of Red algae, occur sporadically.

5. Several authors (e.g. STIRN, 1964; THIENEMANN, 1933) use the term "Chironomide tufa" for algal tufa containing abundant living and feeding burrows of chironomides. In August, the larvae leave the eggs; they live from algae and microorganisms and build living and feeding burrows, lining them with threads from their spinning glands. By the end of April, the larvae attain a length of 5 mm, their feeding structures are 15 mm long. In May they change into pupas and leave the tubes in June. In creeks with abundant bicarbonate, calcite is deposited along the threads, thus strengthening the living and feeding tubes.

In the area, they are found only in the spring layer of Oscillatoriaceae-tufa. In autumn and winter the activity of the larvae seems to be unimportant, thus no visible traces are left.

II. Moss Tufa

The bulk of Recent creek tufa is made up of moss tufa. In the area, only the class of Musci forms calcareous tufa. Cratoneurum and Eucladium are particularly abundant, Cratoneurum mostly occurring in the spray zone. "Surface enlargement of the water favours incrustation of the mosses from bottom to top" (STIRN). Eucladium verticillatum and Bryum ventricusum show clear seasonal growth. Layers of Bryum may be up to 3 cm thick.

III. Calcareous Sinter

In the headwaters of the Schwäbische Alb, rocks, leaves and wood fragments are rapidly covered by thick calcite layers.

The locations of the investigated samples are shown in Fig. 1.

C. Chemical Composition of the Samples

I. Methods

The samples were reduced to grain size of $< 40\,\mu$ and dissolved in dilute HCl. Ca-, Mg-, Sr-, Mn-, and Fe-content of the solutions were determined with an Atomic-Absorption-Spectrometer (Perkin-Elmer). The error of the Ca-determination is less than 1%, that of the Mg-, Sr-, Mn-, and Fe is $\pm$ 20%, due to their small amounts.

II. Results

1. $CaCO_3$ makes up more than 99.2% of the HCl-soluble part of the samples. Its portion — related to the whole of the sample — is between 89.4 and 100%, the average being 96.2%.

2. Mg, Sr, Mn and Fe together make up at most 0.6% (sample 16) of the HCl-soluble portion. Sample 1 has the lowest value: 0.032%.

3. The Mg-content of the samples varies between 240 ppm and 1,060 ppm. Most of the samples have Mg-contents between 300 ppm and 600 ppm. The average is 462 ppm.

4. The Sr-content varies between 17 and 234 ppm, only 3 samples show Sr-contents higher than 60 ppm; the average is 54 ppm.

5. The Mn-contents vary strongly, they are between 6 and 1,600 ppm, the average being 184 ppm. Only 4 samples (4, 8, 14, and 16) show values higher than 43 ppm.

6. The Fe-contents also vary strongly, they lie between 46 and 3,900 ppm. Only samples 4, 8, 14, and 16 show values higher than 215 ppm. The average is 849 ppm.

	Mg	Sr	Mn	Fe
	content in ppm			
Maximum	1060	234	1600	3900
Minimum	240	17	6	46
Average	462	54	184	849

III. Discussion

1. As was to be expected, no relationship between the trace-element portion of the samples and the type of plants incrusted in the tufa could be found. However, a relationship between Fe- and Mn-content and the amount of HCl-insoluble residue was demonstrated: with increasing insoluble residue, Fe- and Mn-contents increase. Probably Fe and Mn are primarily derived from oxide- or hydroxide-compounds.

2. Low Sr-contents and the corresponding low Sr/Ca ratios[1] of the investigated tufas are suprising (Average 0.065).

According to Kulp et al. (1952) as well as Thompson and Chow (1955), the average Sr/Ca-ratios of limestones are 0.71 and 0.63 respectively.

It should be noted that these values are primarily derived from marine limestones deposited in water with extremely high Sr/Ca-ratios.

These low Sr/Ca-ratios, however, are reasonable for the tufas of the Schwäbische Alb, as the Sr/Ca-ratio of the creeks are very low.

According to Knoblauch (1963), the carbonate rocks of the Malm of the Schwäbische Alb — where the waters come from — have an average Sr-content of 430 ppm and an average Sr/Ca-atomic ratio of 0.496. If the above-mentioned Sr/Ca-ratio is used for the creek water of the Schwäbische Alb, a Sr/Ca-ratio of 0.0694 may be calculated for the tufas and calcareous sinter when using Holland's distribution coefficient (Holland et al., 1964) $D = 0.14 \pm 0.02$ for precipitation of calcite from sea water. This value closely agrees with the calculated average value of 0.065.

D. Mineralogy and Organic Matter

I. Thin Sections

More than 99% of the crystals consist of calcite. Their size varies between some micron and 5 mm. The smaller crystals are mostly euhedrally-shaped, the larger ones are prismatic or feathery. The length axis of the long-prismatic crystals is mostly identic with the optic axis. Occasionally, the calcite crystals are brown, probably due to organic matter.

Small amounts of quartz, biotite and feldspars generally occur in cavities.

[1] In the following, the Sr/Ca-ratio will be expressed as atomic ratio Sr/Ca $\times$ 1000.

II. X-Ray Study

X-ray investigations revealed that the minerals are almost exclusively calcite. Small amounts of quartz occur in every sample.

The residue of two tufa samples largely consists of quartz. Also kaolinite, montmorillonite and illite were detected in oriented specimens.

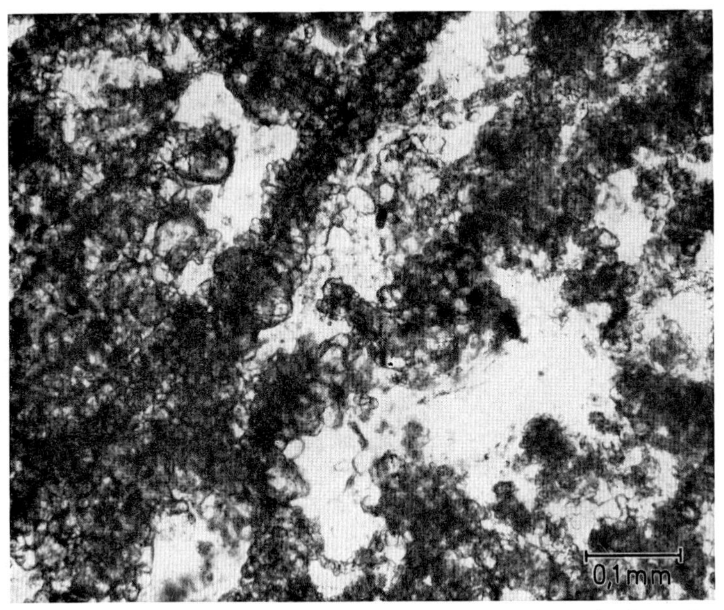

Fig. 2. Oscillatoriaceae tufa from Jakobsbrunnen near Urach. Detail of a spring-summer layer

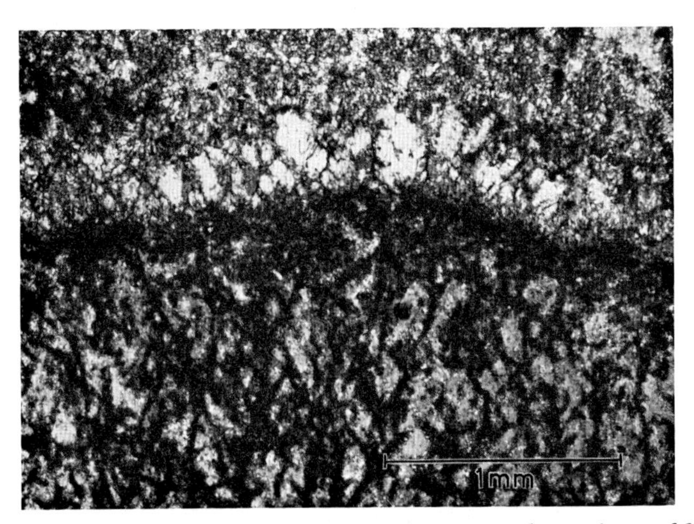

Fig. 3. Oscillatoriaceae tufa, Aubach near Geislingen. Lower and central part of figure show part of a spring layer (the tufa is very porous). This is overlain by calcareous sinter with much larger crystals deposited in winter

All samples contain varying amounts of organic substance, always below 3%. Slightly transformed stems and threads, as well as fragments and finely-divided organic substance were observed.

Fig. 4. Oscillatoriaceae tufa, Aubach near Geislingen. The dark layers were deposited during the winter. The thickness of the white layers, formed during the summer, varies strongly. The elongate cavities are cross-sections of Chironomidea-tubes; they are generally formed in the spring layer

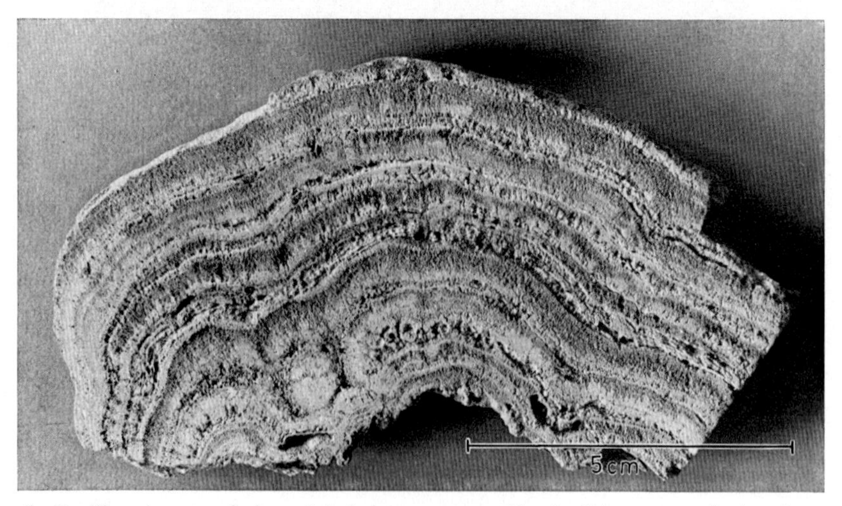

Fig. 5. Oscillatoriaceae tufa from Jakobsbrunnen near Urach. Calcereous tufa showing well developed seasonal growth-layers

Fig. 6. Schizothrix tufa near Gönningen. Threads of Schizothrix showing arrangements in tubes

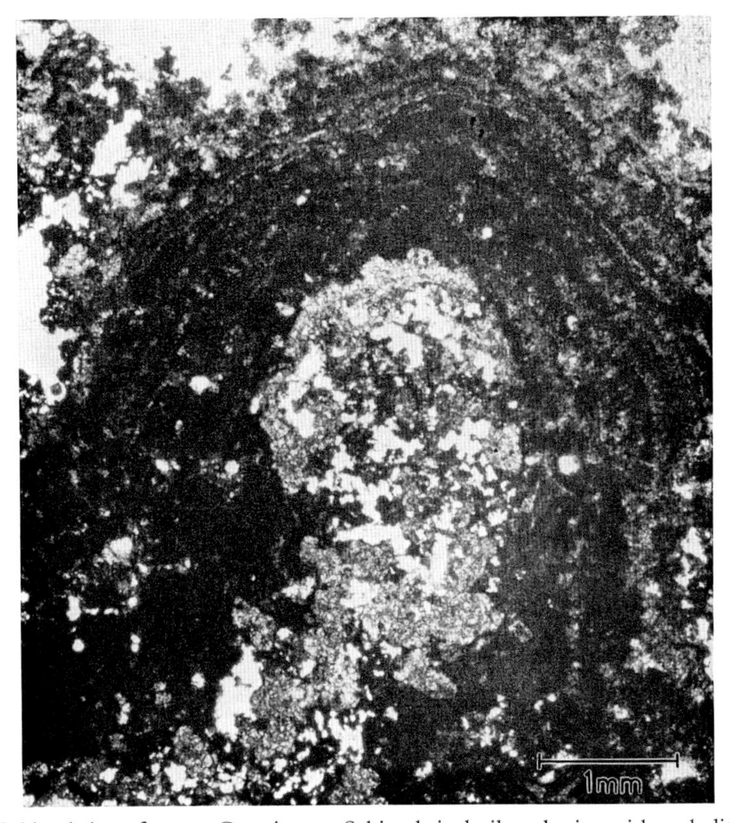

Fig. 7. Schizothrix tufa near Gönningen. Schizothrix built colonies with onkolitic fabric

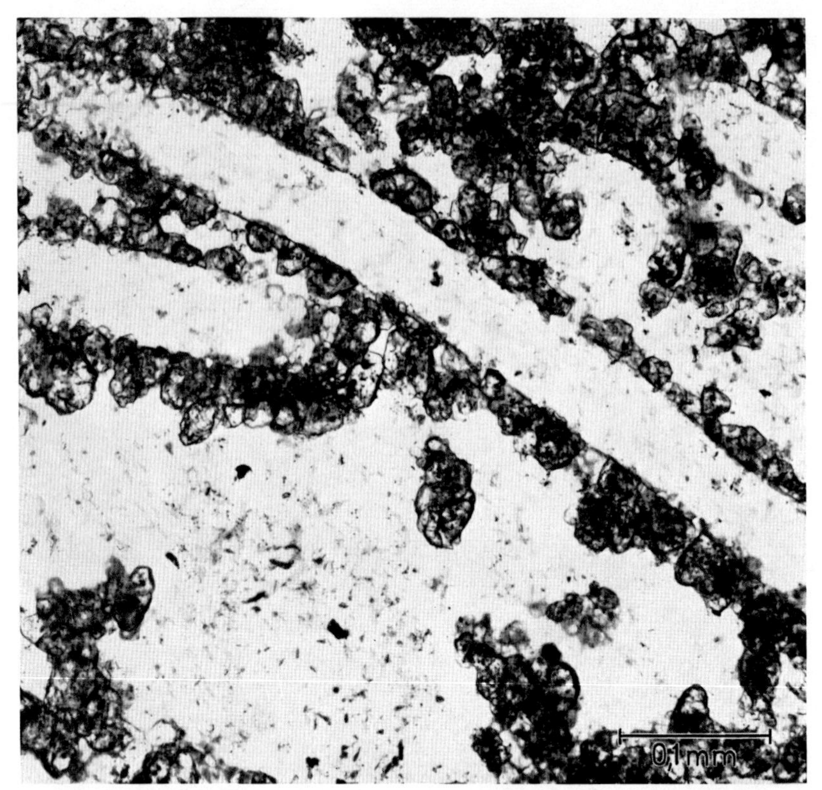

Fig. 8. Vaucheria tufa from Ditz. Vaucheria threads incrusted by relatively small crystals. The threads themselves are caused by decomposition, leaving cavities

Fig. 9. Moss tufa from Bärenthal. Complex moss plants showing an inhomogeneous fabric after incrustation

III. Discussion

All minerals except calcite are detritus and derived from the Malm limestones. Knoblauch (1963) has identified the same minerals in the Malm near Urach.

E. Tufa Fabric

The incrusted plants determine the fabric of the tufa; thus the tufas are classified according to the families and genera of the incrusted plants.

I. Algal Tufas

1. Cyanophyceae (Blue Algae)

a) Oscillatoriaceae (without Schizothrix). The threads of the Oscillatoriceae have a diameter of 5 μ; they are incrusted by isometric crystals with a rounded surface.

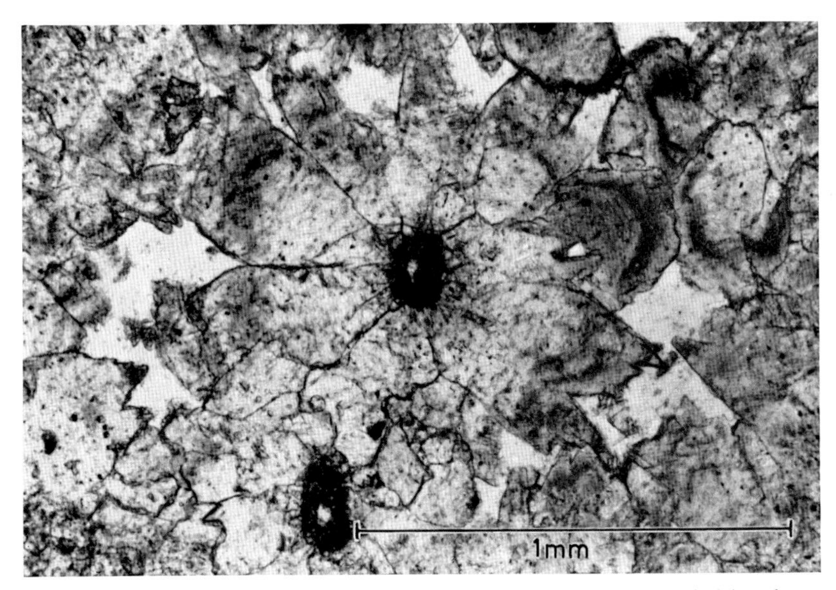

Fig. 10. Moss tufa near Gönningen. An individual moss stalk, surrounded by almost euhedral relatively large crystals. The different growth stages are clearly seen within several crystals

The average diameter of the crystals is 15 μ. The growth rate of the algae changes seasonally. In spring, the algal mats are dense, thus solid tufa may be formed. In summer, the growth rate decreases and the space between the algal threads increases, thus relatively spongy tufa is formed (Fig. 2). As the algae do not grow in winter, pure layers of sinter are formed during this period (Fig. 3). The thickness of layers varies strongly. Sample 2 (Fig. 4) shows layers which have been formed in one year; they are between 0.6 and 7 mm thick. The pores of the summer layer have a diameter of 0.5 mm; those of the layer formed in spring have a diameter of about 0.05 mm. Tubes of chironomides are common in the spring layer. They form long cavities up to 5 mm long and 0.5 mm wide (Figs. 4 and 5). They are always parallel to the annual layers. The Oscillatoriaceae tufas have high porosity; in one of the samples — with a specific surface of 2.4 m^2/gr — a porosity of 73 vol-% was measured.

b) Schizothrix. Subrecent Schizothrix tufa forms nodules with uneven surface (Fig. 6). The cross-section of the nodules shows different layers like that of stromatolites; they are also found in the onkoides of Lake Constance (SCHÖTTLE, 1967). The average size of the crystals is 5 μ.

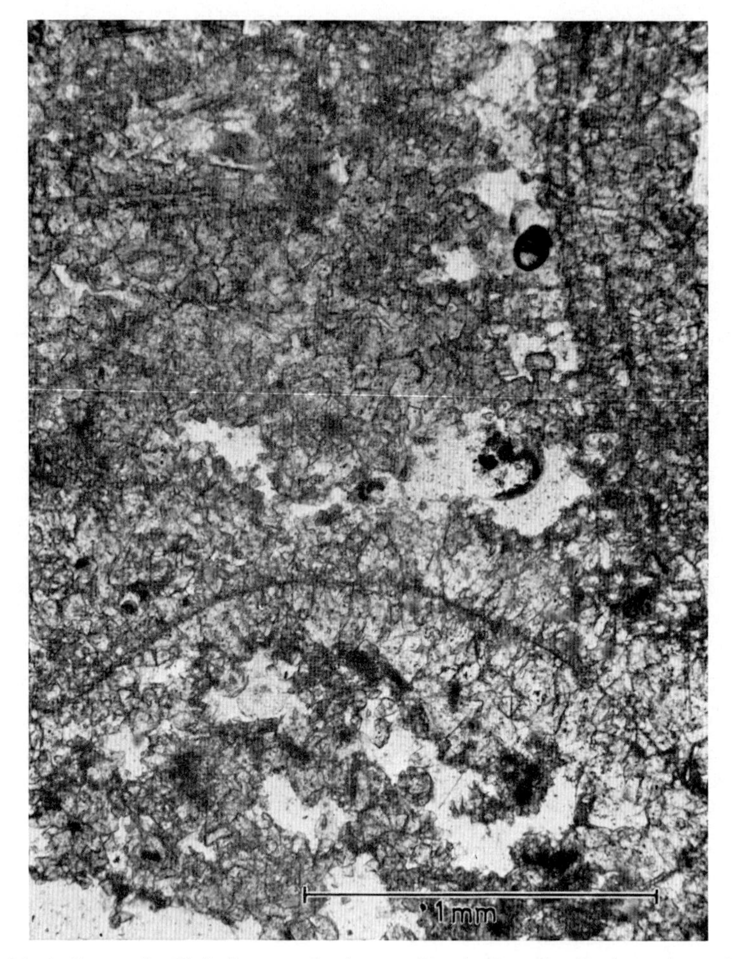

Fig. 11. Eucladium tufa, Fleinsbrunnenbach near Urach. Longitudinal section of a moss stalk showing palisade-like growth of calcite crystals

The Schizothrix tufa of sample 10 has a porosity of 43 vol-% and a specific surface of 0.704 m^2/gr.

2. Xantophyceae (Vaucheria only)

The threads of Vaucheria are decomposed and left oblong cavities; the cross-section of the cavities is round with a diameter of 80μ.

II. Moss Tufa

A typical moss tufa is shown in Fig. 9. The size of the different parts of mosses, e.g. stalk, leaf, etc. varies greatly; some of the stalks are only a few microns thick, large leaves may be several millimetres long (Fig. 9). Essentially, the size of the incrusting crystals depends on the diameter of the plant fragments. Thus the grain sizes of moss tufa are variable; the size of the crystals varies between several microns and several millimetres. The crystals are mostly euhedral, and they are grown on the moss stalks in a palisade-like fashion; in a section perpendicular to the axis of the stalk, the crystals are arranged like rosettes (Fig. 10).

Fig. 12. Sinter-crust on a moss tufa, near Gönningen. Palisade-like growth of calcite crystals on moss tufa. Crystals show plane interfaces. Dark horizontal layers represent organic matter incorporated during growth of the crystals

III. Calcareous Sinter

In the samples calcareous sinter only occurs as crust. The surface layer of these crusts mostly consists of long-prismatic crystals which may reach a length of 5 mm and a width of 0.5 mm (Fig. 12).

The length axis of these crystals mostly runs parallel to the direction of growth. The deeper the crust, the more intergrown the crystals become; they become smaller and the fabric more unoriented, the orientation is no longer uniform (Fig. 13). The layers of some banded sinters are coloured; they correspond to distinct phases of growth, marked by a layer supply of organic matter.

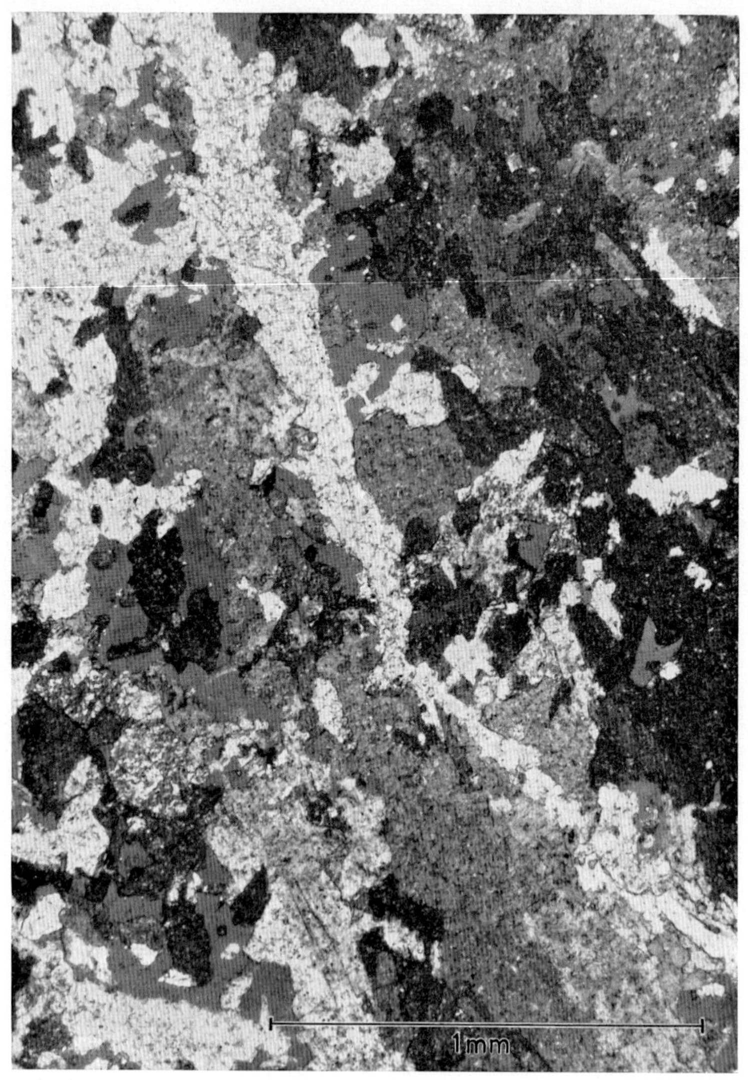

Fig. 13. Calcareous sinter from waterfall near Urach. The internal fabric of calcareous sinter commonly shows xenomorphic texture. There is no preferred orientation of c-axes of calcite crystals

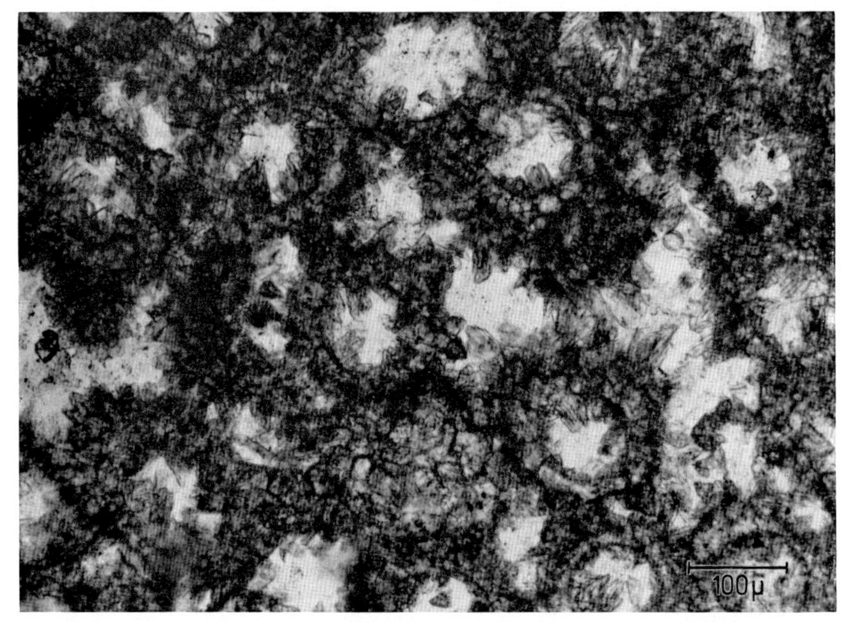

Fig. 14. Fossil moss tufa, near Gönningen. Cross-section through moss-stalks. A second generation of euhedral calcite crystals is grown inside the decomposed stalks

Fig. 15. Moss tufa near Gönningen. Drusy calcite growing in cavity of older moss tufa

F. Diagenesis

As a result of the peculiar growth of moss plants, cavities of several centimetres are formed in moss tufa. Following formation of moss tufa calcite can be deposited again in these cavities from circulating waters; the walls of the cavities are then

Fig. 16. Calcareous sinter from waterfall near Urach. Dark layers of organic matter in calcareous sinter. The originally even layers were later deformed by differential growth of calcite crystals

covered with euhedral calcite (Fig. 14) or completely filled, forming "drusy mosaic" (Fig. 15). These modifications are not observed in Oscillatoriaceae tufa, probably due to the smaller pore diameter and decreased permeability and circulation.

The above-mentioned dark bands are disrupted by recrystallization (Fig. 16). Similar structures are described by Kirchmayer et al. (1968) in cave pearls.

G. Formation of Calcareous Tufa

Inorganic precipitation of $CaCO_3$ from springs and creeks has always be considered the cause for formation of calcareous sinter: Bicarbonate is abundant in the spring waters of calcareous formations; due to a change of CO_2 partial pressure, a certain amount of bicarbonate is transformed into carbonate at the surface of the earth, thus precipitating the poorly soluble $CaCO_3$.

It has long been assumed that assimilation ("biogenic decalcification") played an important rôle in the precipitation of carbonate in plant tufa. With his investigations of the "Uracher Wasserfall", however, GRÜNINGER (1965) has proved that the biogenic-inorganic formation of $CaCO_3$ is only of slight importance and may be neglected as against the purely inorganic precipitation.

References

GRÜNINGER, W.: Rezente Kalktuffbildung im Bereich der Uracher Wasserfälle. Abh. Karst- u. Höhlenk. München, Reihe E (Bot.) **2**, 113 S (1965).

HAHNE, C., M. KIRCHMAYER u. J. OTTEMANN: „Höhlenperlen" (Cave Pearls), besonders aus Bergwerken des Ruhrgebietes. Neues Jahrb. Geol. u. Paläontol. Abhandl. **130**, 1, 1—46 (1968).

HOLLAND, H. D., T. KIRSPU, S. HUEBNER u. U. OXBURGH: On some aspects of the chemical evolution of cave water. J. Geol. **72**, 1, 36—67 (1964).

KNOBLAUCH, G.: Sedimentpetrographische und geochemische Untersuchungen an Weißjurakalken der geschichteten Fazies im Gebiet von Urach und Neuffen. Inaug. Diss., 106 S., Tübingen 1963.

KULP, J. L., K. TUREKIAN, and D. W. BOYD: Strontium content of limestone and fossils. Bull. Geol. Soc. Am. **63**, 701—716 (1952).

SCHÖTTLE, M.: Die Sedimente des Gnadensees; ein Beitrag zur Sedimentbildung im Bodensee. Inaug.-Diss., 104 S., Heidelberg 1967.

STIRN, A.: Kalktuffvorkommen und Kalktufftypen der Schwäbischen Alb. Abh. Karst- u. Höhlenk., München, Reihe E (Bot.) **1**, 92 p. (1964).

THIENEMANN, A.: Mückenlarven bilden Gesteine. Natur und Museum **63**, 370—378 (1933).

THOMPSON, T. G., and T. J. CHOW: The strontium-calcium ratio in carbonate-secreting organisms. Papers in Marine Biology and Oceanography. Suppl. to v. 3 of Deep-Sea Research, 20—39, 1955.

Tidal Flat Deposits in the Ordovician of Western Maryland*

ALBERT MATTER**

With 2 Figures

Various authors have pointed out that the Cambro-Ordovician limestones of the Central Appalachian area have been deposited in warm and shallow seas.

To pinpoint as closely as possible how shallow these seas were at a given time, the Middle Ordovician New Market limestone was studied as a representative small portion of the Cambro-Ordovician sequence of Western Maryland. The New Market limestone measures here about 125 m. Because it is lithologically uniform, emphasis was laid on the bedding and other sedimentary structures, their aspect, variation and origin.

In the well exposed section of the Wilson Quarry (Clear Spring, Md.) the New Market limestone could be subdivided into six structurally different rock types: laminated, lumpy, ribboned, stromatolitic, intraclastic and bioclastic dolomitic limestones shown in Fig. 1. Although these rock types show different kinds of stratification and minor variations in texture and composition, we do not know precisely where the different types formed, except for the stromatolites.

However, on a few bedding planes shrinkage polygons were observed. Planar and sectional views of the same mud cracked beds were compared to recognize the typical mud crack features in sectional views alone. By this method mud cracks were found in many beds which did not seem to contain any shrinkage features (Fig. 2). Two different types of mud cracks are distinguished 1) sediment filled, and 2) spar filled. The first type is mostly present in ribboned limestones and laminated dolomites, whereas the second type was usually observed in stromatolites. Using the experimental results of BURST (1965) it can be shown that these cracks are true mud cracks and not subaqueous shrinkage cracks.

Moreover it was found that the lumpy limestones which rarely show a polygonal pattern on the bedding planes are the product of intense desiccation. The flat fragments (Fig. 1 D) are pieces of polygons torn loose and transported during storm flooding.

Stromatolites occur in the New Market limestone as laterally continuous undulous beds or as Collenia like heads. Studies of recent stromatolites show that they grow most frequently above mean low water level. Recent and ancient stromatolites are interbedded with laminated mud cracked sediments and associated with flat pebble conglomerates. The algal layers of the ancient stromatolites also contained spar filled desiccation cracks, occasionally associated with birdseye structures.

---

* This is a short version of a paper published in the Journal of Sedimentary Petrology vol. 37/2, 1967.

** Department of Geology, University of Berne, Switzerland.

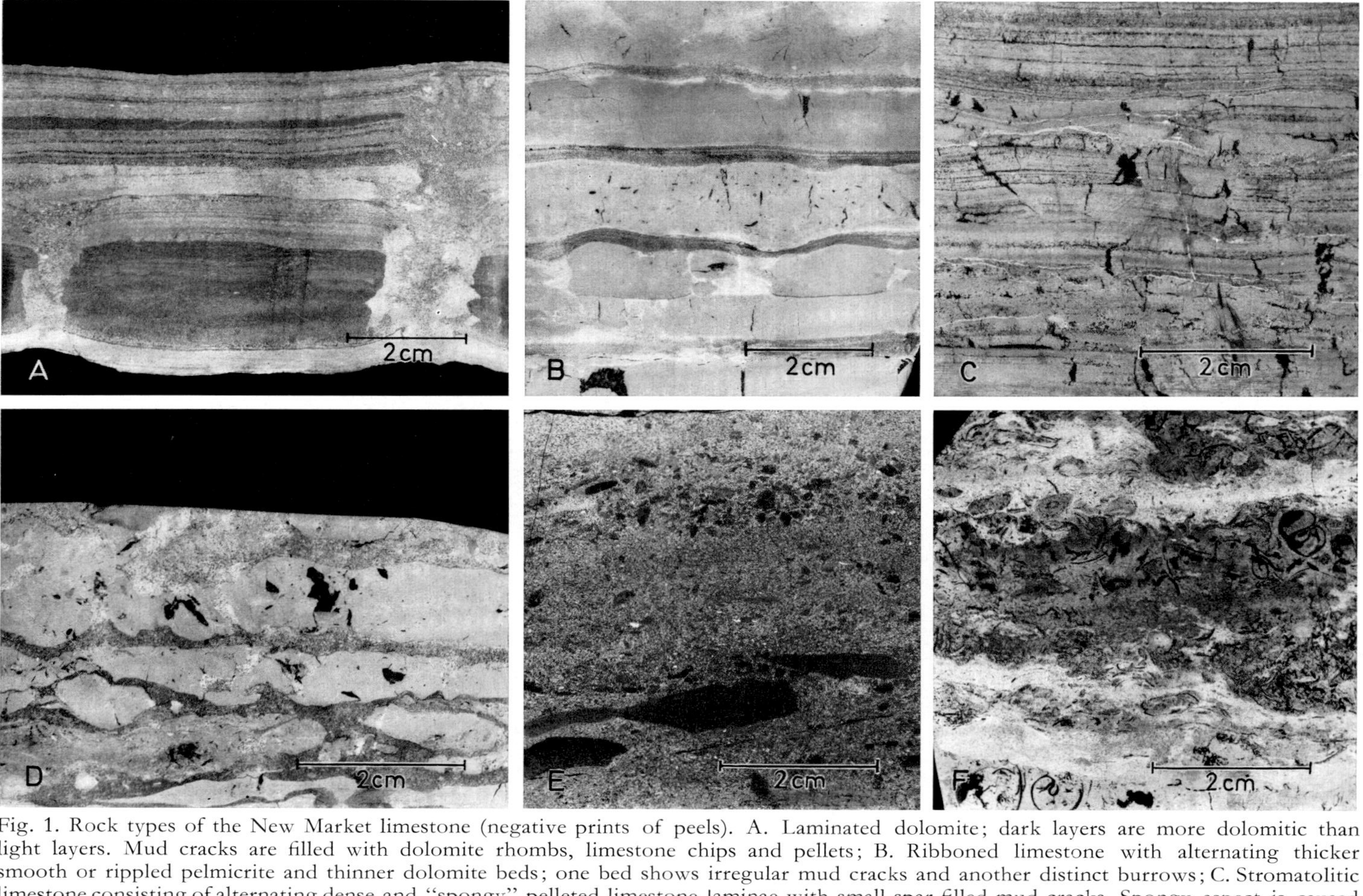

Fig. 1. Rock types of the New Market limestone (negative prints of peels). A. Laminated dolomite; dark layers are more dolomitic than light layers. Mud cracks are filled with dolomite rhombs, limestone chips and pellets; B. Ribboned limestone with alternating thicker smooth or rippled pelmicrite and thinner dolomite beds; one bed shows irregular mud cracks and another distinct burrows; C. Stromatolitic limestone consisting of alternating dense and "spongy" pelleted limestone laminae with small spar filled mud cracks. Spongy aspect is caused by spar filled molds of algal filaments; D. Lumpy limestone; discontinuous and partly burrowed limestone beds and intraclasts in a dolomite "matrix"; E. Poorly bedded intraclastic limestone; intraclasts and pellets with dolomite rhombs in interstices; F. Bioclastic limestone; irregular dolomitic limestone beds with fragments of brachiopods, gastropods, ostracods and other unindentified shells alternating with dolomite beds (light). Note oncolite in upper left

If we take all the evidence collectively: mud cracks, birdseyes, stromatolites and paucity of fossils (disarticulated shells of marine organisms) it seems highly probable that the laminated dolomites, ribboned and lumpy limestones and the stromatolites were formed in the inter — supratidal environment. Because the intraclastic and bioclastic limestones are so closely interbedded with the other four rock types representing nine tenth of the section (Fig. 2) they must have formed within the same environment.

Comparing the rock types of the New Market formation with modern tidal flat sediments it is possible to relate each type to a special sub-environment within the tidal flat gross-environment. The two most important differences between the various rock types are the different thickness of the beds and the frequency of mud cracks. Similar differences are known from modern tidal flats.

The laminated, mud cracked dolomites show all the features of supratidal sediments of Carribbean tidal flats. Here, repetitive flooding of the supratidal flat by

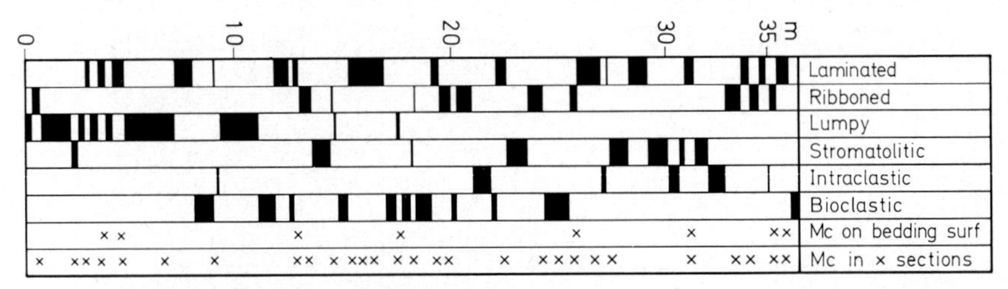

Fig. 2. Measured section of the lower part of the New Market limestone, Wilson Quarry locality, showing vertical distribution of rock types and mud cracks (MC)

heavy winds followed by times of exposure leads to a laminated mud cracked sediment similar to the ancient laminated dolomites. The ribboned limestones drape over stromatolites and small coral bioherms and contain rare desiccation features. They are similar to sediments found in the intertidal zone of well zoned modern mud flats. The lumpy limestones were formed by deposition of thick storm layers which were heavily cracked during long subaerial exposures. They are believed to represent a high marsh deposit. The bioclastic limestones which show occasional algal coated fragments and oncolites were probably formed in the intertidal-subtidal zone, whereas the intraclastic limestones cannot be attributed with certainty to anyone of the sub-environments.

Each of these rock types, correlated with a special sub-environment of a tidal flat, is not unique to the New Market limestone. The same rock types are found in most Paleozoic limestone formations of the Appalachians.

Origin and Diagenesis of the Oriskany Sandstone (Lower Devonian, Appalachians) as Reflected in its Shell Fossils

ADOLF SEILACHER *

With 7 Figures

Abstract

The quartzitic Oriskany Sandstone contains an abundance of calcitic fossil fragments and is cemented by calcite.

A high energy (beach) depositional environment is indicated by (a) rounding and sorting of quartz grains, (b) faunal restriction to few types of fossils, (c) absence of stem columnals of crinoids, indicating drift of the preserved fragile, buoyant calices into this environment, and (d) sturdy form of autochthonous brachiopods.

Diagenesis includes (a) early solution of aragonitic skeletons and redeposition as calcite cement, (b) local solution of calcitic fossils during a period of emergence, and (c) strong corrosion of quartz grains along contacts with shell surfaces during late diagenesis.

A. Introduction

What can the paleontologist contribute to carbonate petrology? — The obvious answer would be that he determines fossils and bioclastic constituents and may thereby back or refine a purely petrographic classification of carbonate rocks. While this is certainly true, the present paper should show that paleontological analysis can go farther than the mere determination. Fossils preserve traces of diverse processes in which they were involved at some time or other. While some of these have little bearing upon petrogenesis, the following processes should be considered in this context.

I. *Phylogeny*, *Ontogeny*, *Physiology* only as far as they determine 1. shell morphology; 2. shell structure; 3. shell chemistry.

II. *Ecology* relates the living organism to the lithotope. It is recorded in 1. functional morphology; 2. biological burial positions.

III. *Sedimentary processes* act mainly on the dead shells. They leave their traces in: 1. regular fragmentation and abrasion; 2. abiological burial position.

IV. *Diagenesis* has affected the buried shell by 1. early diagenetic solution and recrystallisation; 2. late diagenetic solution and recrystallisation.

B. The Oriskany Sandstone

I. Oriskany Lithology

High energy sediment in the Appalachian depositional cycle

The Oriskany Sandstone sticks out in the Appalachian section not only as a prominent ridge-maker, but also as an unusually coarse sediment (0.3 mm median

* Geologisch-Paläontologisches Institut, Universität Tübingen, Germany.

grain size). One would expect such a sand body to be only of local importance, but this sandstone has extensive distribution not only along the Appalachian trough but also in the adjoining shelf area, where under- and over-lying sediments change considerably in thickness and lithology. Even there, the Oriskany sandstone retains its character both in lithological character and fossil content, although the thickness may vary from place to place. In other words, this sandstone does not easily fit familiar environmental models. Neither its lithology nor its rare sedimentary structures provide the necessary clues.

What has this sandstone to do with a colloquium on carbonate rocks? In some places the sand is pure enough to be quarried for glass fabrication. But this is clearly caused by fairly recent leaching that has turned the rock into a friable sandstone. Originally, not only the fossils but also the cement must have been calcareous. At least during part of its history the Oriskany Sandstone would have been classified as a carbonate rock.

II. Shell Chemistry

Original or residual assemblage?

Oriskany fossils differ from other sandstone assemblages not only in abundance and preservation, but also in diversity:

Many brachiopod, crinoid, trilobite, and some gastropod species have been reported.

Still this imposing record is likely to be rather incomplete. Marine shell assemblages normally include calcitic as well as aragonitic forms. In the Oriskany shell beds, however, pelecypods, otherwise common in Devonian shallow marine deposits, are lacking as well as *Tentaculites* and cephalopods. The only gastropods that occur with some frequency are platyceratids that grew on crinoid calices and differed from other snails by having a calcitic shell.

It is reasonable to assume that the aragonitic fraction of the shell assemblages was eliminated by differential solution at an early stage of diagenesis, when the sand could still collapse without leaving a cast. The dissolved carbonate was not lost, but was re-precipitated in the form of calcite cement between the sand grains. In this way the aragonitic shells, though not leaving a determinable record, have "shielded" the calcitic remains from the same fate of diagenetic destruction.

Result: *Diagenesis seems to have started with a phase of aragonite solution which eliminated part of the original shell assemblage and provided the calcareous cement.*

III. Ecology

Allochthonous versus autochthonous faunal elements

While comparable in shell chemistry, the two major groups of Oriskany fossils — crinoids and brachiopods — differ profoundly in their morphology.

1. *Oriskany crinoids* are famous for their perfect preservation. No arms, no pinnules are missing from the calices. Still one would not expect these animals to have lived on the sediment in which they are now preserved. Only very specialized species could become attached in loose sand, particularly if it is frequently reworked by waves and currents. Still, quite a number of species are found. Most of them have long arms and very delicate pinnules. Some are reminiscent of birds' feathers when the pinnulae are spread (Fig. 1). Compared to the sturdy forms that we find in reef

communities, they seem to have lived in a low-energy environment rather than on turbulent sandy shoals.

The evidence rendered by functional morphology is also supported by the *biostratinomy* of Oriskany crinoids. In sharp contrast to the completeness of their calices the corresponding stems are missing to the extent that columnals are practically unknown to the quarry men. Since columnals would be more durable, we can reasonably assume that they had already been broken off when the calices reached the Oriskany lithotope. Storms may have torn the calices from their stems and swept them to the shore where they were left with their arms washed by the receeding

Fig. 1. Crinoid calices that have lost their stems are otherwise perfectly preserved in primary burial situation. Alignment (histogram!) indicates paleocurrent toward NW

waves (Fig. 1). A slight buoyancy, originally designed to hold the calices up on their stems, would have assisted their shoreward drift. The fact that they are usually found in groups and in the upper parts of sandstone beds (Fig. 5) would also favor this model.

2. The *Oriskany brachiopods* differ from the crinoids both in environmental adaptation and in biostratinomy, although both groups had a fixobenthonic mode of life. On the other hand the two common brachiopods *Costispirifer arenosus* and *Rensselaeria marylandica* resemble each other completely in both respects although they are genetically unrelated.

Practically all Oriskany species have a deep *muscle scar* that protrudes as a hump on the internal cast of the pedicle valve. This feature is also common among the brachiopods of the sandy Eodevonian in Europe, but it is absent in species of the Upper Devonian Chemung facies which represents a deeper and less turbulent environment in Appalachian history. Its functional significance becomes obvious if we recall that the deep scar is a consequence of excessive shell thickening. Comparable shell thickening is common among immobile pelecypods that have given up an attached mode of life and use differential shell weight to maintain their favorite position on the ground. In a similar way the heavy umboes of the Oriskany brachiopods must have kept the shells in their proper attitude even if the pedicle was too weak, or the substrate too mobile, to provide safe anchorage (Fig. 3).

Fig. 2. Brachiopod communities were caught by sudden burial in life position with their umboes down. Both seen from below. (a) *Costispirifer* (Johns Hopkins Univ. coll.); (b) *Renssel-laeria* (Tübingen museum, cat. no. 1339/1)

An umbo-down attitude can also be derived from the *burial position* of the Oriskany brachiopods. Large blocks in the Berkeley Springs quarry commonly have their lower surface covered with either *Rensselaeria* or *Costispirifer* in this orientation (Fig. 2). Actually it is necessary to remove the lower centimeter or two of the sandstone bed to expose the brachiopods, but since the rock is friable, one can easily collect hundreds of specimens from which the brachidia can be prepared in course work.

One might still ask whether these are truely autochthonous colonies or accumulations of dead shells swept together and buried by a tide. Some details disagree with the second explanation:

(1) Dead shells would have been deposited like a basal conglomerate at the very bottom of the bed, while the umbonal weight might still have controlled their orientation. On the other hand brachiopods inhabiting such an environment may well

Fig. 3. Oriskany brachiopods had their umboes thickened by internal calcite layers in order to maintain their living position in a turbulent environment (diagram). This adaptational feature again became important during postmortal abrasion, resulting in regular roll fragments that have similar shape in unrelated genera like *Costispirifer* (left) and *Rensselaeria* (right)

have been able to dig up to the surface after they had been covered by a few centimeters of sand. If sedimentation was too heavy, they could at least have reached their present level before they died. (2) The fairly even distribution provides another argument against postmortal accumulation which should have resulted in a more patchy distribution, similar to that of the crinoid calices.

Result: *The Oriskany fossil assemblage combines elements of different origin*: *Excessive sedimentation has occasionally buried whole colonies of brachiopods, which were otherwise well adapted to live in highly turbulent sands. Delicate crinoid calices broken off in other, less turbulent environments, drifted in without suffering further abrasion. A shore environment would best explain this unusual association.*

IV. Biostratinomy

Rules of shell fragmentation and distribution

Museum collections give a biased picture of Oriskany fossil preservation, at least for the *brachiopods*. Collectors will always chose the perfect specimens and neglect the fragmentary ones, particularly if, as in this case, they are more poorly preserved and do not occur associated with the complete ones.

Fig. 4. Disarticulated brachiopod valves aligned by tractional currents. They represent autochthonous species although their high grade of abrasion indicates considerable transport. Top: *Rensselaeria* from Woodmont, Md. (Tübingen cat. no. 1339/2); Bottom: *Costispirifer*, with histogram; (Tübingen cat. no. 1339/3)

In fact, disarticulated parts and fragments of brachiopod shells are more common than complete ones in most places, but they are found on different bedding planes. Contrary to a widespread belief, such fragments are not necessarily inferior in scientific value, since they may preserve a record of postmortal processes that is lacking in complete specimens.

Part of this record is contained in the *burial position*. A slight majority of the single brachiopod valves is embedded with the convex side up (Fig. 6a), as is usual under tractional currents. They also show alignment in azimuth orientation, indicating paleo-current direction (Fig. 4).

The other, less obvious piece of biostratinomic evidence consists of the *fragment shape*. In the case of Oriskany brachiopods abrasion was so regular that the fragments might pass unrecognized if seen only from outside. The internal casts, however, make it obvious that most valves, seemingly complete, are only the thickened umboes of pedicle valves or halves of them broken along the median plane (Fig. 3). This indicates that the shells in this lithotope were subject to considerable transport, which kept the valves rolling and gliding over the sand.

While we have so far learned about direction, strength, and kind of water movement, what was its lateral extension? A true current would have imported alien faunal elements. Since this is not the case, we can assume that it was a transport *in place*, such as by waves rolling the fragments to and fro. They would cause a degree of wear indicative of much greater displacement than has actually occurred.

The various effects of secondary reworking are seen only in the brachiopod part of the Oriskany fauna. Beds of disarticulated and worn *crinoid* ossicles that should be expected under this regime, have never been observed. This would indicate that the crinoid remains, due to different skeletal texture and chemistry (high magnesium calcite), were much less durable, and have practically disappeared from the record except in primary burial situations.

Result: *Orientation and abrasion of disarticulated brachiopod fragments point to a shore environment, in which the shells kept rolling about without being displaced from their original habitat. Crinoid ossicles were considerably less durable and are practically missing in reworked assemblages.*

V. Diagenetic Calcite Solution

Leaching in an elevated beach?

While the calcitic shells escaped early diagenetic solution under the shield of mobilized aragonite shells, they were locally exposed to another phase of carbonate solution. Comparing the two studied sections, one would expect this to have happened in the geosynclinal facies because it is poor in carbonates. But in fact the solution phenomena to be described come from Oriskany Falls, where the sandstone is framed by thick limestones above and below (Fig. 5).

Shells dissolved during this second phase of particle solution did not completely disappear from the record. A thin, dark film of insoluble residue remains visible in vertical sections. Bedding plane fractures may even reveal faint outlines of the original shells (Fig. 6b). They are too indistinct to allow specific determination, because external features were lost as the film was finally pressed over sand grains of the internal cast. Still it is clear that they are the familiar Oriskany brachiopod fragments.

How then did this process differ from the assumed solution of aragonitic shells? Fortunately the Oriskany section has preserved different stages. They show that solution in this case had a pronounced geopetal vector: On partly dissolved shells dark films line only the upper surfaces, whether these happen to be internal or external (Fig. 6c). This indicates that the sand was already cemented but was still capable of internal deformation by pressure solution along calcite shells and interstitial calcite.

This solution was also gravity-controlled in another respect. Although the sandstone bed at Oriskany Falls does not show any parting or other internal discontinuity, shell solution is not uniform in all levels of the bed. Film preservation is limited to the upper foot or so. In the lower part the calcite of the shells is still preserved or leached out by recent weathering. The transition between the two levels is only a few centimeters thick.

A satisfactory explanation for this vertical grading of diagenetic solution can not yet be given. The original idea that we deal with a sort of ancient water table, or an interface between marine and non-marine ground water, is incongruent with pressure-solution. To explain the phenomenon by diagenetic processes would also be difficult

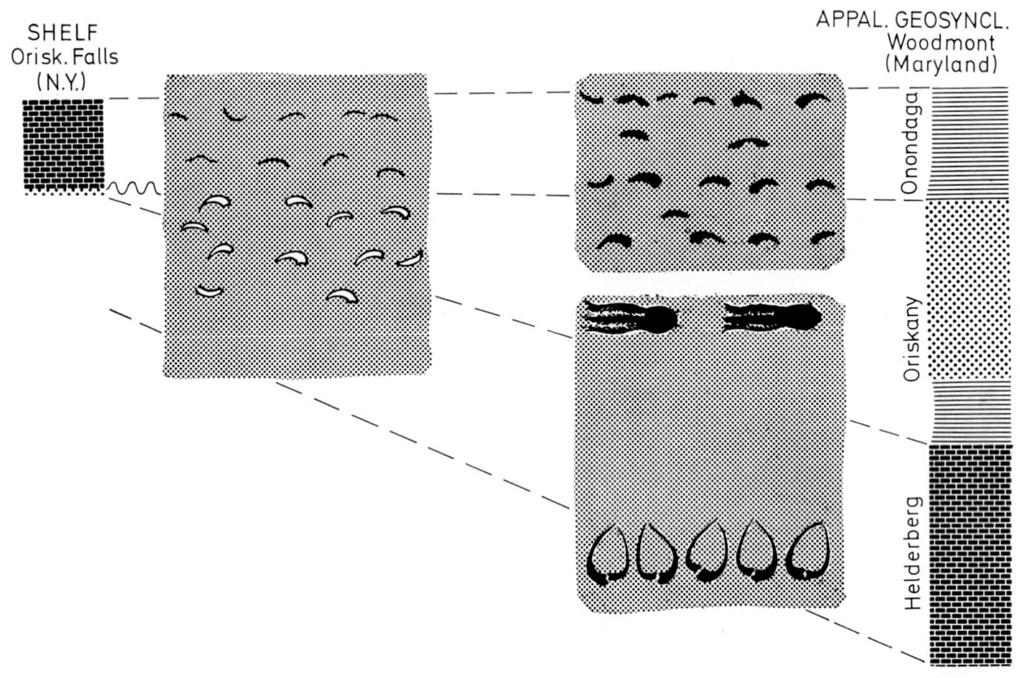

Fig. 5. Lithology and fossil content of the Oriskany sandstone are similar in shelf (Oriskany Falls) and geosynclinal situations (Woodmont near Berkeley Springs, right). All figured specimens unless stated otherwise come from Berkeley Springs

in the rock sequence in which the sandstone is now found. It seems more reasonable to refer it to a stage of emergence in which the sandstone was already cemented and either directly exposed at the surface or covered by shaly sediments of the Springvale stage which are now missing in the section.

Result: *A second phase of carbonate solution, this time affecting the calcitic shells, occurred locally after the sand had already turned into a firm and impermeable rock. This solution was geopetal in direction as well as in degree. It seems to be related to a phase of emergence expressed by an unconformity above the Oriskany.*

VI. Quartz Solution

A later phase of lithification

No further destruction, except for recent weathering, seems to have afflicted the Oriskany fossils after the diagenetic solution processes described in the last chapter.

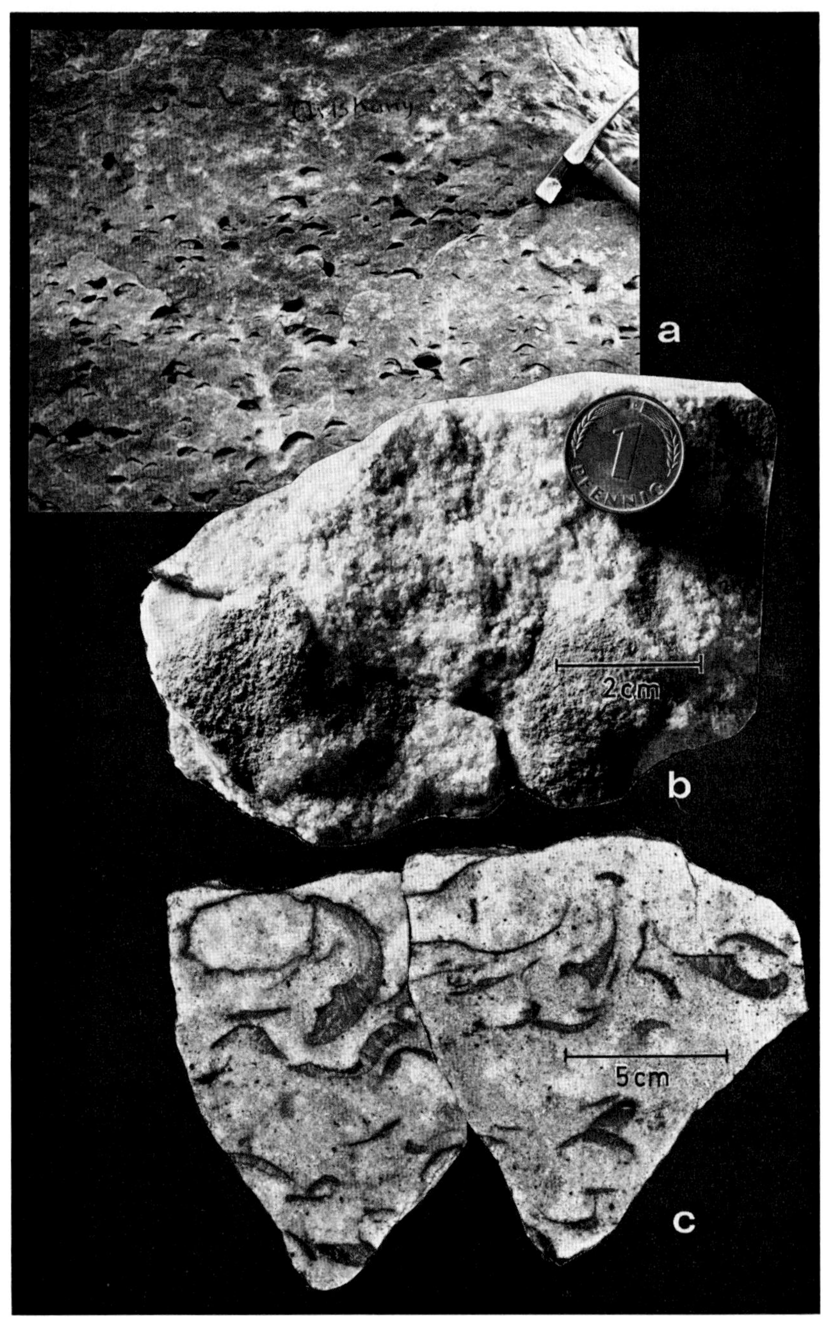

Fig. 6. Preservation of *Costispirifer* roll fragments at Oriskany Falls. (a) Weathered face with obvious shell layers (convex-up position) only in lower part of uniform sandstone bed. In the upper part shells are also present, but reduced to dark films that are too thin to be seen in this photograph. (b) Shells in film preservation from upper part of bed. While they still preserve the shape of the internal cast of the roll fragments (see Fig. 3), all details are lost through early diagenetic pressure solution (Tübingen cat. no. 1339/4; × $^{1}/_{2}$). (c) Vertical section of transition between zones of film and shell preservation. The shells in lower part of specimen show black linings on surfaces facing upwards indicating that shell solution was geopetal (Tübingen cat. no. 1339/5)

Actually solution did go on around the fossils, but it acted on the quartz grains rather than on the calcitic constituents of the rock.

The effect of this third phase of diagenesis is most obvious along the sand/fossil interface. Here the sand grains, otherwise nicely rounded, become heavily corroded. The zigzagged surfaces are reminiscent of stylolitic corrosion. In reality the zigzags

Fig. 7. Thin sections through brachiopod shells showing the effect of late diagenetic solution of quartz grains. (a) From lower part of bed at Oriskany Falls. Sand grains are corroded along their contact with the shell. "Stylolites", however, are not induced by pressure direction, but by the calcite prisms of the shell. (b) Similar corrosion at surfaces between grains and along the surface of un-altered shell cast (upper left), from Berkeley Springs

are not controlled by directional pressure, but by the fibrous structure of the brachiopod shell that acts on the sand grains like a nail bed (Fig. 7a). Similar processes along other interfaces have eventually turned the original accumulation of rounded sand grains into the interlocking texture that is now observed in many thin sections (Fig. 7b).

Evidently this third phase, which was active on the shelf as well as in the geosynclinal section, marks another change in diagenetic regime. We will not try to discuss p^H, pressure and temperature involved. But it can be said that the pressure was not directional, in contrast to the previous phase of calcite solution.

Result: During the final stage of lithification the calcitic shells remained practically unharmed, while quartz grains were corroded along shell surfaces and intergranular faces.

Acknowledgements

Observations on the Oriskany Sandstone were made during the tenure of a visiting lecturership at the Dept. of Geology, Johns Hopkins University, Baltimore. The author wishes to thank his colleagues at the department, Dr. E. Cloos, Dr. H. Eugster, Dr. F. Pettijohn, and Dr. D. Raup for guidance in the field and discussions. Mr. Murphy (Pennsylvania Glass Sand Company) kindly gave permission to make observations in the company's quarry at Berkeley Springs, W. Va. In the Tübingen institute, Dr. W. Lodemann kindly undertook the petrographic analysis of the thin sections. Photographs were made by W. Wetzel.

Facies Types in Devonian Back-Reef Limestones in the Eastern Rhenish Schiefergebirge

WOLFGANG KREBS*

With 12 Figures

Abstract

A study is made of back-reef limestones of Devonian "Massenkalk" in four areas of the eastern Rhenish Schiefergebirge (Dill area, Sauerland, Bergisches Land). The back-reef facies can be divided into eleven subfacies types. Five types belong to the intertidal-, six types to the subtidal facies. The individual subfacies types are characterized by a definite fauna and flora, allochems and matrixes, grain size, colour, bedding, and energy environment. Deformation, recrystallization, and dolomitization have altered the original composition and structure, particularly in the micrite-rich limestones.

A. Introduction

The late Middle Devonian and early Upper Devonian "Massenkalke" of the eastern Rhenish Schiefergebirge consist of a bank-phase and a reef-phase (KREBS, in press). The reef-phase can be divided into fore-reef, reef-core and back-reef. Within

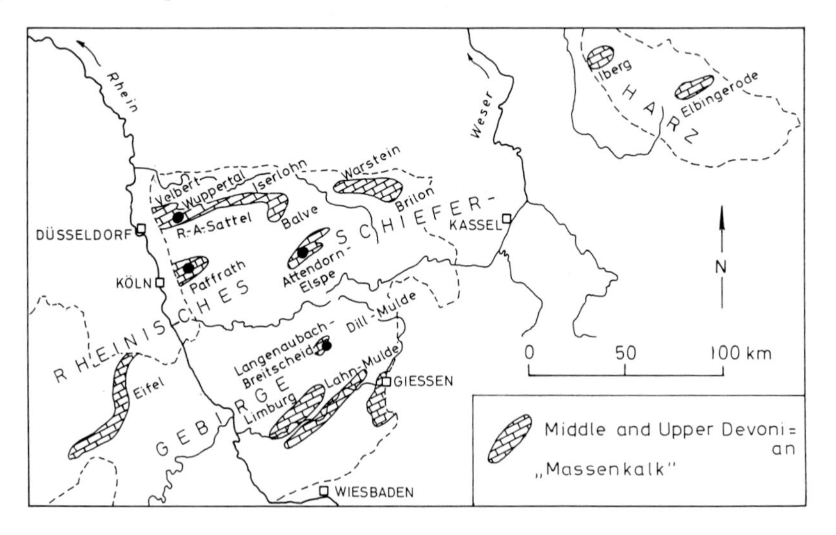

the reef-phase the back-reef limestones take up about 80% in space. So the back-reef facies is very important considering the composition of the Devonian "Massenkalk". To find out the correspondence between and the variations of the Devonian back-reef facies, the author investigated four different areas of the eastern Rhenish Schiefergebirge (see index map):

* Geologisch-Paläontologisches Institut, Technische Hochschule Darmstadt, Germany.

(1) in the early Upper Devonian reef limestones of the Langenaubach-Breitscheid reef complex in the Dill-syncline.

(2) in the late Middle Devonian reef limestones of the Attendorn-Elspe reef complex in the Attendorn-Elspe double-syncline.

(3) in the late Middle and early Upper Devonian reef limestones in the Druiten-Dornap limestone band at Dornap on the northern flank of the Remscheid-Altena anticline.

(4) in the early Upper Devonian "Oberer Plattenkalk" at Bergisch-Gladbach in the Paffrath-syncline. Here, JUX (1964) was the first to recognize the lagoonal nature of these platy limestones.

The facies interpretation of Devonian back-reef limestone is complicated in many cases by the following facts: a) deformation through schistosity as a result of the Variscan orogeny (PLESSMANN, 1966), b) recrystallization of the micrite matrix (micrite enlargement, FRIEDMAN, 1964) and allochems, as well as of the primary fibrous calcite matrix, c) dolomitization of the micrite limestones, which is dominant. The dolomitization quite often follows either the bedding planes or the schistosity plain

The different types of back-reef subfacies, as yet found in the four areas (see index map), shall be described in the following section (limestone classification after FOLK, 1959). An extensive description will be published in another paper.

B. Description

I. Laminated Intrapelsparite (Fig. 1)

Macro: Gray, laminated calcarenites with a) layers of sparry calcite parallel to bedding or b) rounded or oral sparry calcite pores ("birds' eye"). No macrofossils except stromatolites or algal mats. Micro: Intrapelsparite. Irregular wavy layers of micrite intraclasts and pellets interlayered with sparry calcite and rounded or oval patches of calcite parallel to bedding. In the voids drusy fibrous calcite, partly recrystallized to a coarse mosaic. Occurrence: see Table 1. Comparison: laminite (KLOVAN, 1964), microfacies C (KREBS, 1966), laminated "birds' eye" limestones of the *Amphipora*-stromatoporoid beds and of the *Amphipora* beds (MURRAY, 1966).

II. Laminated Pelmicrite (Fig. 2)

Macro: Thinly laminated, dolomitic limestones consisting of a sequence of convex-upward warped, light and dark layers, several mm thick. No macrofossils. Micro: Irregularly and crinkly bedded, often convex-upward algal (?) laminae of dolomitic micrite to dolomite and pelmicrites; interbedded with thin layers of pelsparite and dark brown, bituminous films. Micrite matrix consists in some cases of fine grained dolomite. Locally, detrital quartz grains. No fossils except for algal mats and probable charophytes. Occurrence: see Table 1. Comparison: laminite (PERKINS, 1963), microfacies 2 (BLUCK, 1965), laminated dolomite muds of the supratidal facies (LAPORTE, 1967).

III. Intrasparite (Fig. 3)

Macro: Light gray, calcite-rich, unbedded calcarenites with irregularly distributed sparry calcite pores, patches and cavities. Commonly no macrofossils. Micro: Poorly

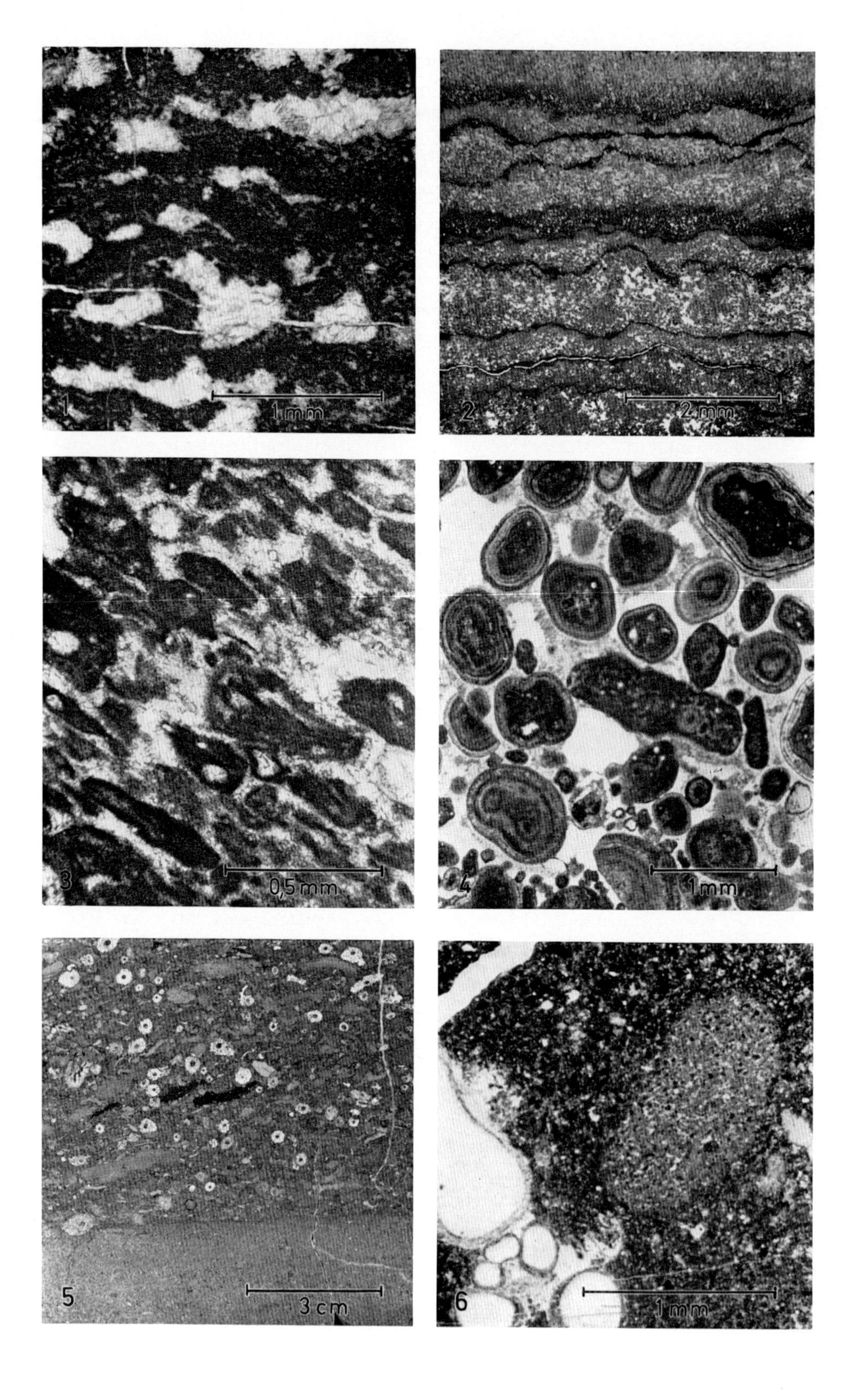

1
1 mm
2
2 mm
3
0,5 mm
4
1 mm
5
3 cm
6
1 mm

rounded and moderately sorted intraclasts (0.3 to 2 mm in diameter) and a few pellets (0.1 to 0.25 mm) in a drusy fibrous calcite matrix. Often recrystallization of the sparry matrix. Few fossils including calcispheres, ostracods, echinoderms, stromatolites. Occurrence: see Table 1. Comparison: non-skeletal calcarenite (KLOVAN, 1964), intrasparite (BEALES, 1965), microfacies C (KREBS, 1966).

IV. Oösparite (Fig. 4)

Macro: Gray oölitic limestone, diameter of oöids up to 2.5 mm, no macrofossils. Micro: Rounded to oval, rarely oblong, superficial oöids, partly well sorted. Nuclei consist of intraclasts, pellets, oömicrite or oösparite, calcispheres and echinoderms or other skeletal fragments. Concentric layers of radial fibrous calcite surround the nuclei. Thickness of single laminae: 0.01 to 0.05 mm; thickness of the whole laminae: 0.05 to 0.9 mm. Oöids with laminae over 0.5 mm are mostly broken. The laminae or the nuclei and the laminae are recrystallized to cryptocrystalline calcite. In some cases oömoldic porosity with former nuclei filled with calcite occurs. A brown, fine drusy calcite cement fills the interstices between the oöids. In the interstices recrystallization to clear, coarse mosaic calcite. Occurrence: see Table 1. Comparison: Hitherto no examples from the Rhenish Schiefergebirge.

V. Conglomeratic Beds (Figs. 5 and 6)

Macro: Beds and layers of calcirudites with rounded and angular, partly dolomitized re-worked sediments (up to 3 cm in diameter), *Amphipora*, dasycladacian algae, brachiopods, gastropods. Micro: Biointramicrudite to -sparudite. Mostly rounded, re-worked sediments and intraclasts of dolomitic pelsparite and pelmicrite; dolomite and fossils in a primary pelmicrite matrix. Micrite matrix shows all stages of recrystallization to coarse calcite mosaic and dolomitization. Settling and compaction

---

Fig. 1 to 6 (Plate 1). Intertidal facies

Fig. 1. Laminated intrapelsparite with cavities parallel to bedding. In the cavities, former drusy fibrous calcite recrystallized to coarse mosaic calcite. Thin section. Mag. 27×. Heggen near Attendorn

Fig. 2. Laminated pelmicrite. Crinkly bedded laminae of dolomitic micrite and pelmicrite interbedded with dark brown, bituminous films. Thin section. Mag. 17×. Marienhöhe, Bergisch-Gladbach

Fig. 3. Intrasparite. Intraclasts deformed by schistosity. Former, drusy fibrous calcite cement recrystallized. Thin section. Mag. 68×. Heggen near Attendorn

Fig. 4. Oösparite. Superficial oöids in drusy fibrous calcite cement, which is partly recrystallized to clear, coarse mosaic calcite. Thin section. Mag. 27×. Erdbach near Langenaubach

Fig. 5. Conglomeratic bed. Dolomitized, rounded, oblong pebbles, current controlled amphiporoids and dasycladacian algae in a dolomitized matrix. Lower part: dolomitized Echinoderm-Pelmicrite. Hand specimen. Marienhöhe, Bergisch-Gladbach

Fig. 6. Conglomeratic bed. Rounded pelmicrite intraclasts in a pelmicrite matrix. Left: serpulid worm. Thin section. Mag. 42×. Marienhöhe, Bergisch-Gladbach

Fig. 7 to 12. Subtidal facies

Table 1. *Distribution of back-reef types in the investigated areas and distribution of allochems and primary matrix in the back-reef types of eastern Rhenish Schiefergebirge.* ( ) *minor quantities*

Langen-aubach-Breitscheid	Attendorn-Elspe double syncline	Dornap near Elberfeld	Bergisch-Gladbach		predominant fossils	oöids and oncoids	intra-clasts	pellets	primary sparite	primary micrite
				intertidal						
×	×			Laminated intrapelsparite			×	×	×	
			×	Laminated pelmicrite				×		×
×	×			Intrasparite			×		×	
×				Oösparite		×			×	
			×	Conglomeratic beds	×	×	×	×		×
				subtidal						
×	×			Biopelsparudite	×			×	×	(×)
×	×			*Stachyodes*-Sparudite	×		(×)		×	
×	× ?	×		*Amphipora*-Pelsparite	×			×	×	
×	×	×		*Amphipora*-Pelmicrite	×			×		×
×	×		×	Biomicrite-Micrite	×			(×)		×
×	×	×	×	Echinoderm-Pelmicrite	×			×		×

phenomena also observed. Occurrence: see Table 1. Comparison: probably non-skeletal calcirudite (KLOVAN, 1964), limestone pebble conglomerate of the intertidal facies (LA PORTE, 1967).

VI. Biopelsparudite (Fig. 7)

Macro: Gray, coarse grained skeletal limestones consisting of allochthonous and fragmented fossils including algal-coated dendroid stromatoporoids of *Amphipora*- and *Stachyodes*-type, echinoderms, tabulate and rugose corals. Cavities filled with calcite and/or red micrites ("Rotpelit"). Micro: The above named fossils as well as brachiopod shells, ostracods, calcispheres and *Solenopora*, together with structure-less pellets (0.1 to 0.5 mm in diameter) and intraclasts (0.5 to 1.1 mm) in a sparite matrix. Strong recrystallization including micrite enlargement, grain diminution, syntaxial replacement rim, euhedral dolomite rhombs, authigenic quartz. Occurrence: see Table 1. Comparison: microfacies E (KREBS, 1966).

VII. Stachyodes-Sparudite (Fig. 8)

Macro: 10 to 30 cm thick detrital beds of denroid stromatoporoids (type *Stachyodes*, *Amphipora*) in a drusy fibrous calcite matrix. Locally reworked sediments up to 4 cm in diameter. Micro: Micrite coatings (algal?) around stromatoporoids. Drusy fibrous calcite matrix partly recrystallized. Occurrence: see Table 1. Comparison: *Stachyodes* calcirudite (KLOVAN, 1964), not Microfacies J on the fore-reef side (KREBS, 1966) and dendroid stromatoporoid beds (MURRAY, 1966).

VIII. Amphipora-Pelsparite (Fig. 9)

Macro: Light gray, massive to bedded *Amphipora*-rich limestones with rounded or oblong calcite-filled cavities. Frequent algal-coated amphiporoids. Micro: Algal-coated dendroid stromatoporoids in a primary sparite or pelsparite matrix. Cavities filled with micrite internal sediment, and/or drusy fibrous calcite. Often recrystallized. Frequent microfossils. Occurrence: see Table 1. Comparison: pelsparite of the *Amphipora*-zone (PERKINS, 1963), microfacies B1, near reef types (KREBS, 1966), *Amphipora*-spar beds (MURRAY, 1966).

IX. Amphipora-Pelmicrite (Fig. 10)

Macro: Dark- to black gray, well bedded, fossil-rich calcilutites with frequent *Amphipora* and calcisponges; also *Stachyodes*-type stromatoporoids, thamnoporoids, rugose corals, burrows. Micro: *Amphipora*-Pelmicrite. Fossils in a primary pelmicrite matrix. Pellets: 0.05 to 0.2 mm in diameter, occasionally up to 0.35 mm. Frequent microfossils including Foraminifera, ostracods, calcispheres, echinoderm fragments. Incipient micrite enlargement around allochems ranging in grain size up to coarsely mosaic. Locally micrite envelopes enclosing former microfossils. Dolomitization of micrite matrix. Dolomite-filled burrows or cracks. Occurrence: see Table 1. Comparison: *Amphipora* facies (KLOVAN, 1964), microfacies B1, dark types (KREBS, 1966), some types of *Amphipora* beds (MURRAY, 1966).

X. Biomicrite-Micrite (Fig. 11)

Macro: Dark to black gray, well bedded, homogeneous calcilutites or dololutites lacking, or with only locally scattered fossils similar to *Amphipora*-Pelmicrite. On the

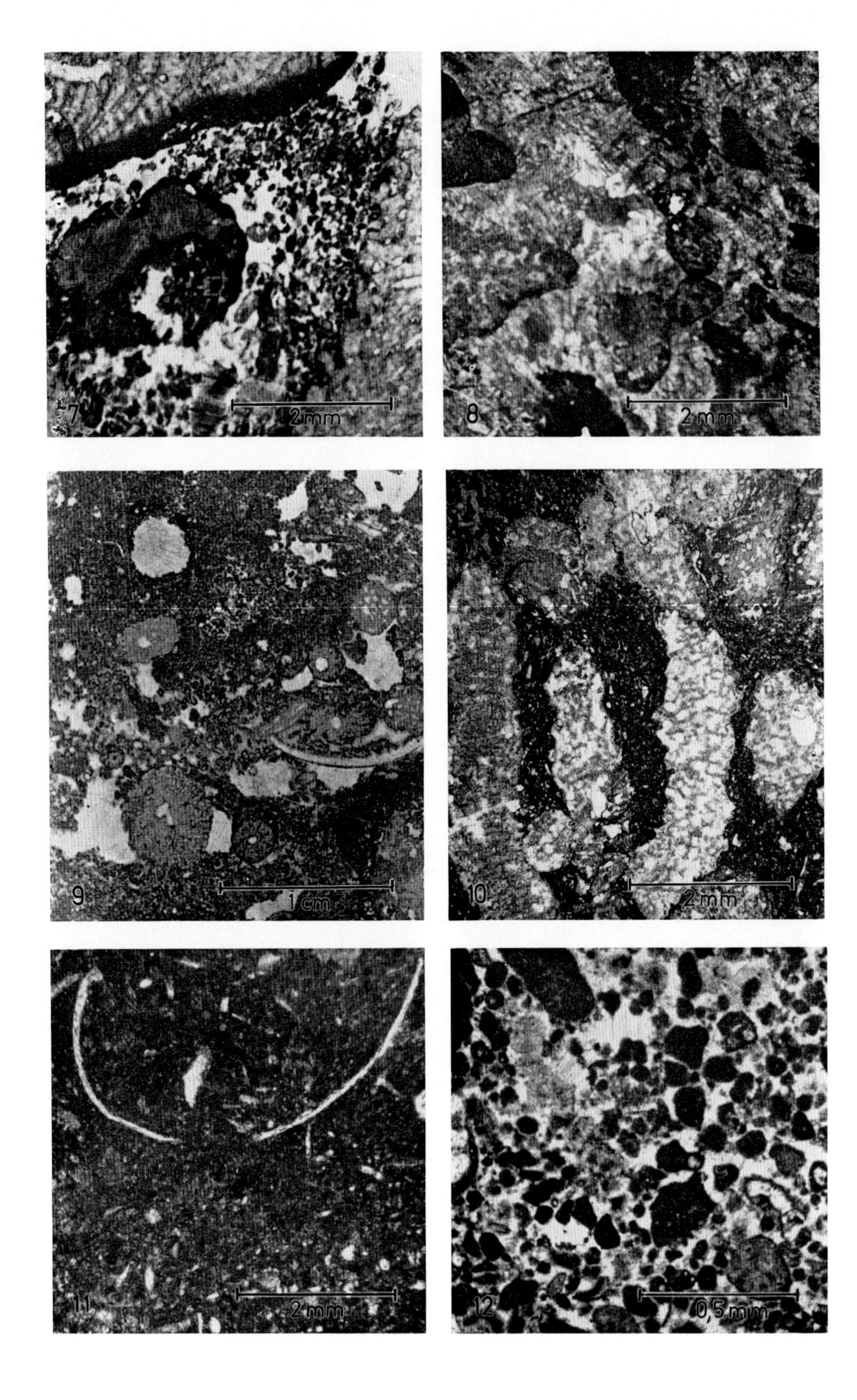
7
2 mm
8
2 mm
9
1 cm
10
2 mm
11
2 mm
12
0,5 mm

bedding planes locally irregular dolomitization, which can simulate re-sedimentation and bioturbation (KREBS, 1966). Micro: Microfossil-rich to microfossil-poor micrite with initial micrite enlargement. Locally transition to pelmicrite or pelsparite. Frequently dolomitization both of the matrix and of the fossils. Interbedded with *Amphipora*-Pelmicrite, Echinoderm-Pelmicrite or pelmicrite. Occurrence: see Table 1. Comparison: microfacies B 2 (KREBS, 1966).

XI. Echinoderm-Pelmicrite (Fig. 12)

Macro: Light to dark gray, coarse to fine-grained, well bedded calcarenites and calcilutites with many echinoderms. Other macrofossils rare. Locally graded bedding. Minor to strong dolomitization. Micro: Well sorted echinoderm fragments (up to 3.5 mm in diameter) and mostly rounded, structure-less pellets (0.05 to 0.25 mm) in a primary micrite matrix. Primary syntaxial rim cement only in coarse echinoderm fragments. Various types of recrystallization and diagenesis: a) micrite enlargement, b) syntaxial replacement rims around echinoderms, c) grain diminution of echinoderms and formation of pseudo-pellets and -intraclasts (BRAITHWITE, 1966), d) replacement fibrous calcite around echinoderm fragments (ORME and BROWN, 1963), e) xenotopic dolomite, f) authigenic quartz. Occurrence: see Table 1. Comparison: pelsparites (BRAITHWITE, 1966), microfacies A (KREBS, 1966).

C. Interpretation (Table 2)

All the intertidal facies types are characterized by the extensive lack of marine fossils (see Table 1). By the intertidal sediments, the laminites are especially conspicuous. At some time, within the lagoons of the atolls of Langenaubach-Breitscheid and Attendorn-Elspe, the nondolomitized, laminated intrapelsparites probably occurred behind bars or above shallow reef-debris flats in the intertidal zone. The calcite layers or pores parallel to bedding originated through desiccation (shrinkage pores

---

Plate 2.

Fig. 7. Biopelsparudite. Above: algal-coated dendroid stromatoporoid (type *Stachyodes*). On the right: *Amphipora*. Pelsparite recrystallized matrix. Thin section. Mag. 17×. Heggen near Attendorn

Fig. 8. *Stachyodes*-Sparudite. Algal-coated, dendroid stromatoporoids in former, drusy fibrous calcite cement with incipient recrystallization to coarse mosaic calcite. Thin section. Mag. 17×. Heggen near Attendorn

Fig. 9. *Amphipora*-Pelsparite. *Amphipora*, brachiopod shells, echinoderms in a pelsparite matrix with sparry calcite filling in primary cavities. Thin section. Mag. 3.7×. Dornap near Elberfeld

Fig. 10. *Amphipora*-Pelmicrite. *Amphipora* in pelmicrite matrix. Thin section. Mag. 17×. Dornap near Elberfeld

Fig. 11. Biomicrite. Brachiopods, ostracods, calcispheres and a few pellets in micrite matrix. Thin section. Mag. 17×. Medenbach near Langenaubach

Fig. 12. Echinoderm-Pelmicrite. Pellets, few intraclasts and echinoderms in a strongly recrystallized and dolomitized matrix. Echinoderms show replacement fibrous calcite. Thin section. Mag. 68×. Marienhöhe, Bergisch-Gladbach

A. G. Fischer, 1965), gas blisters through the decay of organic material (algal mats?) or burrows. These laminated carbonates with intercalated stromatolites correspond in many details to back-reef types recently described by Tebutt et al. (1965) and A. G. Fischer (1965). The laminated pelmicrites from Bergisch-Gladbach, which show probably early dolomitization, were deposited in more restricted lagoons protected from currents. The intrasparites probably resulted from the continuous erosion and re-deposition of periodically aerated algal coated carbonate muds. Shrinkage structures and stromatolites also show intertidal deposition. Great tidal current agitation, probably in tidal channels, and supersaturation of seawater with calcium carbonate favoured the formation of sparite cemented, superficial oöids. The intraformational conglomerate beds at Bergisch-Gladbach, which already contain flat dolomitized pebbles, originated from reworking of tidal flats and skeletal material during periodical storm tides in the lagoon.

Table 2. *Interpretation of the horizontal distribution of the various back-reef types in the Devonian reef limestones of the eastern Rhenish Schiefergebirge*

intertidal		Conglomeratic beds →		
	→ Oösparite	→ Intrasparite →	Laminated intrapelsparite	→ Laminated pelmicrite
subtidal		↗ *Stachyodes*-Sparudite		
	Reef → Biopelsparudite	→ *Amphipora*-Pelsparite →	*Amphipora*-Pelmicrite ↘	Biomicrite-Micrite
		↘ Echinoderm Pelmicrite →	Pelsparite Pelmicrite ↗	

The biopelsparudites consist of coarse-grained, reworked, current controlled skeletal components, which were transported from the leeward reef margin into the adjacent back-reef areas. They occasionally converge into coarse, skeletal reef debris, like fore-reef debris. Strong currents within the lagoon concentrated thin layers of dendroid stromatoporoids; the fine matrix between the stromatoporoids was almost completely washed out. The light *Amphipora*-pelsparites represent still stronger current agitation. In this facies type, the frequent cavities originated from washing-out effects or burrows. On the other hand, the dark, well-bedded *Amphipora*-pelmicrites were deposited in a quiet environment, in which pellets and microfossils frequently occur. They converge into macrofossil-poor, dark coloured biomicrites and micrites, which are situated in the central or restricted parts of a lagoon. Finally, the well-bedded Echinoderm-pelmicrites consist of well-sorted echinoderm fragments representing an extensive facies type within the back-reef areas. This type, which is characterized by the abundance of stenohaline echinodermata, is absent in the Canadian back-reef areas described by Edie (1961), Klovan (1964), and Murray (1966).

While the relationship of the back-reef limestones in Langenaubach-Breitscheid and Attendorn-Elspe is quite close, according to the present investigations, only limestones of subtidal facies exist at Dornap. Divergent from these three areas is the Oberer Plattenkalk at Bergisch-Gladbach, which was deposited in a much restricted, shallow lagoon, as Jux (1964) has shown. Similar conditions exist on the broad shelf-lagoon of the Old-Red-Continent at Aachen and in the Sötenich syncline, northern Eifel.

References

BEALES, F. W.: Diagenesis in pelletted limestones. In: PRAY, L. C., and R. C. MURRAY: Dolomitization and limestone diagenesis. Soc. Econ. Palaeontol. Mineral., spec. publ. **13**, 49—70 (1965).

BLUCK, B. J.: Sedimentation of Middle Devonian carbonates, south-eastern Indiana. J. Sediment. Petrol. **35**, 656—681 (1965).

BRAITHWITE, C. J. R.: The petrology of Middle Devonian limestones in South Devon, England. J. Sediment. Petrol. **36**, 176—192 (1966).

EDIE, R. W.: Devonian limestone reef reservoir, Swan Hills oil field, Alberta. Trans. Canad. Inst. Min. and Metall. **54**, 278—285 (1961).

FISCHER, A. G.: The Lofer cyclothems of the Alpine Triassic. Kansas Geol. Survey Bull. **169**, 107—149 (1965).

FOLK, R. L.: Practical petrographic classification of limestones. Am. Assoc. Petrol. Geologists Bull. **43**, 1—38 (1959).

FRIEDMAN, G. M.: Early diagenesis and lithification in carbonate sediments. J. Sediment. Petrol. **34**, 777—813 (1964).

JUX, U.: *Chaetocladus strunensis* n. sp., eine von *Spirorbis* besiedelte Pflanze aus dem Oberen Plattenkalk von Bergisch-Gladbach (Devon, Rheinisches Schiefergebirge). Palaeontographica, Abt. B, **114**, 118—134 (1964).

KLOVAN, J. E.: Facies analysis of the Redwater reef complex, Alberta, Canada. Canad. Petroleum Geologists Bull. **12**, 1—100 (1964).

KREBS, W.: Der Bau des Langenaubach-Breitscheider Riffes und seine weitere Entwicklung im Unterkarbon (Rheinisches Schiefergebirge). Abhandl. senckenberg. naturforsch. Ges. **511**, 1—105 (1966).

— Reef development in the Devonian of the eastern Rhenish Slate Mountains, Germany. Internat. Symposium on Devonian System, Calgary, Alberta (In press).

LAPORTE, L. F.: Carbonate deposition near mean sea-level and resultant facies mosaic: Manlius Formation (Lower Devonian) of New York State. Am. Assoc. Petrol. Geologists Bull. **51**, 73—101 (1967).

MURRAY, J. W.: An oil producing reef-fringend carbonate bank in the Upper Devonian Swan Hill Member, Judy Creek, Alberta. Canad. Petroleum Geologists Bull. **14**, 1—103 (1966).

ORME, G. R., and W. W. M. BROWN: Diagenetic fabrics in the Avonian limestones of Derbyshire and North Wales. Proc. Yorkshire Geol. Soc. **34**, 51—66 (1963).

PERKINS, R. D.: Petrology of the Jeffersonville limestone (Middle Devonian) of southeastern Indiana. Bull. Geol. Soc. Am. **74**, 1335—1354 (1963).

PLESSMANN, W.: Lösung, Verformung, Transport und Gefüge. Z. deut. geol. Ges. **115**, 650—663 (1966).

TEBUTT, G. E.: Lithogenesis of a distinctive carbonate rock fabric. Wyoming Geol. Survey. Contrib. Geology **4**, 1—13 (1965).

Carbonate Sedimentation and Subsidence in the Zechstein Basin (Northern Germany)

HANS FÜCHTBAUER*

With 4 Figures

Abstract

The rock distribution in the evaporitic Zechstein formation is described. In the "Ca 2" series, the following sequence towards the shore is developed:

1. dark, laminated limestones in the basin,
2. dark limestones and dolomites covering the slope, with isolated forams and ostracods,
3. oncolite shoal (early diagenetic dolomites),
4. light grey dolomites with few bryozoan-stromatolite bioherms in the lagoon.

Subsidence and sedimentation rates are discussed briefly. Two alternatives are formulated.

A. Introduction

The erosion of the Variscian chains has been completed during Lower Permian time, leaving sandstones and shales of the "Rotliegend" red beds. At the same time, evaporation began in the North. The subsequent Upper Permian "Zechstein" evaporites are different from those of the "Rotliegendes" by their very small admixture of clay and sand. This possibly is caused by an extremely low morphology in the southern mainland, combined with a high rate of subsidence in the basin which was compensated by evaporitic sedimentation. Four evaporation cycles are developed, most of them grading from claystones through limestones/dolomites and anhydrites to rock salt and potassium salts, in several cases with minor recessive branches (RICHTER-BERNBURG, 1955, p. 632).

B. Sedimentation

The facies relationships in the Zechstein formation are described, beginning from the bottom.

"T 1" = "copper shale"

In the major part of the basin, the Zechstein was ingressive. The first sediment, not more than about 0.3 m in thickness, is a black, ± pyritic, flasery marlstone which contains copper near the southeastern margin.

The carbonate content of about 30%, being calcite in the interior of the basin, grades into dolomite towards the shore (WEDEPOHL, 1964, p. 305). The dolomite $(Ca_{0.57}\,Mg_{0.43})\,CO_3$ is enriched in Ca, a fact indicative for a more or less normal salinity. This is in contrast to the stöchiometric dolomites in the later Zechstein (FÜCHTBAUER, 1964, pp. 504, 514).

* Geologisches Institut, Ruhr-Universität, Bochum, Germany.

A similar relationship exists between the clay minerals in the copper shale (which is free of rock salt) and the higher Zechstein "Salztone" (clays with salt): Whereas the latter show about equal amounts of chlorite and illite, the copper shale is poor in chlorite and contains even traces of kaolinite, a composition typical to normal marine claystones.

"Ca 1" = "Zechstein Limestone"

The differences between basin and margins increase. The basin facies consists of about 4 m of dark limestone, which is dolomitic in places. Its thickness and dolomite content increase towards the margins: Near the Rhenanian Massif, only dolomites are developed. The bryozoan and "Stromaria" bioherms in Thuringia are well known (MÄGDEFRAU, 1956; HECHT, 1960). The "Stromaria", according to their texture and growth attitudes, are supposed to be stromatolites of the cryptozoon type (LOGAN et al., 1964). A seperate bryozoan-stromatolite bioherm (Fig. 4) has been found recently within the basin (well "Schale Z 1", FABIAN et al., in preparation).

Large, tender oncoïds of a special kind (Fig. 2a) are typical of the basin facies. No reworking occurs. These oncoïds, which are in striking contrast to the shallow water oncoïds found in the "A 1", "Ca 2" and "Ca 3", suggest deeper water. Echinoderms, foraminifera and ostracods are found occasionally. The dolomite composition is about $(Ca_{0.55}\ Mg_{0.45})\ CO_3$. This together with the oncolites, may point to a slightly increased salinity.

"A 1" = "Werra Anhydrite"

In the northern basin, this member is a layered anhydrite (anhydrite with dolomite layers) of 45 m thickness ("6" in Fig. 4) which is fringed to the South by a 200 to 300 m thick rampart of massive to flasery anhydrite (RICHTER-BERNBURG, 1955, p. 598, 1960; FÜCHTBAUER, 1964, Fig. 1) ("5" in Fig. 4). In many cases, the "A 1" rests upon the "Ca 1"-limestone with a sharp contact (RICHTER-BERNBURG, 1955, p. 631), whereas the upper contact of the anhydrite is gradual, especially in the basin. Here, laminae of anhydrite grade into such of limestone even laterally, whereas the dolomite laminae grade into silt laminae (FÜCHTBAUER, 1964, p. 502). The near-shore facies of the "A 1" is a brown-grey dolomite, about 30 m thick with isolated oncoïds of the shallow water type (Fig. 2b).

"Ca 2" = "Staßfurt Carbonate"

The anhydrite rampart mentioned above divided the basin into four different environments, which are, from North to South (Fig. 1):

1. Basin facies: "stinking shale" (Stinkschiefer), 2 to 8 m of black limestone (occasionally dolomite), which is laminated by bituminous silt layers (Fig. 3). It has no fossils (Fig. 4, "10").

2. Slope facies: "stinking limestone and dolomite" (Stinkkalk, -dolomit), 10 to $>$ 100 m of dark grey limestones with a few ostracods and foraminifera and early diagenetical dolomites $(Ca_{0.51-0.52}\ Mg_{0.49-0.48})\ CO_3$ without fossils (calcitized in places) (Fig. 4, "9").

3. Ridge facies: "main dolomite" (Hauptdolomit), a 30 to 100 m sequence of medium grey dolomites, especially in the upper part rich in oncoïds, i.e. blue-green algal nodules, 0.05 to 5 mm in diameter, occasionally with well preserved laminae. Cross bedding and reworking, shown by multiple oncoïds, indicate considerable

movement in shallow water. There is a clear connection between the occurrence of oncolites and the position of the mentioned anhydrite "1" rampart (Richter-Bernburg, 1960; Füchtbauer, 1964, Fig. 1) (Fig. 4, "8").

The dolomites are suggested to be early diagenetical in origin: Their mineralogical composition — mostly $(Ca_{0.50}\,Mg_{0.50})\,CO_3$ (44 x-ray analyses) — the neighbour-

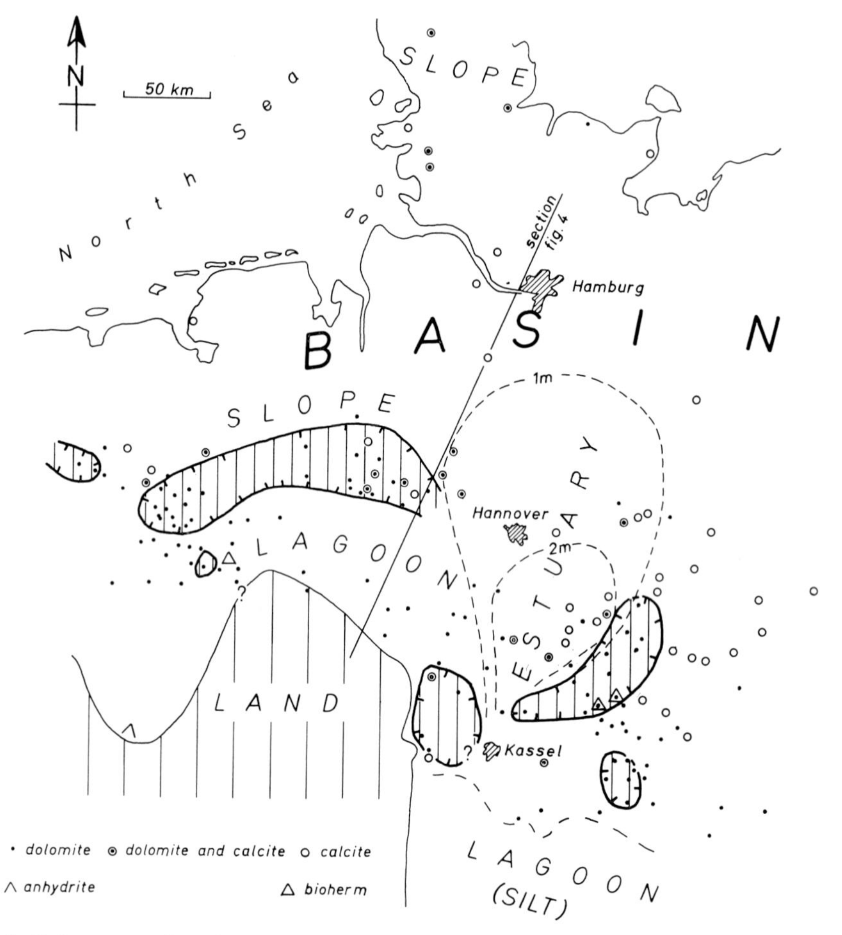

Fig. 1. Paleogeography of Ca2-Staßfurtkarbonat time. Thick lined with vertical hatching: algal shoals (early-diagenetical dolomites, calcitized in places). "1 m, 2 m" = calculated cumulative thickness in meters of the insoluble residues in the "Ca 2" carbonate rocks

hood of anhydrites, the occurrence of very few gastropods and ostracods as well as the blue-green algae, which are known to be indifferent even to extreme chemical environments — all these facts point to an increased salinity, typical for early diagenetical dolomites. This is supported by the low crystal size (< 0.01 mm) and the good preservation of the oncoïds and other fossil remains, showing a high magnesium concentration and a high nucleation rate during the dolomitization (Füchtbauer and Goldschmidt, 1965).

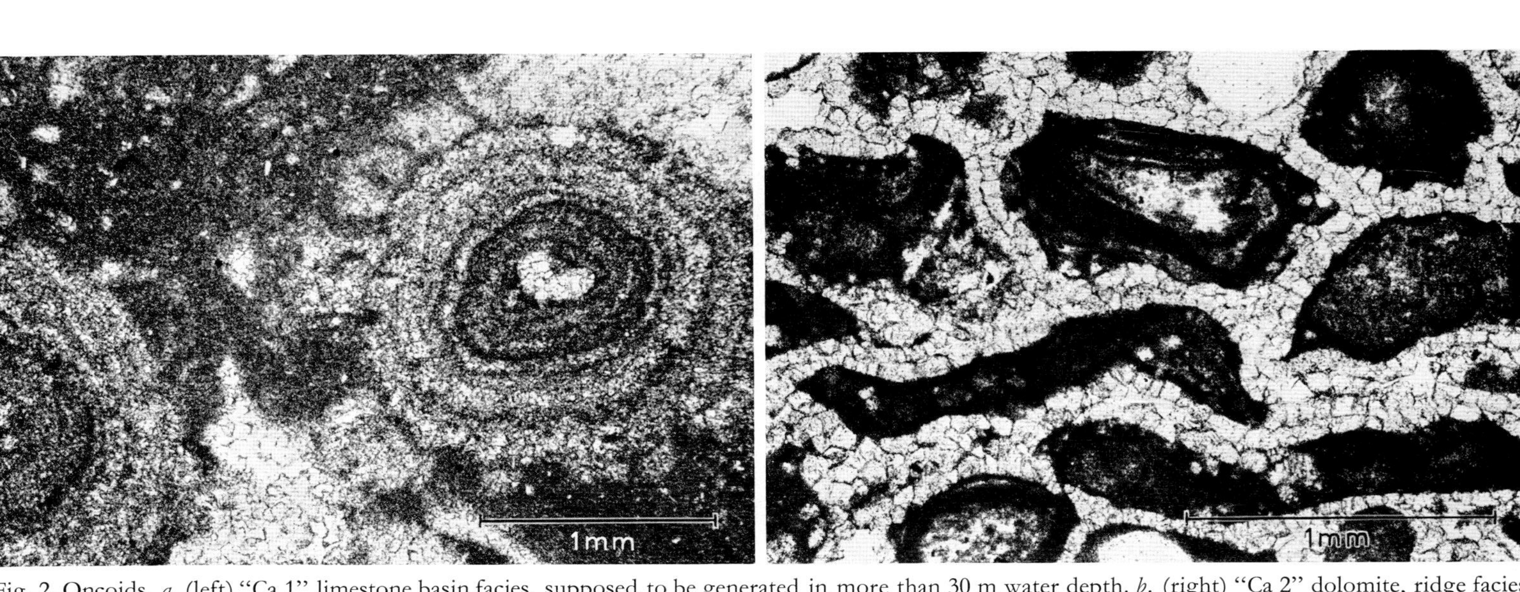

Fig. 2. Oncoids, *a*. (left) "Ca 1" limestone basin facies, supposed to be generated in more than 30 m water depth. *b*. (right) "Ca 2" dolomite, ridge facies. Oncoids presumably generated in a few meters depth. Secondary solution porosity

In general, secondary solution processes are more effective in carbonate rocks with wide spread than with uniform crystal size. Owing to their 1 to 5 micron crystals, the oncoïds are more soluble than the 30 to 60 micron matrix (Fig. 2b). It is for this reason, that the porosity of the uniform limestones and dolomite lutites is generally below 1%, whereas the dolomite oncolites are porous (2 to 30%). Intercrystalline and secondary solution intraparticle porosity prevail.

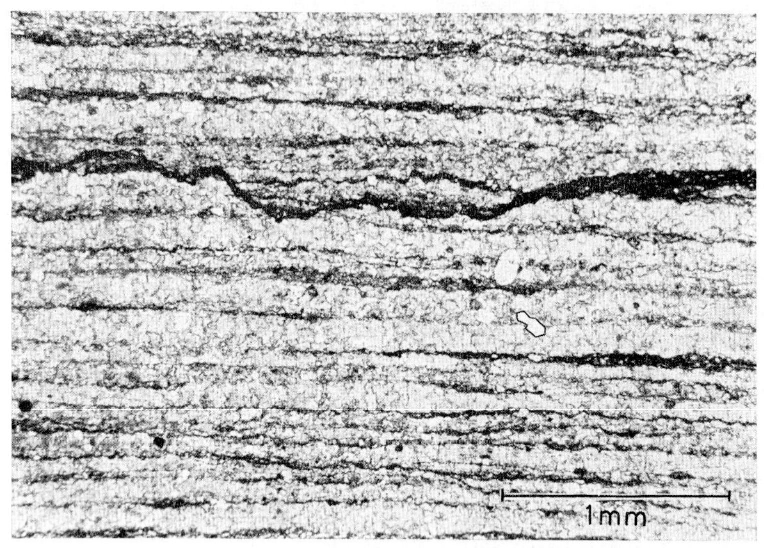

Fig. 3. "Stinking shale" = Ca 2 basin facies. Calcite with bituminous silt laminae. White spots are authigenic albites

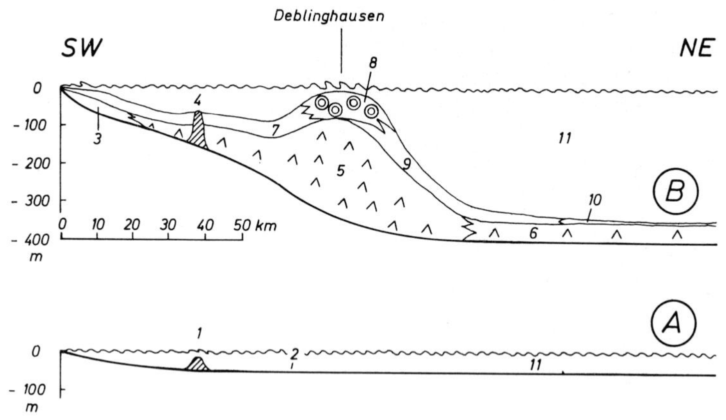

Fig. 4. Hypothetical sections through the Zechstein basin, *A* after deposition of Ca 1, *B* after deposition of Ca 2. 1 = bioherm (bryozoans, stromatolite). 2 = basin facies of Ca 1; 4 m of dark, oncolitic limestone. 3 = nearshore facies of A 1; light dolomites, sporadically with oncoïds. 4 = bioherm continued during A 1 and Ca 2 time. 5 = massive anhydrite "1" rampart. 6 = basin facies of A 1: laminated anhydrite. 7 = lagoonal facies of Ca 2: anhydritic light dolomites. 8 = ridge facies of Ca 2: dolomite oncolites. 9 = slope facies of Ca 2: limestones and dolomites. 10 = basin facies of Ca 2: laminated limestone (dark "stinking shale"). 11 = water

4. Lagoonal facies: light grey dolomites with anhydrite and sporadic oncoïds (Fig. 4, "7").

Isolated stromatolite-bryozoan bioherms occur on the southern rim of the oncolite bank S of the Harz mountains as well as in the lagoon farther to the West, where the bioherm of Schale Z 1 possibly continued to grow (Fig. 4, "1", "4").

"A2" = "Basal Anhydrite"

Similar to the "A 1", the anhydrite "2" is laminated only in the basin, where its thickness is about 2 to 3 m (RICHTER-BERNBURG, 1955, p. 606). Nearer to the coast, its thickness increases considerably. The boundary to the "Ca 2" on the ridge is sharp; dissolutions may be involved.

"Na 2" = "Staßfurt Rock Salt"

In the basin it is about ten times thicker ($\geqq$ 200 m) than above the ridge.

"T 3" = "Grauer Salzton"

A more or less plastic, grey claystone, very unlike the copper shale, not only by view but also according to x-ray analyses, which proved about equal contents of illite and chlorite, the latter mainly diagenetic, presumably (FÜCHTBAUER and GOLDSCHMIDT, 1959), but nevertheless indicating primary enviroment.

"Ca 3" = "Leine Carbonate"

Dolomites, in places oncolitic like those of the "Ca 2", are covering an area 30 to 150 km wide in front of the southern coast (FÜCHTBAUER, 1964, pl. 19; HECHT, 1960, Fig. 48). Occasional beds of limestone contain green algae ("Tubulites articulatus", BEIN; HECHT, 1960) and sporadically other fossils. To the North, the dolomites become dark, platy, and anhydritic ("platy dolomite") and merge to anhydrite laterally.

"A 3" = "Main Anhydrite"

In consequence, both stages, "Ca 3" and "A 3", are represented by anhydrite in the basin, whereas in the shallow water environment the anhydrite is confined to the "A 3" stage. It may be for this reason, that the anhydrite "3" increases in thickness towards the basin, opposite to "A 1" and "A 2".

Na 3—T 4—A 4—Na 4

These stages which partially are not developed, will not be discussed in this article.

C. Subsidence

It is intriguing to estimate the rate of subsidence during Zechstein, since lamination is abundant in the basin facies of the "Ca 2"-limestone and of several anhydrites, as well as in diverse rock salt units. In the limestones, these laminae are formed by silt, in the anhydrites by dolomites, and in the rock salt by anhydrite and other sulfates. Their distance in limestones is frequently 0.05 mm, in anhydrites 0.5 mm, and in rock salts 50 mm (RICHTER-BERNBURG, 1955, p. 621). The laminae have been interpreted as annual layers. BRAITSCH explained that this is possible for the limestones and the basin anhydrites only. Here, it is even supported by the resulting rates of deposition, which correspond to slow sedimentation on platforms (5 cm/1000 years). But the laminae in the rock salt probably represent more than one year each (JUNG, 1959; BRAITSCH, 1962, p. 415).

Provided that the laminae in limestones and anhydrites are annual layers, in the rock salt, however, about 10 year's layers, and that a constant rate of subsidence and equal sedimentation rates for equal facies types are assumed, then the net sedimentation time of Zechstein period would amount to about 0.6 million years only. This is a very short time, compared with the length of the Permian era, 50 million years. LOTZE (1938, p. 165) and RICHTER-BERNBURG (1950) both already realized this difficulty. One must, however, admit major periods of non-sedimentation. These omissions may be inserted mainly at the end of the carbonate units, which is suggested by visible facies breaks as well as by the requirement of major subsidence after the deposition of the "Ca 1" oncolite and the shallow water "Ca 2" oncolite, in order to provide the place for up to 250 m "A 1" anhydrite and up to 70 m "A 2" anhydrite, which have been deposited in a short time, as suggested by the laminations in the contemporary basin facies.

But even with these omissions, the Zechstein period would not exceed 1 million years too much. Since, however, the correlation between the Zechstein basin, and regions with continuous sedimentation during the Permian is poor, this is difficult to prove.

Under the above assumptions, the subsidence rate would be as high as 50 cm/1000 a, a figure which is supported by the following calculations:

a) The basin facies of "Ca 1" is characterized by tender oncoids differing markedly from the "Ca 2" shallow water oncoids. Therefore, a higher depth of deposition may be assumed. Approximately 30 m are suggested as a minimum figure. On the other hand, the "Ca 1" developed after only 0.3 m of "T 1" which was the first sediment of the Zechstein ingression. The minimum sedimentation rate of the "T 1" copper shale may be 0.5 cm/1000 years (porosity always eliminated). This rate is found in present oceans as well as in platform rocks with very slow sedimentation (STRAKHOV, 1962/67, Tables 27, 28). From these figures, a maximum of 60.000 years results for the "T 1". The subsidence then would be 30 m/60.000 years = 50 cm/1000 years.

b) The shallow water oncolites of "Ca 2" are characterized by local cross bedding and reworking. They attain 50 m in thickness and have been deposited in the same time interval as the 5 m basin limestone with its 100.000 laminae which probably are annual layers. Supposed that the depositional depth of the "Ca 2" oncolites was constantly very shallow, a subsidence rate of 50 m/100.000 years = 50 cm/1000 years would result.

This figure suggests a very high sedimentation rate in order to keep step with subsidence. Most of the units display normal sedimentation rates either in the basin or on the ridge. Since, however, the time correlation between basin and ridge facies is probably good, a high mean sedimentation rate results.

Therefore, the subsidence should have been high, too. Otherwise, very long omissions had to be inserted mainly above the carbonate rocks (see above). In this case, however, one would expect a clay layer, for at least in "Ca 2" time, according to Fig. 1, a river provided the basin with small quantities of detrital material, which should have accumulated during longer periods of non-sedimentation.

Even if compared with regions of major subsidence in present time or in the geological record, a subsidence rate of 50 cm/1000 years is high: Epeirogenic movements of the earth's crust of 100 to 1000 cm/1000 years are not unusual in the present (STRAKHOV, 1962/1967, p. 121). In older sediments, however, a relatively high subsidence rate is 70 cm/1000 years, displayed by the folded Lower Molasse, S-Germany.

On the other hand, the diagram of subsidence, plotted from the thickness of major rock units at the locality of Deblinghausen (see Fig. 4), is nearly a straight line between the Permian and the Upper Cretaceous, corresponding to a subsidence rate of 2 cm/1000 years only (FÜCHTBAUER, 1964, p. 520, Fig. 10). This suggests a similar rate of subsidence for Permian time. Consequently, as much as 25 million years would result for this era, a figure, which is much in contrast to 1 million years suggested above.

This discrepancy leaves only the following alternatives:

1. The laminae are annual layers. Then the subsidence rate in Zechstein time was much higher than during the following periods.

2. The Zechstein subsidence rate did not differ from the rates during the following periods. Even in this case, the laminations may be annual layers, but longer omissions must be assumed within the dark laminae.

References

BRAITSCH, O.: Die Entstehung der Schichtung in rhythmisch geschichteten Evaporiten. Geol. Rundschau **52**, 405—417 (1962).

FÜCHTBAUER, H.: Fazies, Porosität und Gasinhalt der Karbonatgesteine des norddeutschen Zechsteins. Z. deut. geol. Ges. **114**, 484—531 (1964).

—, u. H. GOLDSCHMIDT: Die Tonminerale der Zechsteinformation. Beitr. Mineral. u. Petrog. **6**, 320—345 (1959).

— — Beziehungen zwischen Calciumgehalt und Bildungsbedingungen der Dolomite. Geol. Rundschau **55**, 29—40 (1965).

HECHT, E.: Über Kalkalgen aus dem Zechstein Thüringens. Freiberger Forschungsh. **C 89**, 127—168 (1960).

JUNG, W.: Das Steinsalzäquivalent des Zechstein 1 in der Sangerhäuser und Mansfelder Mulde und daraus resultierende Bemerkungen zum Problem der „Jahresringe". Ber. geol. Ges. (Berlin) **4**, 313—325 (1959).

LOGAN, P. W., R. REZAK, and R. N. GINSBURG: Classification and environmental significance of algal stromatolites. J. Geol. **72**, 68—83 (1964).

LOTZE, F.: Steinsalz und Kalisalze, Geologie. In: „Die wichtigsten Lagerstätten der Nicht-Erze", Bd. 3, Teil 1, 936 p. Berlin: Borntraeger 1938.

MÄGDEFRAU, K.: Paläobiologie der Pflanzen. 3. Aufl., 443 p. Jena: VEB Gustav Fischer 1956.

RICHTER-BERNBURG, G.: Zur Frage der absoluten Geschwindigkeit geologischer Vorgänge. Naturwissenschaften **37**, 1—8 (1950).

— Über salinare Sedimentation. Z. deut. geol. Ges. **105**, 593—645 (1955).

— Paläogeographische Position der gashöffigen Dolomite im Zechstein. Vortr. deutsch. Ges. Mineralölwiss. Kohlechemie 18. 10. 1960. Erdöl u. Kohle **13**, 827 (1960).

STRAKHOV, N. M.: Principles of lithogenesis, vol. 1, translated by FITZSIMMONS, TOMKEIEFF and HEMINGWAY, New York. Consultants Bureau, 245 p. Edinburgh and London: Oliver and Boyd 1962/67.

WEDEPOHL, K. H.: Untersuchungen am Kupferschiefer in Nordwestdeutschland; ein Beitrag zur Deutung der Genese bituminöser Sedimente. Geochim. et Cosmochim. Acta **28**, 305—364 (1964).

References not quoted in the text, but used for the calculations:

Andres, J., E. Brand, W. v. Engelhardt u. H. Füchtbauer: Die Erdgaslagerstätten im Zechstein von Nordwestdeutschland. Giacimenti gassiferi dell' Europa occidentale, vol. 1, Accad. Nazion. dei Lincei, Milano, 101—136 (1959).

Fabian, H.-J.: Die Zechsteinprofile im Gebiete Barnstorf-Diepholz-Rehden-Wagenfeld. Fazielle Eigenheiten in der Staßfurtserie. Erdöl u. Kohle **10**, 741—747 (1957).

— Die Faziesentwicklung des Zechsteins zwischen Bielefeld und Hameln. Geol. Jahrb. **73**, 127—134 (1957).

— Das Jungpaläozoikum zwischen Diepholz und Twistringen (Konzession Ridderade) und seine Erdgasführung. Erdöl-Ztschr. 3—20, Wien, Hamburg, Juni 1963.

Philipp, W.: Zechstein und Buntsandstein in Tiefbohrungen zwischen Harz und Lüneburger Heide. Geol. Jahrb. **77**, 711—740 (1960).

The Bioclastic Limestones of the "Trochitenkalk" (Upper Muschelkalk) in SW Germany

(Preliminary report)

Manfred P. Gwinner, Gerhard Bachmann, Karlheinz Schäfer, and Klaus Skupin*

With 12 Figures

Abstract

Rock-forming components of the "Trochitenkalk" (Upper Muschelkalk) and their diagenetic modifications are described. Several types of bioclastic limestones are illustrated.

A. Introduction

The Upper Muschelkalk ("Hauptmuschelkalk", Middle Triassic) consists of a sequence of dense, micritic limestones, and of marls and bioclastic limestones. Some beds of bioclastic limestones can be regionally traced and are thus lithostratigraphic marker horizons. The lower part of the Hauptmuschelkalk is called "Trochitenkalk", because of the abundance of fragments of Encrinus liliiformis. Particularly dissociated columnals ("Trochiten") occur as rock forming components together with shell debris of brachiopods and molluscs.

Several stratigraphic sections of the "Trochitenkalk" in Baden-Württemberg were studied in detail in order to analyze the depositional environment of these rocks. The present preliminary report discusses the rock-forming components.

The work was supported by the Deutsche Forschungsgemeinschaft.

B. Rock-Forming Components and Their Diagenetic Modifications

The bioclastic limestones of the Trochitenkalk consist of the following essential rock-forming components:

I. Matrix and Cement

1. Micrite

The average grain size of the fine-grained matrix of the Trochitenkalk is about 7 μ, but may be $> 10\,\mu$, and is thus larger than the maximum value of micrites (Folk 1959). Various types can be distinguished:

a) equigranular crystallites with a diameter of 2 to 8 μ, grain boundaries always straight, nearly no dolomite,

b) coarser crystals up to 10 μ, otherwise like a),

c) matrix of unequal grain size with dolomite "phenocrysts", grain boundaries irregular. Obviously more pronounced diagenetic modification,

* Geologisch-Paläontologisches Institut, Universität Stuttgart, Germany.

d) micrite with crumbly to cloudy texture. Alternating dark- and light-colored, irregular areas. The color difference is due to differences in grain size of the crystallites and only to a small extent to coloring admixtures. It is unknown, if this fabric results from differential recrystallization or if it is due to pods of fine-grained micrite, later cemented by sparry cement. This type of micrite is confined to dense limestones free of allochems.

2. Sparite (Pore-Filling Cement)

Sparite crystallized in the intergranular space of washed out bioclastic components. Crystals generally increase in size from the walls of the intergranular spaces towards the center, the crystals are oriented and of the palisade-type on skeletal surfaces without a micrite envelope and on coarsely crystalline pseudo-ooids, in contrast to shells with a micritic envelope. The central large crystal of cement-filled pores is commonly dolomite.

In the "Trochitenkalk", sparite only occurs in bioclastic limestones, whose clastic constituents are in contact with each other as best shown by rocks composed of well-sorted shell-fragments or rounded constituents (columnals, pseudo-ooids).

Calcite envelopes around echinoderm fragments should be distinguished from pore-filling cement. These envelopes form single crystals of calcite or dolomite together with the echinoderm particles. Such overgrowths can form prior to final deposition because they were abraded during transport.

II. Shells and Shell-Fragments can be Subdivided According to their Internal Structure in:

1. Shells with a coarse, sparry mosaic (partly dolomite), with larger inner and coarser outer crystals. This fabric results from the dissolution of aragonite and redeposition of calcite or dolomite.

2. Shells with retained prismatic layer. The interior of mollusc shells, originally aragonite, has recrystallized or is dissolved. The outer layer primarily composed of calcitic prisms remained stable.

3. The fibrous brachiopod shells were relatively stabel during diagenesis and show few changes.

4. Micritic envelopes around shells.

Shells and shell fragments commonly show a dark rim, 40 to 80 μ wide, which becomes drusy outwards. This envelope is commonly broken, the fragments being piled on top of each other. Only such envelopes may remain after the shells were dissolved. Broken shell surfaces commonly show no micritic envelope. Possibly the envelopes are the result of overgrowths.

III. "Trochiten" and other Fragments of Encrinites

IV. Pseudo-Ooids

In contrast to rare "true" ooids, pseudo-ooids are very common. They are made up of a tabular commonly dolomitic crystal mosaic even where the host rock is calcite.

Well-sorted and enriched pseudo-ooids occur mostly in well washed-out and sparry cemented rocks. They are distributed irregularly in limestones. Relic-textures in the pseudo-ooids indicate that some formed from ooid-like enveloped shell fragments, others from rounded micrite particles.

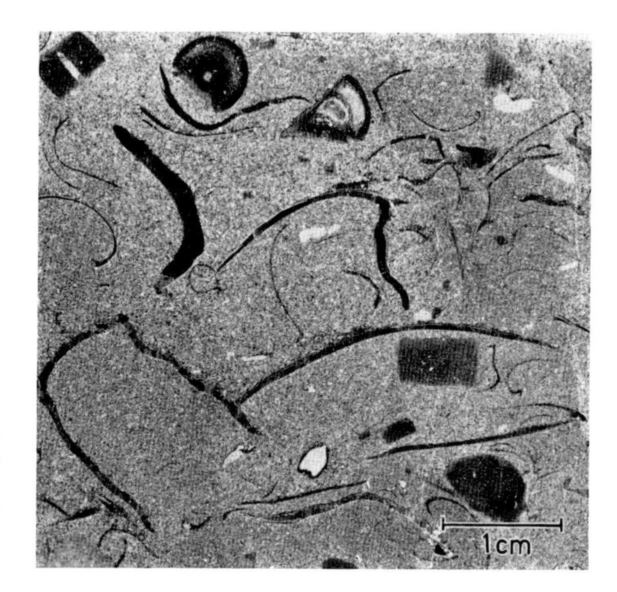

Fig. 1. Schalen- und trochitenführender Kalkstein (encrinitic limestone with shell debris). Matrix (M) = 79%, shell debris (SD) = 13%, crinoidal columnals (CC) = 8%

Fig. 2. Schalentrümmerkalkstein (coquina). M = 67%, SD = 31%, CC = 2%

Fig. 3. Schalentrümmerkalkstein (coquina). M = 52%, SD = 48%

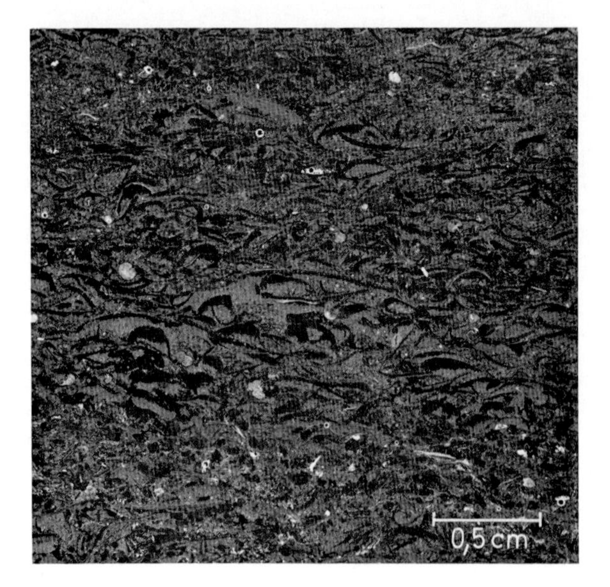

Fig. 4. Schalentrümmerkalkstein (coquina). M = 35%, SD = 65%

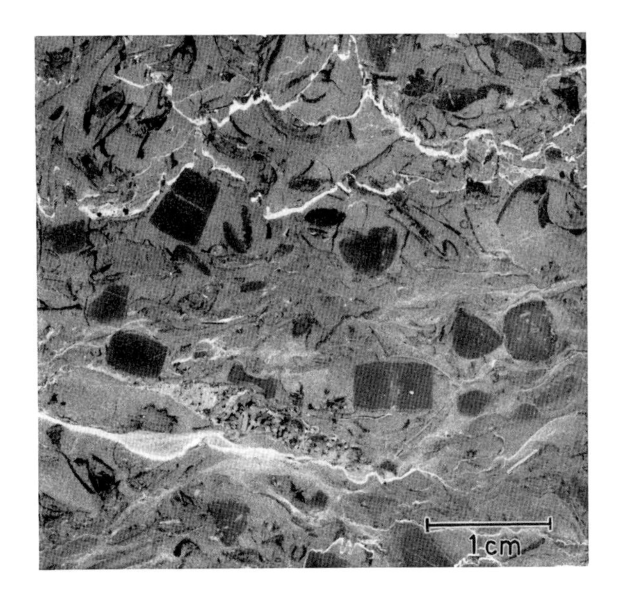

Fig. 5. Trochitenführender Schalentrümmerkalkstein (encrinitic coquina). M = 67%, SD = 22%, CC = 11%

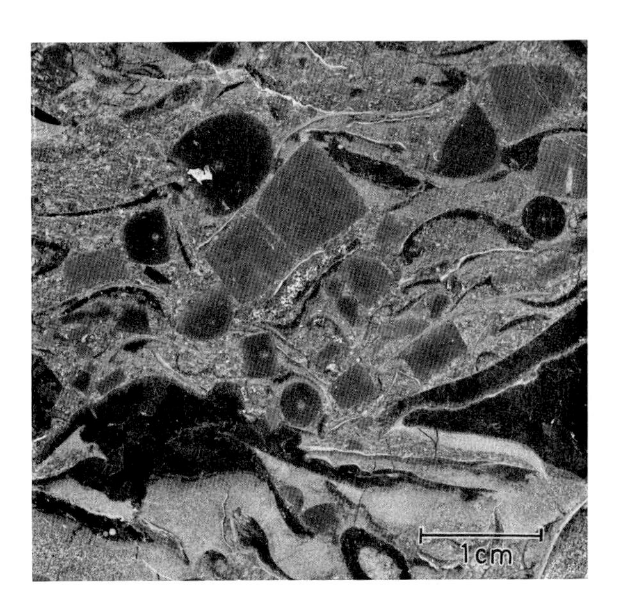

Fig. 6. Trochiten-Schalentrümmerkalkstein (encrinite-coquina). M = 44%, SD = 32%, CC = 24%

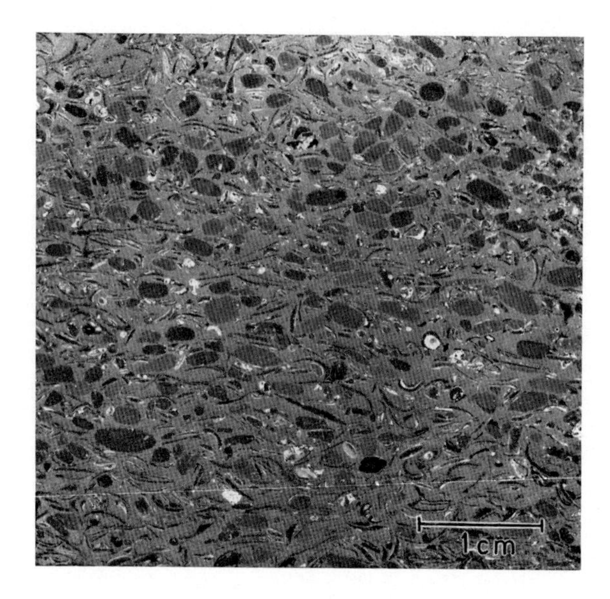

Fig. 7. Trochiten-Schalentrümmerkalkstein (encrinite-coquina). M = 50%, SD = 26%, CC = 24%

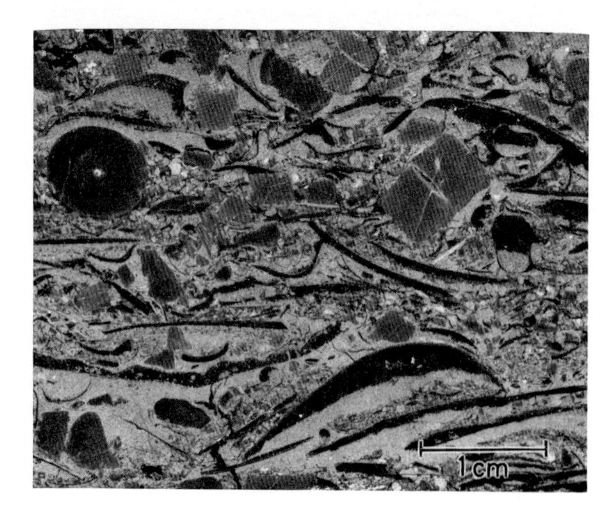

Fig. 8. Schalentrümmer-Trochitenkalkstein (coquina-encrinite). M = 30%, SD = 43%, CC = 27%

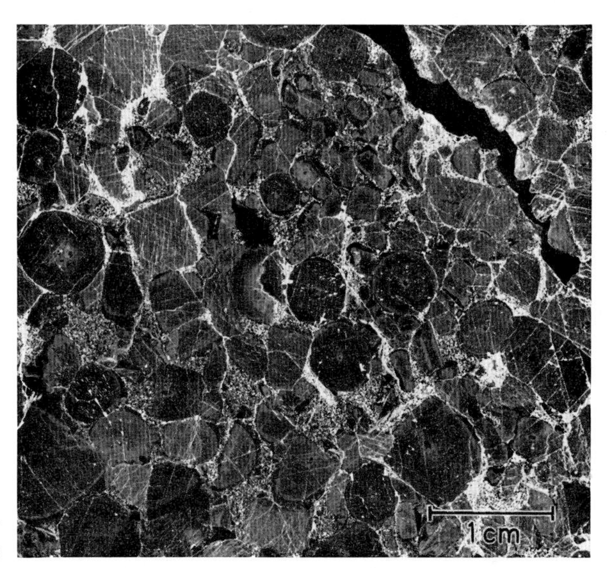

Fig. 9. Trochitenkalkstein (encrinite). M = 19%, CC = 81%

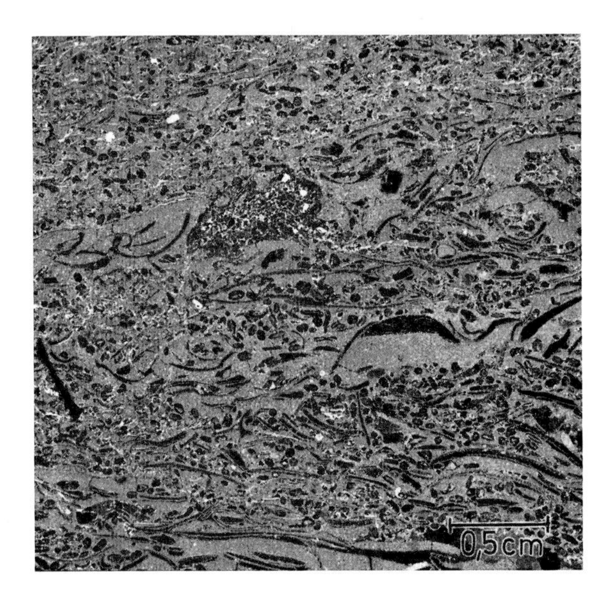

Fig. 10. Schwach pseudoolithischer Schalentrümmerkalkstein (coquina, weakly pseudoolitic). M = 58%, SD = 33%, pseudo-ooids (PO) = 9%

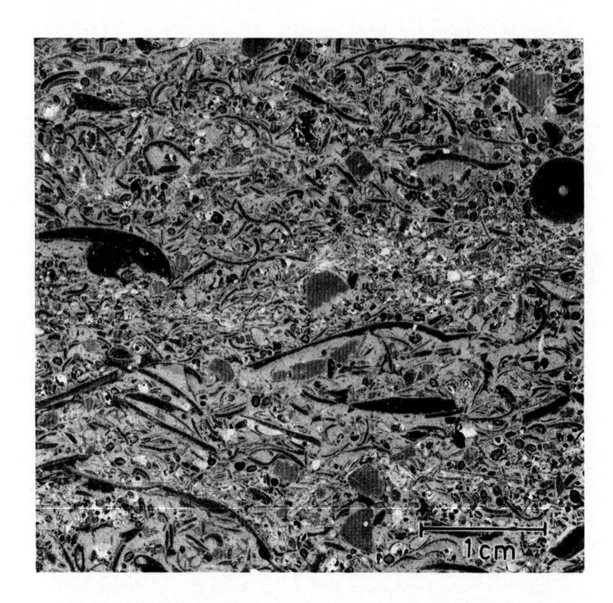

Fig. 11. Schwach pseudoolithischer, trochitenführender Schalentrümmerkalkstein (coquina, encrinitic and weakly pseudoolitic). M = 46%, CC = 40%, PO = 7%

Fig. 12. Pseudoolithischer Schalentrümmerkalkstein (coquina, pseudoolitic). M = 62%, SD = 25%, PO = 13%

Table 1. *Bioclastic rocks of the "Trochitenkalk"*

Grundmasse (Matrix)	Schalen-trümmer (Shell debris)	Trochiten (Columnals)	Pseudooide (Pseudooides)	Gesteinstyp	Rock type
90—100		0—10		reiner, mikritischer Kalkstein	micritic limestone
70—90		0—30		allochemarmer Kalkstein	micritic limestone poor in allochems
30—70	20—70	0— 5	0—10	(schwach pseudoolithischer) Schalentrümmerkalkstein	(Weakly pseudoolitic) coquina
	10—65	10—15		trochitenführender Schalentrümmerkalkstein	encrinitic coquina
	10—55	15—25		Trochiten-Schalentrümmerkalkstein	encrinite-coquina
	10—45	20—55		Schalentrümmer-Trochitenkalkstein	coquina-encrinite
	0—15	10—70		schalentrümmerführender Trochitenkalkstein	encrinite with shell debris
	—60	—10	10—30	pseudoolithischer trochitenführender Schalentrümmerkalkstein	Pseudoolitic encrinitic coquina
	—50	10—20		Trochiten-Schalentrümmerkalkstein	encrinite-coquina
	—40	10—50		Schalentrümmer-Trochitenkalkstein	coquina-encrinite
	—10	10—60		schalentrümmerführender Trochitenkalkstein	encrinite with shell debris
	0—40		30—70	Pseudoolith	Pseudoolite
0—30	0—100			allochemischer Kalkstein	Limestone rich in allochems

V. Pellets

They occur in two grain size classes. Coarser pellets have a diameter of 0.7 to 1.5 mm. These are fecal pellets, in part with an internal structure. The small pellets have a diameter of 20 to 200 μ and consist of finely crystalline calcite which is finer grained than the enclosing micritic matrix. Pellets and pseudo-ooids are irregularly distributed in coquinas. They generally lack in washed-out and bioclastic limestones cemented by sparry cement. They do occur in micritic limestones void of, or poor in shells, where they are well sorted and may cause lamination characterized by an alternation of layers of pellets and micrite.

Accessory components are autigenic quartz and glauconite.

C. Classification of the Bioclastic Limestones

The diversity of rock-forming components causes a diversity of rock types in the "Trochitenkalk". This diversity is not reflected in common rock classification. We have, therefore, set up a special scheme, by avoiding conflict with commonly used classifications, in order to more precisely delineate different types of depositional environments. Our classification is based on the modal composition of rocks, counted in thin-sections or acetate peels:

Groundmass (micrite or sparite)	up to 100%
Shells and shell fragments	up to 65%
(Molluscs and brachiopods cannot be exactly distinguished)	
Columnals and other crinoidal fragments	up to 30%
rarely	up to 80%
Pseudo-ooids	up to 55%
Pellets	up to 25%

This classification was so chosen that individual rock-types can be easily distinguished on polished surfaces. The most important bioclastic rocks are shown in Figs. 1 to 12. Table 1 summarizes the rock types.

Sedimentological and Biological Characteristics of a Dachsteinkalk Reef Complex in the Upper Triassic of the Northern Calcareous Alps

HEINRICH ZANKL*

With 2 Figures

The massive Dachsteinkalk (Norian-Rhaetian) of the Hohe Göll area in the Berchtesgaden Alps was already known to the geologists of the past century as a reef complex. But untill now there was only little knowledge about the structure of the

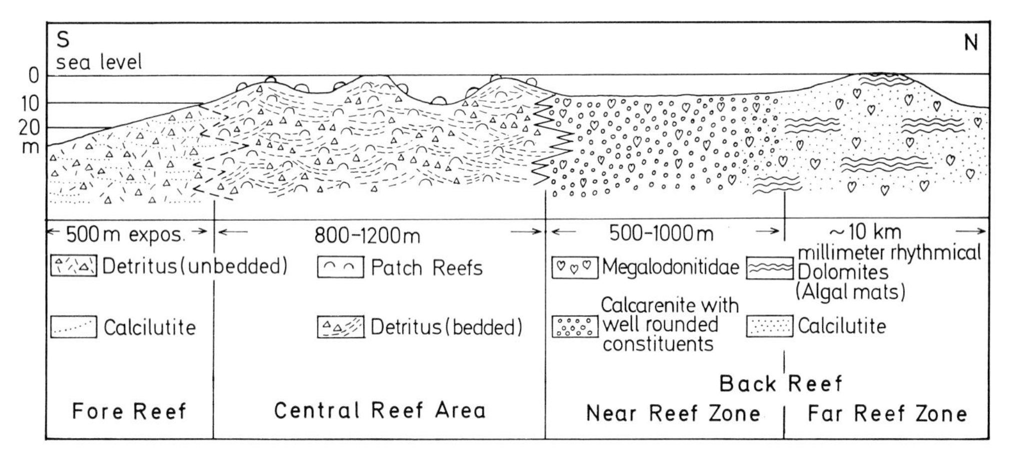

Fig. 1. Schematic profile of the Dachsteinkalk reef complex at the Hohe Göll area (Berchtesgaden Alps)

reef and about the reef-building organisms. Modern investigations started in the Dachstein area by H. ZAPFE (1960) and E. FLÜGEL (1962); a small reef complex was studied by E. FLÜGEL and E. FLÜGEL-KAHLER (1962) near Gußwerk in Styria.

A recent reconnaisance (ZANKL, 1967, 1968) in the Hohe Göll area shows a reef complex subdivided in different zones (Fig. 1). In the *central reef area* there are the relicts of the original reef framework with the reefbuilding organisms in growth position. This area strikes E-W, is 1200 m thick and 800 to 1200 m wide. In the south towards a basin the *fore reef* joins the central reef. The deposits consist mainly of reef detritus which interfingers on a low talus slope (max. 5°) with the basin sediments of the Hallstatt facies.

In the north the *back-reef sediments* are deposited in a *near-reef zone* which is characterized by well rounded reef detritus and in a *far-reef zone* of shallow-water sediments

* Institut für Geologie und Paläontologie, Technische Universität Berlin, Germany.

with an alternation of laminated dolomites — evolved from supratidal algal mats — and massive beds of lutitic limestones which contain large colonies of the pelecypode family Megalodontidae (A. G. Fischer, 1964). The near-reef zone is about 500 to 1000 m wide. The joining far-reef zone extends several kilometers to the north on a platform.

In the *central-reef area* the reef structure is erected by the activity of the organisms which include reefbuilders and reefdwellers. I want to avoid the term "reef core" because this term is too much connected with the idea of a cohering organic framework. The Hohe Göll reef convincingly shows that there is more reef detritus than primary reef framework. The proportion of reef framework and reef detritus amounts to 1:9. Taking in consideration 50% of detritus of finer grain sizes within the reef framework, this proportion changes in favour of the part of the detritus.

The *reef framework* is restricted to patch reefs of limited extension. The patch reefs are distributed irregularly in the central reef area. They are no relicts of erosion but the original areas occupied by the biological community of reef-building organisms. Each patch reef has a characteristic community whose individuals show zonal arrangements according to their forms of growth or their systematic classification. An uninterrupted reef ridge could not be observed.

The organic framework in the patch reefs is constructed by a restricted number of generations of reefbuilders (Fig. 2). Optimally five generations follow each other. The first generation contains corals, calcisponges, calcareous algae (Solenoporacea). The second generation with sessile foraminifers shows the dying out of the first generation. A third generation follows with sessile and incrusting calcisponges, bryozoans, Cheilosporites and very seldom Solenoporaceas. The fourth generation contains again sessile and incrusting foraminifers and crusts of calcareous algae. The fifth and last generation is formed by crusts of Spongiostromata. This succession of generations either is completely developped or is interrupted before by sedimentation. If one or more than one generation is omitted, the succession always remains the same. The succession of generations may be considered as a biogenous cycle.

Calculating the average volume of the reefbuilders referring to systematic groups, a value of 75% is found for calcisponges and corals, distributed in equal parts. Among the calcisponges the unsegmented Inozoa predominate the segmented Sphinctozoa. The most important coralls are represented by certain species of the genera Astraeomorpha and "Thecosmilia". The other coralls with the genera "Montlivaltia", Thamnasteria and Palaeastraea are reduced. The remaining 25% of reefbuilders are distributed among the calcareous algae, hydrozoans, bryozoans, foraminifers and some problematica.

The *reef detritus* represents, with more than 90%, the main mass of the central reef area. The detritus is surprisingly well bedded which is not recognizable by the morphology of the massive rock faces. The bedding is evident by a change in the grain size of the constituents as well as by the changing matrix.

As constituents besides the broken reefbuilders the detritus contains lithified and reworked material derived either from the patch reefs or the reef detritus. The patch reefs are the inexhaustible source of the reef detritus. As the distribution of the constituents shows, fine-grained detritus of the reef framework is supplied by biological and mechanical erosion. This material is quickly lithified, reworked by erosion and resedimented. This cycle may be repeated up to four times, the constituents of

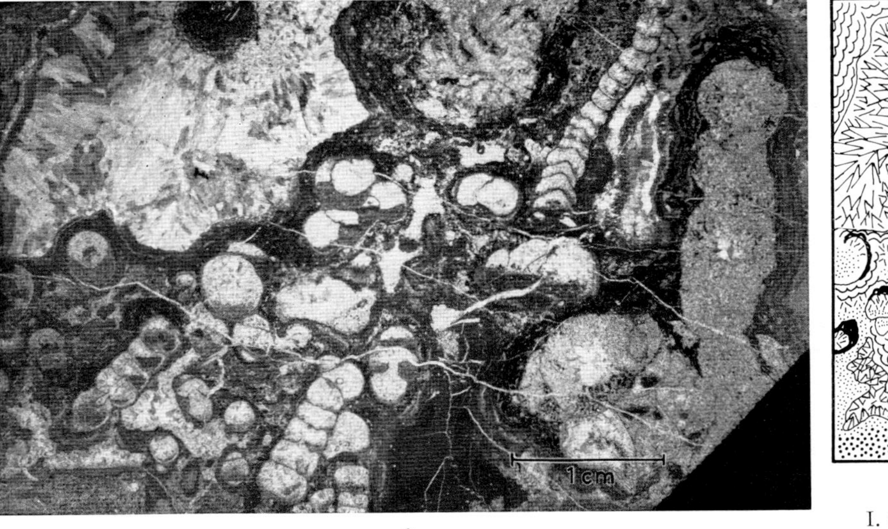

a

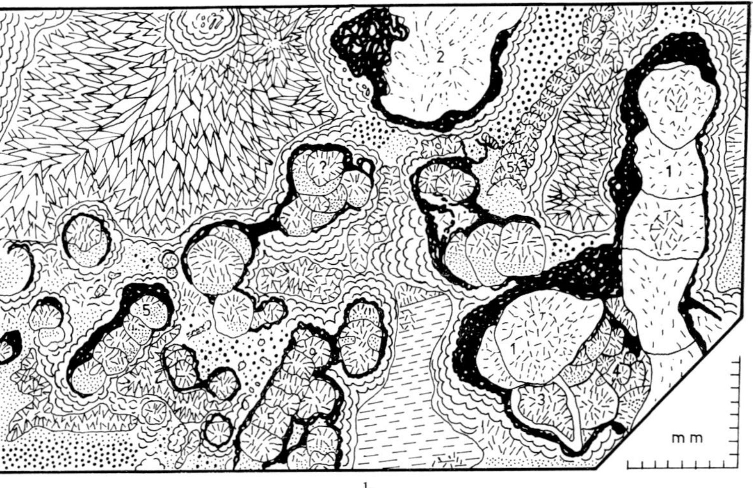

b

I. Generation
1 Calcisponge
2 Coral

II. Generation

encrusting Foraminifers

III. Generation
3 attached Pelecypod
4 Calcisponge
5 Cheilosporites tirolensis

IV. Generation

encrusting Foraminifers

V. Generation

Spongiostromata

calcisiltite
fine grained Calcarenite

Micrite

sparry Calcite

Fig. 2 a and b: The microphotograph and the explanatory sketch demonstrate a sequence of 5 generations of reef-building organisms and the cavity-filling by micrite and spary calcite. The encrusting foraminifers are the same in the II. and IV. generation — Microtubus, Calcitornella and Nubecularia (seldom) —; the two generations are not distinguishable around the calcisponges and the coral. Cheilosporites tirolensis (a dubious foraminifer) always starts growing when the first generation of reefbuilders is already incrusted, therefore the foraminifers attached around Cheilosporites belong to the IV. generation. Spongiostromata-crusts (V. generation) are formed by Blue-Green-Algae following at the end of skeletal growth. The production and sedimentation of detritus is also stopped at this point. The remaining cavities are filled at first partly by micrite and than by spary calcite cement detritus getting coarser. The rapid lithification of the reef detritus is the consequence of an syngenetic internal crystallisation of calcite in the pores between the constituents. Thus the reefbuilders always find rigid substrate for settling.

The great thickness of the reef of 1200 m is the consequence of a subsiding underground which is compensated by the growing of the reef. The biotope of the central reef area thus always keeps close to the sea level. The compensation of the subsidence is not caused by building up of a huge reef framework but by rich supply of detritus from the patch reefs.

Finally a note concerning the changing opinions of the term "reef" shall be given. The different definitions shall not be discussed neither shall a new one be added. The term reef is derived either from an isolated calcareous structure rising above the sea bottom or from a wave-resistant reef framework. These definitions do not fit for the type of a Dachsteinkalk-reef, because here the organic framework with the patch reefs has only a very small extension. The balance between the constructive and destructive activities is of significant influence. The main constructive activity is the skeletal organic growth; but the resedimentation and the rapid lithification of the detritus are important constructive factors as well. Certain boring organisms and mechanical erosion are the main destructive factors.

If skeletal growth prevails and other factors are unimportant, only small reefs originate as we can see in some contemporary Oberrhätkalk reefs in the Northern Calcareous Alps (F. Fabricius, 1966). In contrast, if the destructive activities prevail, the reef is covered by its own detritus. The balance between the constructive and destructive activities and a steadily subsiding underground leads to the growth of the hugest reefs known in the Earth's history. The Dachsteinkalk reef of the Hohe Göll represents a good example for such a reef.

References

Fabricius, F.: Beckensedimentation und Riffbildung an der Wende Trias/Jura in den Bayerisch-Tiroler Kalkalpen, 143 S., 53 Mikrophotogr., 24 Abb. Leiden: E. J. Brill 1966.

Fischer, A. G.: The Lofer Cyclothems of the Alpine-Triassic. Kansas geol. Survey Bull. **169,** 107—149 (1964).

Flügel, E.: Untersuchungen im obertriadischen Riff des Gosaukammes (Dachsteingebiet, Oberösterreich). III. Zur Mikrofazies der Zlambach-Schichten am W-Ende des Gosaukammes, Taf. 5, 1 Abb., Verhandl. geol. Bundesanstalt (Wien) **1962,** 1, 138—144 (1962).

—, u. E. Flügel-Kahler: Mikrofazielle und geochemische Gliederung eines obertriadischen Riffes der nördlichen Kalkalpen (Sauwand bei Gußwerk, Steiermark, Österreich). Mitt. Mus. Bergb. Geol. Techn. **24,** 1—129 (1962).

Zankl, H.: Die Karbonatsedimente der Obertrias in den nördlichen Kalkalpen. Geol. Rundschau **56,** 1, 128—139 (1966).

— Der Hohe Göll — Aufbau und Lebensbild eines Dachsteinkriffes in der Obertrias der nördlichen Kalkalpen. Abhandl. senckenberg. naturforsch. Ges. (Im Druck).

Zapfe, H.: (2) Untersuchungen im obertriadischen Riff des Gosaukammes (Dachsteingebiet, Oberösterreich). IV. Bisher im Riffkalk des Gosaukammes aufgesammelte Makrofossilien (exkl. Riffbildner) und deren stratigraphische Auswertung. Verhandl. geol. Bundesanstalt (Wien) **1962,** 2, 346—352 (1962).

Sedimentological Investigation of the Ladinian "Wettersteinkalk" of the "Kaiser Gebirge" (Austria)

PETER H. TOSCHEK*

With 11 Figures

Summary

The Middle Triassic "Wettersteinkalk" of the "Kaiser Gebirge" (Nördliche Kalkalpen, Austria) can be divided into several types of carbonate facies, each of which can be assigned to the different areas of a reef complex.

A. Introduction

Dr. W. HEISSEL, Professor of Geology at the University of Innsbruck suggested a sedimentological investigation of the Ladinian "Wettersteinkalk" of the "Kaiser Gebirge" (Austria).

The following account represents a summary of the preliminary results.

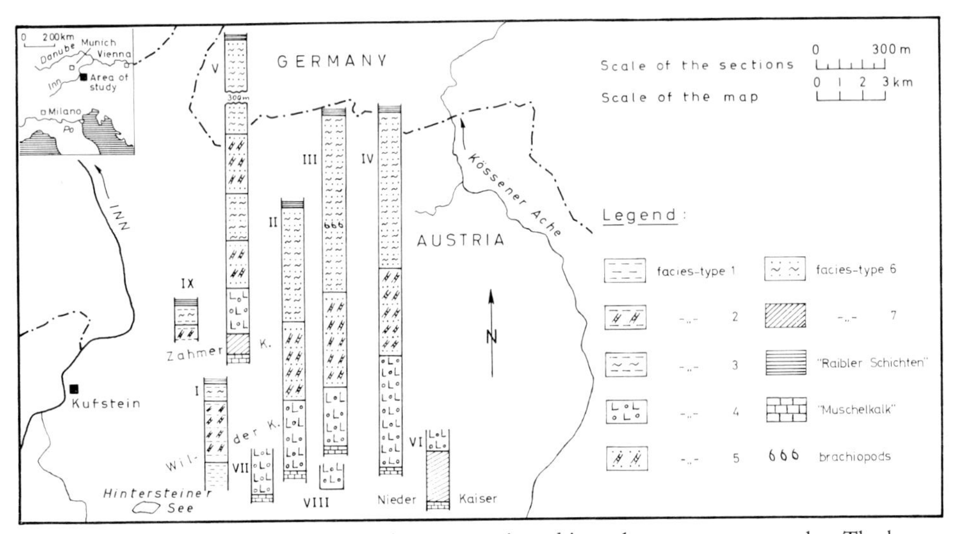

Fig. 1. Area of study. The cross sections are projected into the recent topography. The base of each columnar section corresponds to the proper beginning of the section in the field

The area of study is situated east of the river Inn where it cuts through the Nördliche Kalkalpen (Fig. 1). About 6000 m of vertical stratigraphic sections were examined in the northern part of the area ("Zahmer Kaiser") and in the southern part ("Wilder Kaiser"), comprising nine sections with nearly 700 oriented samples.

* Institut für Geologie und Paläontologie, Universität Innsbruck, Austria.

Most of the stratigraphic sections occur above dark-gray limestones with chert nodules ("Hornsteinknollenkalke") and above stratified calcarenites which are rich in crinoids; both horizons belong to the "Anisian Alpine Muschelkalk Series". The lower shale member of the "Raibler beds" (Carnian) forms the top of all the sections. Observations in the field showed that the entire area of the Ladinian "Wettersteinkalk" investigated can be lithogenetically and stratigraphically grouped into seven types of carbonate facies that comprise a general facies scheme. They will be described briefly and assigned to the different areas of a reef complex (for a definition of a reef complex, see Sarnthein, 1967: 119).

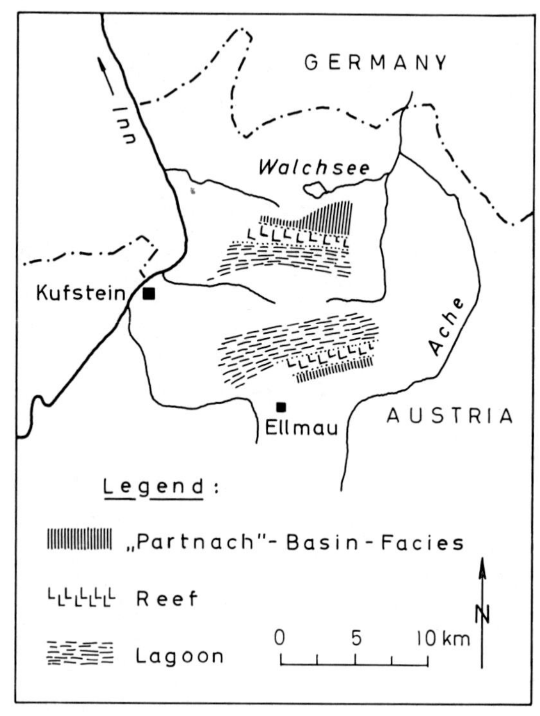

Fig. 2. Attempt of reconstruction of the palaeogeography of "Kaiser Gebirge" in the Lower Ladinian stage

B. Description of the Stratigraphic Sections

I. Section I

Section I is located at the eastern slope of the "Scheffauer". It includes a unit of layers with a thickness of 436 m and can be subdivided into 3 types of carbonate facies.

1. Carbonate Facies Type 1

Dark gray, thick-bedded (0.4 to 0.6 m) limestones, bioarenomicrites with detritus of crinoids, algae (dasycladaceae), and gastropods. The sedimentation took place in a calm (micrite), intermittently agitated (biodetritus) environment.

2. Carbonate Facies Type 2

Gray, stratified limestones, overly type 1 in a section of 50 m thickness. Beds, varying from 0.4 to 0.7 m in thickness, of a "sterile" and partly dolomitic calcimicrite (early diagenetic and belteroporic dolomitization, Fig. 9) alternate with biospararenites (which are also partly dolomitized, Figs. 10, 11), beds of 0.3 to 0.6 m thickness. The grains (up to 60 per cent) of these biospararenites are: biogenes, detrital grains (rock fragments, intraclasts), coated grains, fecal pellets, and lumps. Among the biogenes, the dasycladaceae predominate (mainly longitudinal- and

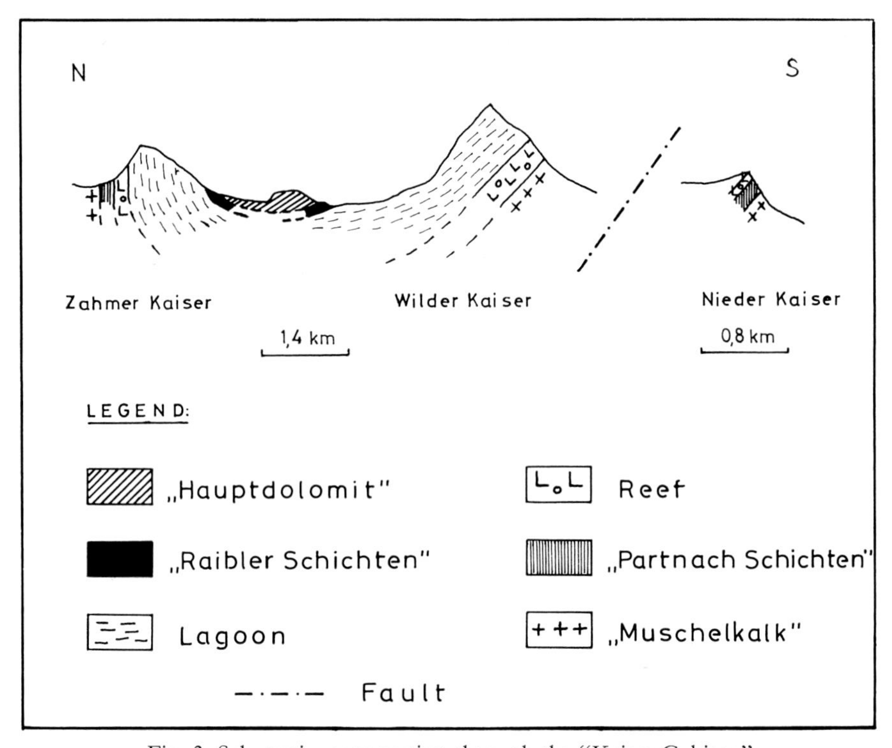

Fig. 3. Schematic cross section through the "Kaiser Gebirge"

cross-sections of Poikiloporella duplicata (PIA), sensu OTT, 1963). During deposition, periods of quiet, silty sedimentation ("sterile" micrites) alternated with phases of constant water turbulence (biospararenites).

3. Carbonate Facies Type 3

Thinly stratified layers (0.01 to 0.3 m) of nearly constant thickness are found in calcimicritic beds with a thickness of 0.5 to 1.5 m. The former are characterized by crust-like laminae which consist of micritic carbonate particles and are outlined by laminoid fenestrae, filled with sparry calcite (laminoid fenestral fabric, "Wyoming", Type „A" according to G. E. TEBBUTT et al., 1965: 6, Fig. 6). Parts of these crusty laminae, the thicker layers, are most probably algal stromatolites of the geometrically defined type LLH (B. W. LOGAN et al., 1964).

Besides the LF-fabric, Type "A", calcarenites with irregularly formed fenestrae occur, the alignment of which can hardly be recognized. These calcarenites are termed LF-fabric, Type "B" (G. E. TEBBUTT et al., 1965: 7, Fig. 7).

This type of fabric is also characterized by arenaceous foraminifera and algae (cross-sections of dasycladaceae) as well as by sequences of reverse graded bedding and the heterogeneity of the grains.

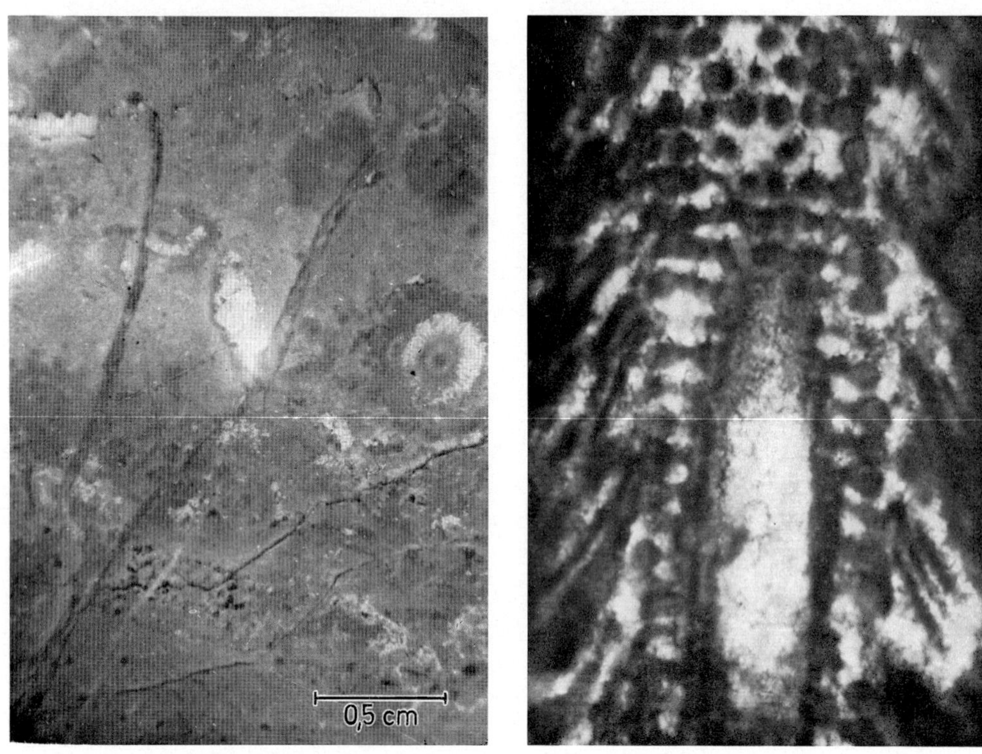

Fig. 4 Fig. 5

Fig. 4. Dasycladaceaespararenite (carbonate facies type 5), polished and etched sample surface

Fig. 5. Microphotograph of a thin-section of the sample shown in Fig. 4 (Poikiloporella duplicata (PIA), sensu OTT, 1963)

During deposition of carbonate facies type 3, relatively short periods of oscillation in sea level changed conditions from mud flats with relatively shallow depth of water or tidal flats (laminoid fenestral fabrics) to an environment with relatively deeper water and calm sedimentation (micrites).

II. Sections II, III and IV

Sections II ("Hoher Winkel — Kopftörl"), III ("Steinerne Rinne — Kübelkar") and IV ("Kreidegrube — Ackerl Spitze") are 1113, 1407, and 1523 m thick and allow the differentiation of three additional carbonate facies types, which appear in all three sections; they are, however, of different thickness (see Fig. 1).

1. Carbonate Facies Type 4

This carbonate facies type at the base of the stratigraphic sections consists of stratified calcirudites with biodetrital grains of, for instance, sponges, Solenoporaceae, gastropods, and lamellibranchiata. In the basal portions, well-preserved remains of sponges, green algae (Codiaceae), Solenoporaceae occur besides gastropods and in some places corals. The pore spaces are either filled with poorly sorted organic

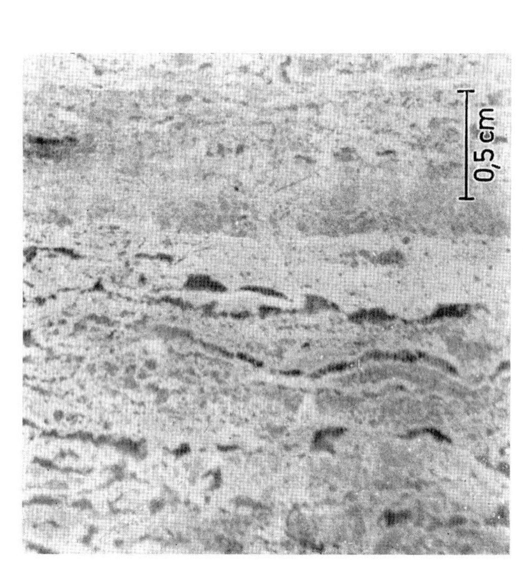

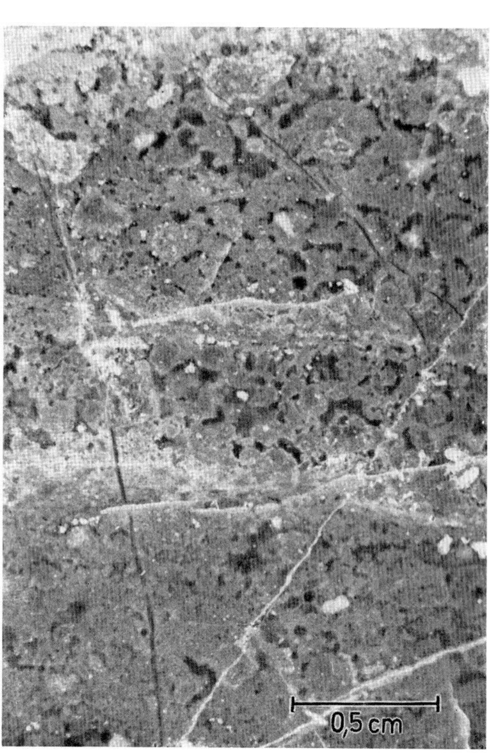

Fig. 6 Fig. 7

Fig. 6. Laminoid fenestral fabric, LF-fabric, Type "A" (G. E. TEBBUTT et al., 1965) of carbonate facies type 3, polished and etched sample surface

Fig. 7. Laminoid fenestral fabric, LF-fabric, Type "B" (G. E. TEBBUTT et al., 1965) in the lower part of the picture; in the upper part, algal stromatolites (hardly recognizable) (carbonate facies type 3; polished and etched sample surface)

fragments or with chemically, internally deposited carbonate ("Grossoolith" fabrics).

The biodetrital sediments of facies type 4 may be called reef débris from a back-reef area. The faunal and floral elements, which appear at the base of the sections, are partly preserved in situ and can be classified most probably as reef core facies (biogenic deposition, B. SANDER, 1936, Fig. 8).

2. Carbonate Facies Type 5

Beds of a Dasycladaceae-spararenite (0.4 to 1.2 m thick) alternate over a vertical distance of 20 m with biospararenite beds of nearly the same thickness, the grains of which consist predominantly of coated grains and tubes of Cyanophyceae.

Periods of relatively deep and well-ventilated water (up to 20 m; conditions for the growth of Dasycladaceae; Figs. 4, 5) on the one hand, and periods of more

Fig. 8. Fragments of sponges (carbonate facies type 4), weathered sample

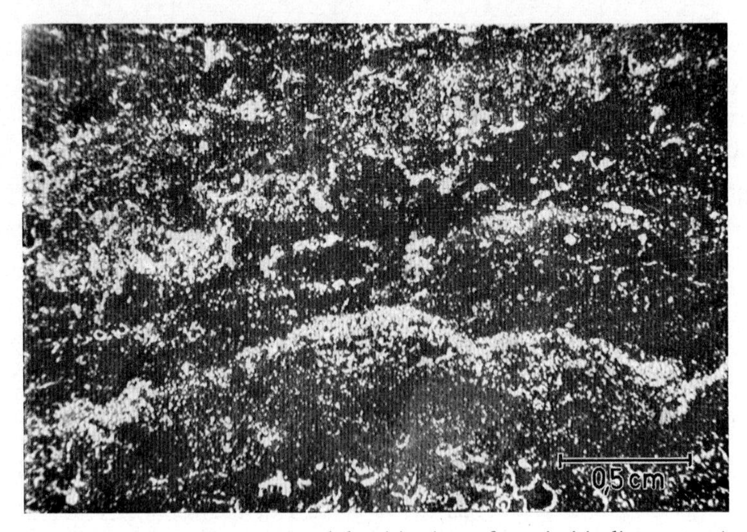

Fig. 9. Early diagenetic (belteroporic) dolomitization of a micritic limestone (carbonate facies type 2); polished, etched and stained sample surface (dolomite: white)

shallow water depths (up to 10 m; conditions for the growth of Cyanophyceae) and slight water movement on the other hand were typical of this depositional environment. In comparison with carbonate facies type 2, the absence of beds of micrite is a striking and distinguishing feature, which reveals that the water action must have been relatively constant during the entire period of sedimentation of carbonate facies type 5.

3. Carbonate Facies Type 6

Thinly stratified limestones with "laminoid fenestral fabrics" of carbonate facies type 3 were found between biospararenites of facies type 5. An unusual horizon of brachiopods (family: Terebratulidae) of 1 m thickness appears in section III. The

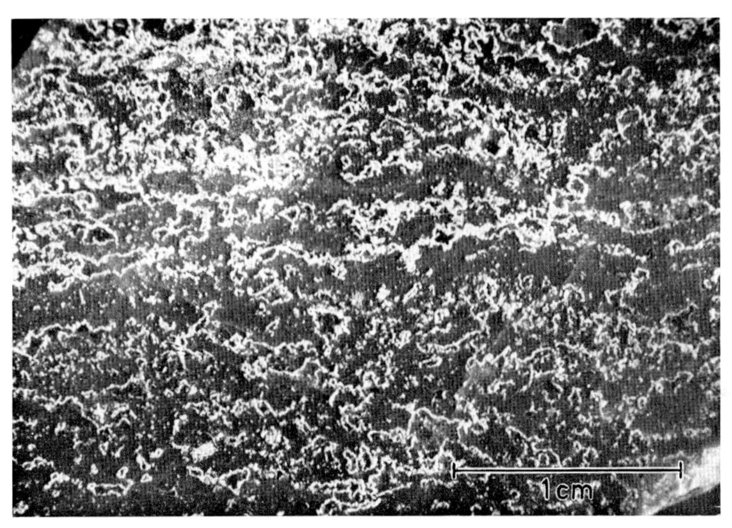

Fig. 10. Early diagenetic (belteroporic) dolomitization of a calcarenite (carbonate facies type 2). The grains, filled with drusy calcitic mosaics, are surrounded by dolomitic crystals (white); polished, etched and stained sample surface

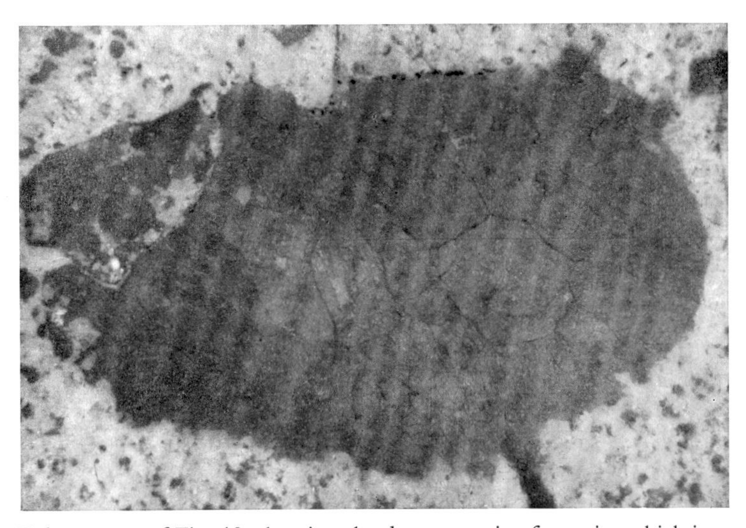

Fig. 11. Enlargement of Fig. 10, showing the drusy mosaic of a grain, which is surrounded by dolomitic crystals. Because of its belteroporic advancement, the dolomitization must have been early diagenetic, but it must have happened so late, however, that the grain growth inside of the components had already been completed

depositional environment of carbonate facies type 6 can be considered as equivalent to that of facies type 3, a mud flat. Periods of shallow water depths and/or drying out (carbonate facies type 3) alternated cyclically with periods of shallow water and slight water turbulence (facies type 5). The absence of micritic beds constitutes a difference with regard to facies type 3.

III. Section V

Section V ("Joven Alm—Pyramiden Spitze—Vordere Kesselschneid") comprises a unit of layers about 1580 m thick, which contains a seventh carbonate facies type at its base.

1. Carbonate Facies Type 7

A complex of 80 m in thickness (increasing toward the south-east and northwest) of grayish-blue limestones with chert nodules occurs above bedded calcarenites (0.3 to 0.9 m) of the "Alpine Muschelkalk Series" which are rich in crinoids. Up to three horizons of brownish-gray shale occur within this complex. At the top of this shale-limestone facies, named "Partnach Schichten" by C. W. Gümbel (1858), carbonate facies type 4 appears, followed by types 5 and 6 as in sections II to IV (see Fig. 1). The depositional environment of carbonate facies type 7 can be considered as continuation of the basin facies of the "Alpine Muschelkalk Series". Thus the "Partnach Schichten" can best be explained as basinal sediments which developed at the same time as the "Wettersteinkalk" reef complex (H. J. Schneider, 1964: 39, and M. Sarnthein. 1965: 150).

IV. Section VI

Section VI in the "Nieder Kaiser" shows a development similar to that of section V. The thickness of the sediments, however, is smaller (about 300 m). The basal „Partnach Schichten" which show a dolomitic development (M. Sarnthein, 1966) in the upper parts of the "Labturkopf" are followed by a thin complex of carbonate facies type 4.

V. Sections VII, VIII, and IX

All three sections are tectonically reduced. They are respectively about 250, 110, and 140 m thick. Section VII ("Treffauer") and section VIII ("Am Fuss") consist almost exclusively of sediments of carbonate facies type 4; section IX ("Naun Spitze") consists of facies type 2 (base of the section) and of facies type 3 (top of the section).

C. Reconstruction of the Paleogeography

The base ("Anisian Alpine Muschelkalk Series") and the sharply delineated top ("Carnian Raibler Beds") of the sections permit correlation of their different carbonate facies types and thus reconstruction of the paleogeography of the area investigated. Reef débris, or remains of a reef core which overlap at the base of sections V and VI to the "Partnach Basin Facies", change into sediments of a relatively deep, well-aired environment (up to 21 m deep; conditions of growth for Dasycladaceae) in the central portions. In the upper parts of the sections the water depth gradually decreased, sometimes even drying out (lagoonal parts of a reef platform, Sarnthein, 1967: 119) (see Fig. 3).

In Lower Ladinian times (see Fig. 2) the paleogeography was as follows under pretectonic conditions:

In the NE and SE of the area investigated, E-W and respectively NE-SW striking areas of reef débris of the back-reef occur, respectively limestones of the reef core; in front of them (N and S) the sediments of the "Partnach Basin Facies" are to be found. Between the two débris areas, sediments of a reef platform appear, the exact limitation of which, in the E and W as well as in the NE and SW, is not yet determined.

References

Bathurst, R. G. C.: (1) Diagenetic Fabrics in some British Dinantian Limestones. Liverpool Manchester Geol. J. **2,** 11—36 (1958).

— (2) The Cavernous Structure of some Mississippian Stromatactis Reefs in Lancashire, England. J. Geol. **67,** 506—521 (1958).

Friedman, G. M.: Early Diagenesis and Lithification in Carbonate Sediments. J. Sediment. Petrol. **34,** 777—813 (1964).

— Terminology of Cristallization Textures and Fabrics in Sedimentary Rocks. J. Sediment. Petrol. **35,** 643—655 (1965).

Gümbel, C. W. v.: Geognostische Karte des Königreichs Bayern. 1. Abth.: Das bayrische Alpengebirge mit einem Theil der südbayrischen Hochebene. München 1858.

Logan, B. W., R. Rezak, and R. N. Ginsburg: Classification and Environmental Significance of Algal Stromatolites. J. Geol. **72,** No. 1, 68—83 (1964).

Sander, B.: Beiträge zur Kenntnis der Anlagerungsgefüge (Rhythmische Kalke und Dolomite aus Tirol). Mineral. Petrol. Mitt. **48,** 1/2, 27—139, 6 Diagr., 36 Abb., 7 Tab.; 3/4, 141—209, 2 Diagr., 10 Abb., 6 Tab. (1936).

Sarnthein, M.: Sedimentologische Profilreihen aus den mitteltriadischen Karbonatgesteinen der Kalkalpen nördlich und südlich von Innsbruck, Austria. Verh. geol. Bundesanstalt (Wien) **1/2,** 119—162 (1965).

— Sedimentologische Profilreihen aus den mitteltriadischen Karbonatgesteinen der Kalkalpen nördlich und südlich von Innsbruck, Austria. 1. Fortsetzung. Ber. Nat.-Med. Ver. **54,** 33—59 (1966).

— Versuch einer Rekonstruktion der mitteltriadischen Paläogeographie um Innsbruck, Austria. Geol. Rundschau **56,** 116—127 (1967).

Schneider, H. J.: Facies Differentiation and Controlling Factors for the Depositional Lead-Zinc Concentration in the Ladinian Geosyncline of the Eastern Alps. Develop. Sediment. **2,** 29—45 (1964).

Tebbutt, G. E., C. D. Conley, and D. W. Boyd: Lithogenesis of a Distinctive Carbonate Rock Fabric. Wyoming Geol. Survey, Contrib. Geol. **4,** No. 1, 1—13 (1965).

Sedimentary Petrologic Investigation of the Upper Triassic "Hauptdolomit" of the Lechtaler Alps, Tyrol, Austria

W.-U. Müller-Jungbluth*

With 14 Figures

Abstract

The initial results of a detailed examination of the "Hauptdolomit" (Norian) are presented. They include the description of the various types of lithology and fossil assemblages which led to the proposed subdivision: *Lower-*, *Middle-*, *Bituminous-*, and *Upper "Hauptdolomit"*. An attempt is made to give an interpretation of the rock types, which represent distinct facies. These are thought to have been deposited very near mean sea-level, in a range between supra- to subtidal. There is much evidence to agree with A. G. Fischer (1964) who suggested that the "Hauptdolomit" developed on an extended, more or less steadily sinking platform, representing the "ultra back-reef facies" of the contemporaneous "Dachstein Kalk".

A. Introduction

The Upper Triassic Alpine "Hauptdolomit" ("HD") within the "Nördliche Kalkalpen" has been studied in the area between Landeck and Imst, north of the river Inn (Fig. 1). The "HD" is a very well-bedded sequence throughout with

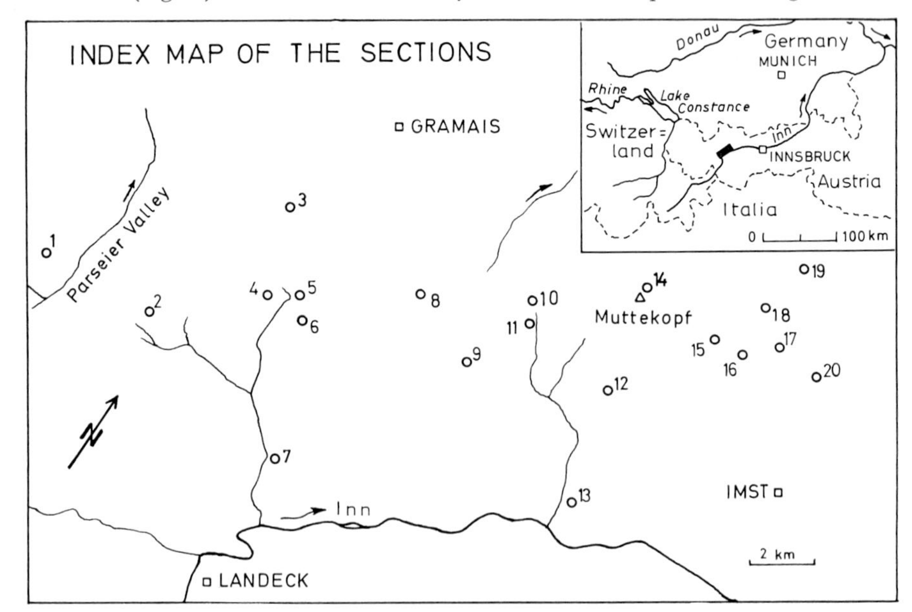

Fig. 1. Index map and location of sections

* Institut für Geologie und Paläontologie, Universität Innsbruck, Austria.

sporadically interbedded shale layers. Its sucrose, brittle weathering has created large talus slopes. Detailed cross-sections of lengths of more than 4 km have been sedimentologically examined (Fig. 2). Tectonic influences have been taken into consideration only, when they considerably interrupted the complete sections. On the other hand, the detailed sedimentologic studies helped to detect doubtful tectonic structures.

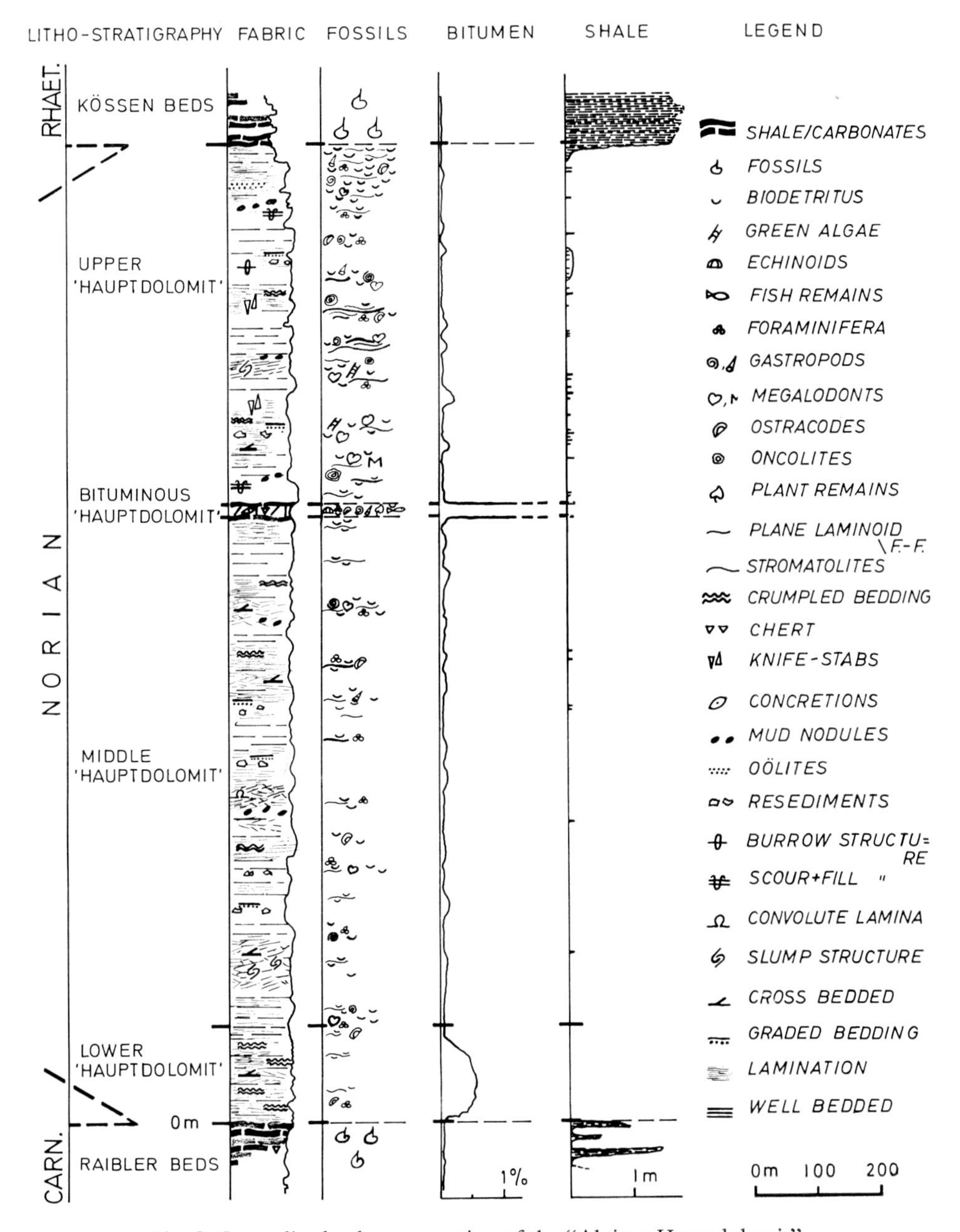

Fig. 2. Generalized columnar section of the "Alpiner Hauptdolomit"

After F. TRUSHEIM (1930) and B. SANDER (1936), only few geologists worked on the "Hauptdolomit". Recently, more has been done, notably by A. BOSELLINI (1965, 1967) and F. PURTSCHELLER (1962).

B. Litho-Stratigraphic Subdivision (Main Sedimentary Features)

I. "Raibler Beds"

The base of the "Hauptdolomit" is gradationally underlain by the "Raibler beds". The uppermost sequence of these consists of medium, to very thinly bedded carbonates with frequent shale partings reaching a thickness of up to 150 cm. Most of the carbonates are micritic (micrite enlargement is common), virtually non-fossiliferous, buff to olive-gray weathering dolomite or dolomitic limestone. From the base till the top, within a thickness of 50 m or more, the bedding is convolute, slightly wavy, continuous and discontinuous, parallel and non-parallel, and even laminated. The gradation into the Lower "Hauptdolomit" starts with the alternating of continuous to discontinuous, crumpled bedding, with more or less parallel, plane to platy bedding. Changes in grain size along with a changing bitumen content (which increases slowly but steadily upward), cause the fine lamination. Chert nodules and pyrite crystals are found nearby shale layers. These are dark-gray, partly green to black and up to 30 to 150 cm thick. Very often they contain carbonate nodules embedded like phacoids parallel to the stratification. Within the transition zone from "pure" shale to carbonate the shale caused convolute bedding and in the uppermost parts tectonically fragmented "Raibler beds", the so-called "Hauptdolomit-Basalbreccien" and a lot of the "Rauhwacken". The real "Rauhwacken" breccia, however, occurs in the lower portions. These may have been formed through sedimentary precesses (H. JERZ, 1966: 60). The last shale appearence defines the contact with the Lower "Hauptdolomit".

II. Lower "Hauptdolomit"

Briefly stated the Lower "Hauptdolomit" is a dark- to yellowish-brown weathering, more or less bituminous, thinly well-bedded dolomite sequence (Fig. 3). Alternation between finely crystalline, well-sorted dolomite and originally pelletal, badly sorted micrite causes fine lamination (Fig. 4). Only a few signs of organic remnants (Foraminifera and Ostracods) can be recognized.

The grain size of the dolomite ranges from < 0.001 to 0.08 mm, (up to 4 mm for the larger dolomite crystals in healed fractures). The finely crystalline strata contain light brown bitumen, diffused or oriented as bitumen stripes, but in lesser amounts than in the pelletal beds, where the bitumen fills in the irregularities of the bedding planes as very thin crumpled films, < 0.01 to 0.1 mm thick. The bitumen was concentrated around tiny grains (0.009 to 3.0 mm) recognized as molds, but is now accumulated on crystal faces, as dolomitization "purified" the dolomite (H. FÜCHTBAUER, 1966). Generally, the bitumen emphasizes lamination. It may have been derived from a thin organic film which likely represents the remains of algal mats that covered the original sediments from time to time above low mean sea-level. Crinkling and curling structures here and there within sets of regularly laminated sediments are thought to have been caused by the drying-out of the mats (Fig. 5). In the outcrop, the single laminae rapidly wedge out laterally indicating diagonal bedding. The whole succession, however, is noted for its wide lateral persistence.

III. Middle "Hauptdolomit"

The monotony of the "mm-Rhythmite" (B. SANDER, 1936) is broken towards the top, where it contains interbeds of very finely crystalline pelletal dolomite which alternatively bear sparry, skeletal relics or a "bird's-eye" fabric (Fig. 6). This is one characteristic indicating transition to the Middle "Hauptdolomit". The other one is the marked colour-changes which depend on crystal size (very fine to finely crystalline) and reduction in magnesium and bitumen content. The weathering, as well as the rock colours become lighter, predominantly buff to light gray.

The main petrologic data deal with detritus. Carbonate rock fragments and faunal relics such as tiny forms of Gastropods, Molluscs, Ostracods, arenaceous Foraminifera, together with a lot of undefinable skeletal debris, occurring as spar-filled molds, are cemented in a differentially recrystallized dolomicritic matrix (Fig. 7). Often, single euhedral dolomite rhombs (0.03 to 0.8 mm) are found floating or oriented parallel to bedding within the otherwise fine, to medium crystalline (up to 0.05 mm) mottled dolomite.

One of the striking features are "bird's-eyes" (as single spar-filled, initial vugs) and "laminoid fenestral fabrics" ("LF-fabric" of G. E. TEBBUTT et al., 1965) (Fig. 6). Such fabrics appear first in lower strata and are discontinuous, filigreed LF-fabrics with a lot of small skeletal remnants in a pelmicritic matrix. These biota-rich beds are about 30 to 80 cm thick. Higher in the sequence, the organic-sedimentary structures become more distinct and continuous and more or less even. As larger curvatures are rare along these thinly laminated (0.08 to 1.0 mm) structures, perhaps they should be referred to as "plane laminoid fenestral fabric" ("PLF-fabric"), a term which may be added to those of G. E. TEBBUTT et al. (1965). Skeletal debris is uncommon, a few tests of Foraminifera or Ostracods are scattered within the laminae. Finally, higher up in the Middle "Hauptdolomit", the laminae become thicker; more irregular, larger spar-filled fenestrae and gentle convex-up warps lead to the LF-fabric, type 'A' (crustiform), subtype "LLH-S" (B. W. LOGAN et al., 1964). These presumably algal stromatolite structures alternate with skeletal, scattered pelmicrites which contain the major parts of the biota-rich sets of beds, about 60 to 150 cm thick. Large fossils, about 3 to 10 cm in diameter, the Megalodonts, occur singley or accumulated in nests.

Up to five or more thin shale layers, possibly existing but not always observable, interfinger the section here and there and often cause irregular bedding planes. A very micritic, nearly structureless dolomite, weathering yellowish-gray and with a lot of scattered pyrite crystals, occurs with the shale or may replace it.

Reworking and transportation account for "protointraclasts" (A. BOSELLINI, 1964), mud pebbles, intraclasts and "inhomogeneous-breccias" (B. SANDER, 1936) (Fig. 8). Generally the zone of resedimentation occupies an interval of 2 to 5 cm above the eroded bottom. The fragments (less than 0.05 up to 20.0 mm) are usually subrounded to well rounded. Larger, flat ones are scattered amongst smaller ones without orientation or oriented with their long axes parallel to the stratification (Fig. 7, top). Sometimes, the whole bed consists of resedimented, tiny fragments, graded and fairly rounded, embedded in a dark-gray dolomicrite. In addition, washouts and geopetal fillings (Fig. 9), channelings ("Priele") and burrow-mottling, proving organic activity, disrupt the well-bedded strata. The skeletal-pelletal dolomicrite, and "algal stromatolite" sets reappear in arbitrary distances from 2 to 20-, and

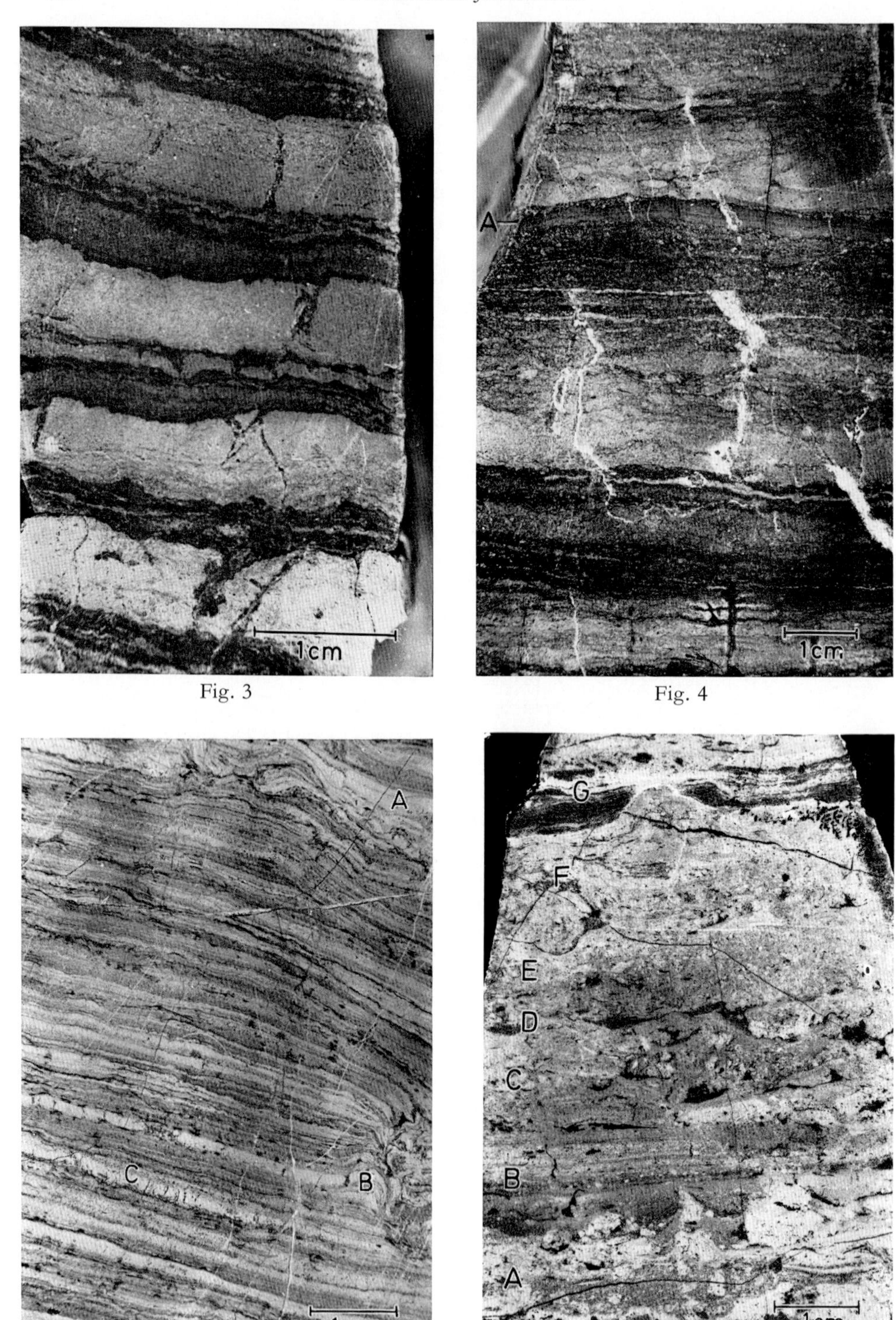

Fig. 3

Fig. 4

Fig. 5

Fig. 6

more meters, getting more abundant towards the top. In the outcrop these rocks can appear as massive, laminated sets, that weather from very light gray to light yellowish-brown, depending on the magnesium content. Etching and staining (for methods refer to G. M. FRIEDMAN, 1959) of samples have yielded only dolomite, but with different magnesium contents. The crystal size of the matrix (less than 0.006 mm), may be the reason for the small amount of recrystallization and 'well'-preserved organic remains.

The generally uniform bedding ranges from 25 to 45 cm in thickness limited by well-developed stylolites. The Middle "Hauptdolomit" is more than 400-, but less than 900 m thick depending upon the development and definition of the base.

IV. Bituminous "Hauptdolomit"

(Equivalent of the "Seefelder" type, B. SANDER, 1921)

The Middle "Hauptdolomit" succession is interrupted by extremely dark-coloured carbonates, containing a higher bitumen ratio (maximum up to 20%, but generally less than 5%) and very little coal (less than 1%), interbedded with black shales. These shales might have prevented some of the calcite layers from being dolomitized. The Bituminous "Hauptdolomit" consists of very finely crystalline, extremely thinly laminated dolomite, as well as detrital dolomite (rock fragments and marine skeletals) (Fig. 10).

Bitumen films surround the crystals and form discontinuous, wedging-out strata. Pigmented rims or areas outline original pellet forms. Pyrite and quartz are scattered throughout the sequence. Chert nodules, elongate or spherical, lined up or without orientation occur near the shale partings. Within the shales a lot of different sized carbonate nodules ($<$ 1 to more than 10 cm in size) are embedded parallel to the stratification. Most of them are convex-concave upwarped laminae only, others are laminae-disturbing interruptions caused by decaying organic matter (Fig. 11). The fauna is relatively undiversified, small-sized and with fragile shells, occurring as abundant individuals, or locally in accumulations (Fig. 10). Several forms of Gastropods, Echinoids, fish remains, Ostracods and Foraminifera, as well as plant remains are found.

Slump structures with convolution especially near contact with shale partings are not rare. The most obvious structure is very thin lamination, but without regular rhythm of deposition and almost without grading.

Fig. 3. Lower "HD" 'mm-Rhythmit': more (dark) and less (light) bitumen containing laminae (loc. 19)

Fig. 4. Lower "HD" laminated bituminous pelletal mudstone (supratidal) interbedded with a micrite (A) (caused by high tide) (loc. 9)

Fig. 5. Lower "HD" laminated dolomite (supratidal); some sets of laminae disturbed (desiccation curling of algal mats) (A); mudcracks "healed" by later mechanical (B) and chemical (C) depositions (loc. 9)

Fig. 6. Middle "HD" doloarenomicrite. (A): algal stromatolites; (B): laminated pelletal micrite; (C): LF-fabric, 'B', shrinkage pores, spar-filled (dark); (D): relief with condensed residue; (E): pelletal skeletal micrite, grading into mat-covered area (F), with current relief fill by delaminated bituminous dolomite (G) (loc. 9)

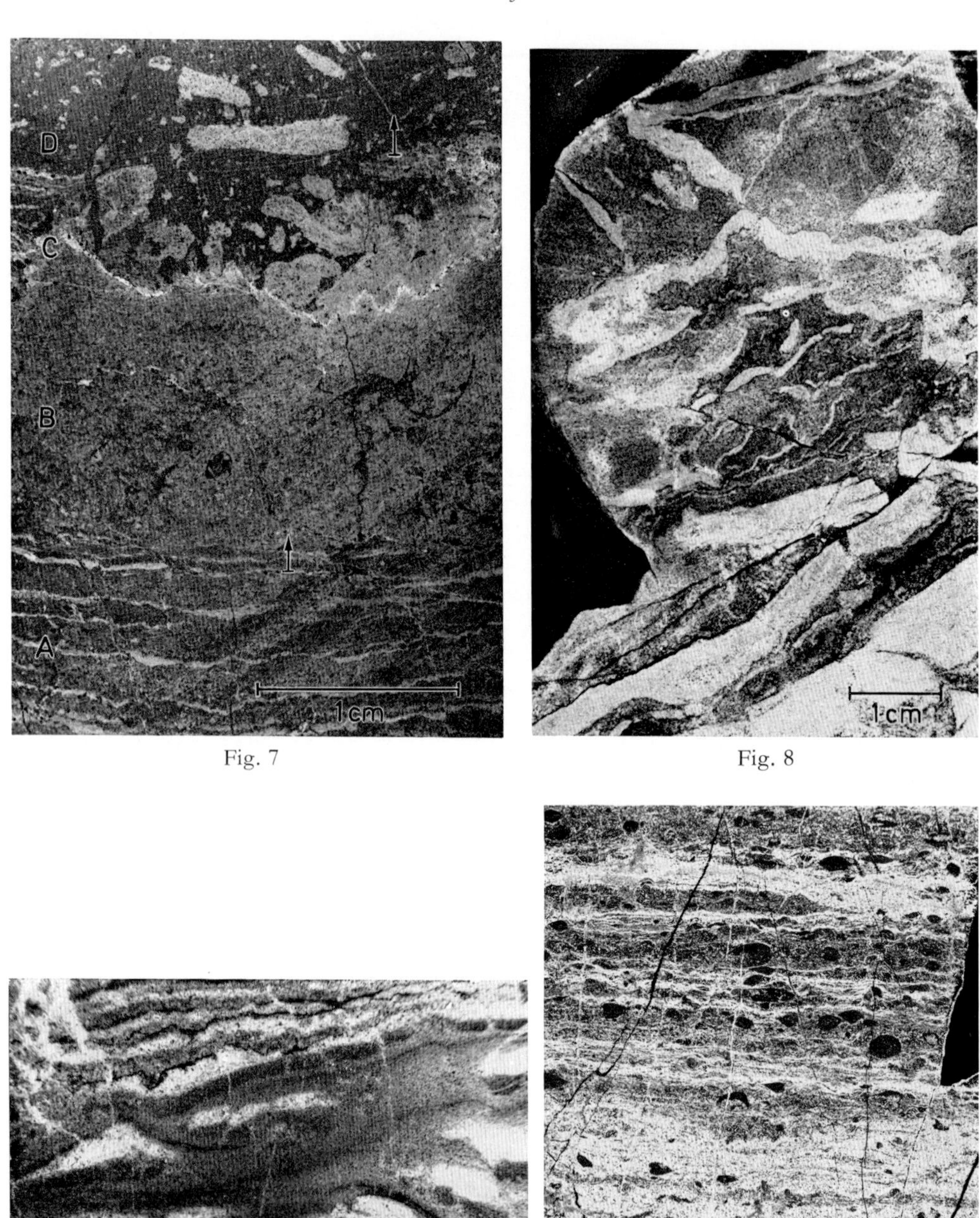

Fig. 7

Fig. 8

Fig. 9

Fig. 10

The Bituminous "Hauptdolomit" sequence, 10 to 60 m thick, 30 m on an average, varies in texture and structure from outcrop to outcrop. This has caused great uncertainty regarding its stratigraphic classification. The present sedimentological study, however, allows the placement of the sequence as the deviding litho-stratigraphic set between Middle and Upper "Hauptdolomit". Many of the "Kössener Schichten" on O. AMPFERER's (1932) geologic maps are now recognized as "Bituminous Hauptdolomit". This too supports the position of the "BIT HD" high up in the stratigraphic sequence.

V. Upper "Hauptdolomit"

The contact between the "BIT HD" and the Upper "HD" is as sharp as the one at the base of the "BIT HD". The Upper "HD" contains all the sedimentary features recognized before, but more distinctly and with some new varieties added. The coarse-fraction constituents have grown in number: the biota-rich beds have developed a more prolific fauna, the individuals have become larger, the shells stronger. Algal stromatolites and oncolites commonly occur. The LF-fabric, type 'A', subtype 'LLH-S' and '-C' have larger, irregularly shaped fenestrae occurring within thicker laminae of pelletal mudstone. Fossil debris and reworked semi-lithified detritus belong to the LLH-mat-covered-deposit. One set of mats is generally about 25 cm, or up to 60 to 120 cm thick. They alternate with pelletal, skeletal micrite or doloarenite. Groups of constituent particles are coated on a large scale into oncoid-like spheres (LF-fabric, 'A', subtype 'SS' and 'SS-I'), locally reaching a maximum diameter of up to 15 cm. More common are the crumbled concentric, multiple laminae around smaller detritus and with fewer layers (Fig. 12). Combined with algal mats, doloarenites could take up the LF-fabric, 'B', subtype II ("New Mexico" as high detrital) or subtype I ("Colorado" as low detrital) partly by carbonate precipitated through photosynthesis (G. E. TEBBUTT et al., 1965) (Fig. 13). Another common fabric is the pelletal ortho-sparite. On weathered surfaces, the sparite stands out as an irregular network, while the sometimes well-sorted and graded pellets have been dissolved. This should not be confused with oolites which are preserved only as pigmented circles in totally recrystallized dolomite. Within this facies evident organic mottling appears in doloarenomicrites. Beds containing Dasycladaceae sp. and Megalodonts (up to 15 cm in size) become apparent (Fig. 14) and serve as examples for a subtidal facies. Quite another facies is represented by non-fossiliferous, fine- to medium-crystalline, low to higher bituminous dolomite, locally with mud-cracks and "knife stabs" ("Messerstiche": M. SARNTHEIN, 1966), presumably initial evaporite-crystal

---

Fig. 7. Middle "HD" skeletal doloarenomicrite. (A): PLF-fabric (light colour: dolomite, dark: less dolomitic); (B): skeletal-pelletal doloarenite (intertidal); (C): stylolite (white): bituminous peel of shale or residue; (D): resediments (clasts: A + B) in doloarenomicrite (loc. 7)

Fig. 8. Middle "HD", reworked coarsely grained argillacious dolomite: "inhomogenous-breccias" (autoclastic breccia) (loc. 7)

Fig. 9. Upper "HD" coarsely crystalline, vuggy dolomite; geopetal intern-sedimentation and spar-filled top (loc. 4)

Fig. 10. "BIT HD", very thin laminations of fragile shells (loc. 6)

molds, a supratidal feature. All the representatives of the supratidal facies are totally dolomitised (penecontemporaneously and later, see L. V. Illing et al., 1965); they retain fabrics and components as "ghosts" and molds only. On the other hand, the

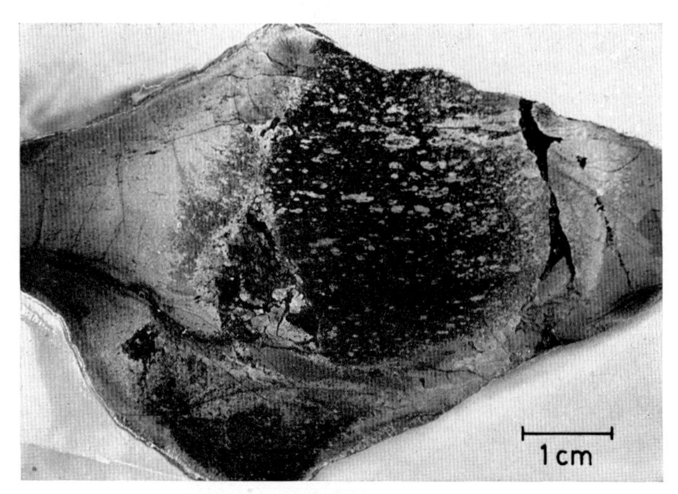

Fig. 11. "BIT HD" carbonate nodule with stratification (white: dolomitic, black: bituminous calcite) (loc. 20)

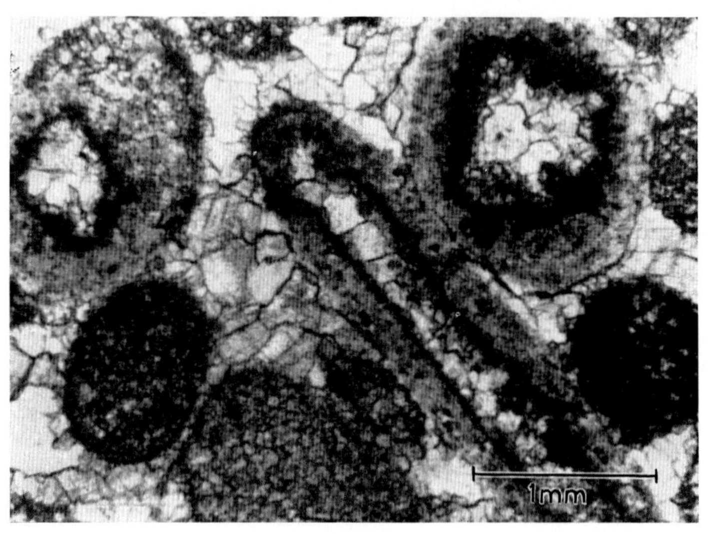

Fig. 12. Thin section of algal coating in pelletal dolosparite; nuclei (biodetritus) are replaced by 'drusy mosaik', moulded by pigmental rims (Upper "HD", loc. 7)

latter are presumably without indigenous marine fossils because of the restricted environment and not the result of differential loss by dolomitisation.

The distinct, well-warped stromatolitic forms are excellently developed up to the middle part of the Upper "HD"; higher up, the PLF-fabric replaces them continuously. In general, the beds of the Upper "HD" are thicker and the development of single features is more persistent. The top region (locally called "Plattenkalk", preponderatingly subtidal facies) grades into the "Kössen beds", marked by the increasing abundance of organic matter (skeletal dolomicrite and lumachellas) and steadily increasing shale partings. In addition, recrystallization has not reached the degree shown above; the magnesium content is much less, while the pyrite

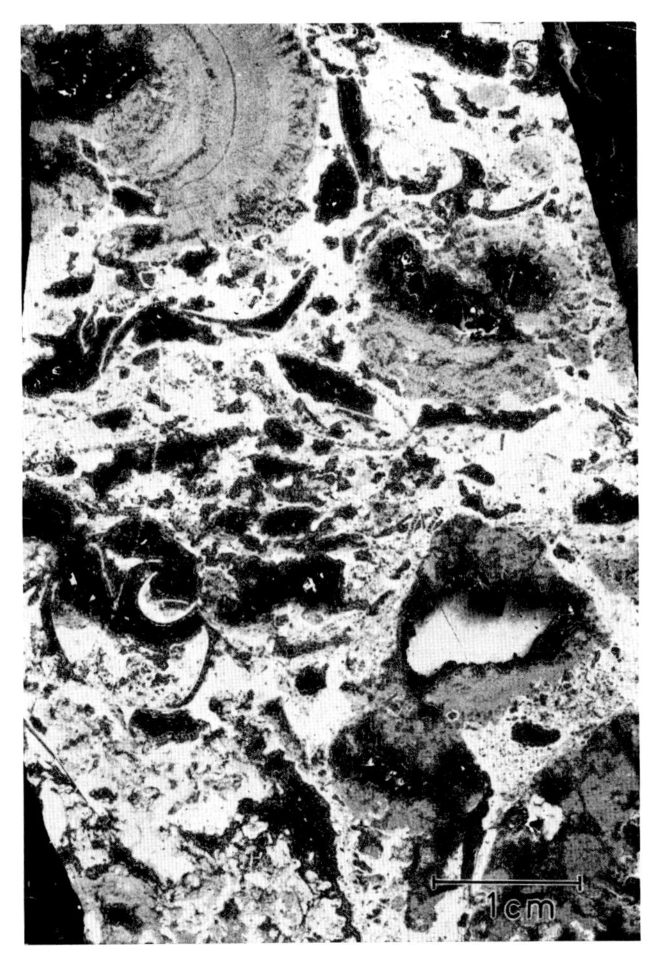

Fig. 13. Upper "HD" algal, skeletal doloarenomicrite with LF-fabric, 'B' (black): as well as the dissolved larger biodetritus; blue-green algal coating (gray) and algal fragments: Solenoporaceae (?) (loc. 9)

content increases. The rock colour changes to dark gray. The "Kössen beds" themselves consist of dark shales interbedded with limestones having an extremely fine, crystalline matrix.

C. Discussion

Recent work on the development of carbonate facies has revealed more and more features within various rock types, which allows their categorization into distinct depositional environments. The rocks described above are generally interpreted as having been deposited in a supratidal-, intertidal-, subtidal- and their respective transitional zones. [Additional, detailed data may also be found in A. Bosellini (1965), A. G. Fischer (1964), L. V. Illing et al. (1965), L. F. Laporte (1967), A. Matter (1967), for example.]

Developing after the subtidal facies of the Upper "Raibler beds", the principal rock types of the Lower "HD" is thought to represent supratidal development. The restricted depositional environment resulted in the rocks characterized above

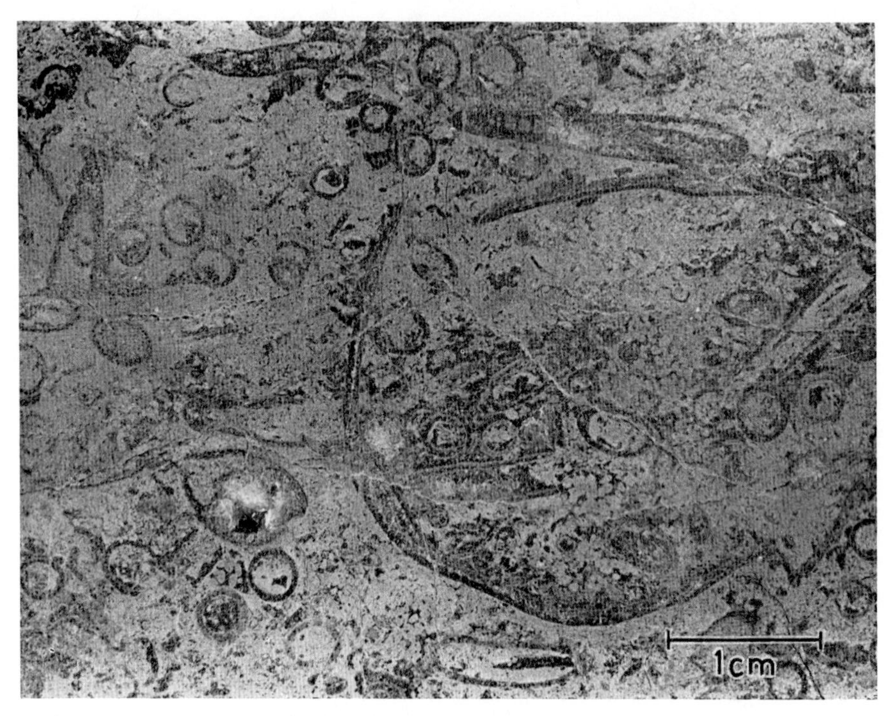

Fig. 14. Upper "HD" green algal, skeletal dolomicrite. Dasycladaceae and Megalodontidae in the same deposition, less dolomitic (subtidal) (loc. 9)

and with abundant desiccation marks. The relatively long lateral persistence of these beds suggests a fairly unique depositional environment at the beginning of the "Hauptdolomit" sequence. Later, both interfingering, thin mudstone layers due to short periods of flooding, and intertidal sediment mark the transition to the "M HD". The "M HD" itself starts with a subtidal facies, indicated by Megalodonts. Fluctuating environmental conditions, which are almost identical with intertidal ranges, explain the varying mechanically—and organically deposited fabrics that represents most of the "M HD". The "BIT HD" is called a subtidal facies because of its faunal elements. The Upper "HD" marks the return to intertidal and supratidal environments of deposition. Higher up in the Upper "HD", a subtidal environment again controlled the sedimentation ("Plattenkalk"); this sequence eventually grades into the overlying "Kössen beds" shale.

Acknowledgements

The author expresses his sincerest thanks to all those persons who gave him indispensable aid, especially to Dr. C. PENDEXTER who introduced him to sedimentology and Dr. M. SARNTHEIN who took care of this work. This paper is rewritten from a portion of a thesis, supervised by Prof. Dr. W. HEISSEL, which will soon be presented to the Geologic Institute of the University of Innsbruck, Austria.

References

AMPFERER, O.: Geologische Karte der Lechtaler Alpen, mit Erläuterungen, 1:25000. Wien: G. B. A. (1932).

BOSELLINI, A.: Stratigrafia, petrografia e sedimentologia delle facies carbonatiche al limite Permiano-Trias nelle Dolomiti occidentali. Mem. Mus. Nat. Ven. Trident., Vol. **XV**, Fasc. 2, 59—160, Trento (1964).

— Analisi Petrografica della "Dolomia Principale" nel Gruppo di Sella (Regione Dolomitica). Mem. Geopal. Univ. Ferrara, **I,** 2, No. 3 (1965).

— La tematica deposizionale della Dolomia Proncipale (Dolomitie Prealpi Venete). Boll. Soc. Geol. Ital. **86**, 133—169 (1967).

FISCHER, A. G.: The Lofer Cyclothems of the Alpine Triassic. Kansas Geol. Survey Bull. **169,** 1, 107—149 (1964).

FRIEDMAN, G. M.: Identification of Carbonate Minerals by Staining Methods. J. Sediment. Petrol. **29,** 1, 87—97 (1959).

— Early Diagenesis and Lithification in Carbonate Sediments. J. Sediment. Petrol. **34,** 4, 777—813 (1964).

FÜCHTBAUER, H., u. H. GOLDSCHMIDT: Beziehungen zwischen Calciumgehalt und Bildungsbedingungen der Dolomite. Geol. Rundschau **55,** 1, 29—40 (1966).

ILLING, L. V., A. J. WELLS, and J. C. M. TAYLOR: Penecontemporary Dolomite in the Persian Gulf. Soc. Econ. Pal. Min., Spec. Pub. **13** (1965).

JERZ, H.: Untersuchung über Stoffbestand, Bildungsbedingung und Paläogeographie der Raibler Schichten zwischen Lech und Inn (Nördliche Kalkalpen). Geol. Bavarica **56** (1966).

LAPORTE, L. F.: Carbonate Deposition near Mean Sea-level and Resultant Facies Mosaic: Manlius Formation (Lower Devonian) of N. Y. State. Am. Ass. Petrol. Geologists, Bull. **51,** 1 (1967).

LEIGHTON, M. W., and CH. PENDEXTER: Carbonate Rock Types. Am. Ass. Petrol. Geologists, Memoir **1,** 33—61 (1962).

LOGAN, B. W., R. REZAK, and R. N. GINSBURG: Classification and Environmental Significance of Algal Stromatolites. J. Geol. **72,** 1, 68—83 (1964).

MATTER, A.: Wattensedimente im Ordovic der Appalachen von Maryland. Vortrag zum Karbonat-Kolloquium, Heidelberg 1967. J. Sediment. Petrol. **37,** 2 (1967).

PURTSCHELLER, F.: Sedimentpetrographische Untersuchungen am Hauptdolomit der Brentagruppe. Tschermak's mineral. u. petrog. Mitt. **8,** 167—217 (1962).

SANDER, B.: Über bituminöse Mergel. Jahrb. Geol. Staats-A. (Wien) **71,** 3 (1921).

— Beiträge zur Kenntnis der Anlagerungsgefüge, I und II. Mineral. u. petrog. Mitt. N. F., Leipzig **48,** 27—209 (1936).

SARNTHEIN, M.: Sedimentologische Profilreihen aus den mitteltriadischen Karbonatgesteinen der Kalkalpen nördlich und südlich von Innsbruck. Ber. Nat.-Med. Ver. Innsbruck **54,** 33—59 (1966).

TEBBUTT, G. E., C. D. CONLEY, and D. W. BOYD: Lithogenesis of a Distinctive Carbonate Rock Fabric. Wyoming Geol. Survey Contrib. Geol, **4,** 1, 1—13 (1965).

TRUSHEIM, F.: Die Mittenwalder Karwendelmulde. Wiss. Veröff. D. Ö. A. V. **7** (1930).

Calcareous Sea Bottoms of the Raetian and Lower Jurassic Sea from the West Part of the Northern Calcareous Alps

FRANK H. FABRICIUS*

With 3 Figures

Abstract

The Raetian and Lower Jurassic marine sediments are described as types of fossil sea bottoms. The marly to limy bottomfacies are classified in regard to their former consistence as (1) "loose or soft bottoms" (mud bottoms, stable mud bottoms, sand bottoms) and as (2) "solid or hard bottoms" (biogenic hard bottoms, and solution bottoms). A "temporary hard bottom" is interpreted as a cyclic change of (1) and (2), probably caused by a tectonic oscillation. Sediments trapped in tectonical fissures and deposited below the general plane of the sea floor are classified as "internal bottoms". — The paleogeographic distribution of the different types of sea bottoms of this area and at that time are outlined.

A. Significance of Sea Bottom Features

I. Introduction

The properties of Recent marine sediments at and immediately below the sediment water interface are especially of geological and ecological interest. These properties are a complex result of the mingling and the various interferences of sedimentological, oceanographic and biological factors.

When we try to interprete a marine environment of the past, we cannot depend on the petrographical and paleontological evidences of the sediment only (LADD, 1957: 47). We also have to regard each fossil bed of the sediment as the once uppermost layer of the lithosphere, which was exposed to the sea water, and which was an integral part of the whole environment. We must study all features and fabrics which can be related to the former and now fossilized sea bottom. This approach is mainly based on the comparison of the fossil bottom features and patterns with those of modern sea floors. But in cases where the Recent complement to the fossil record is not yet well known, the fossil document may be the guide to the research in the Recent sea.

A great variety of facies of Upper Triassic to Lower Jurassic age from the Northern Calcareous Alps of Bavaria and the Northern Tyrol admits the study of a number of different types of marly and especially limy sea floors (Fig. 1). Dolomitic and non-carbonate sediments (with the exemption of clay minerals) are rare and do not form special types of sea bottoms in this area at that time.

* Institut für Geologie, Technische Hochschule München, Germany.

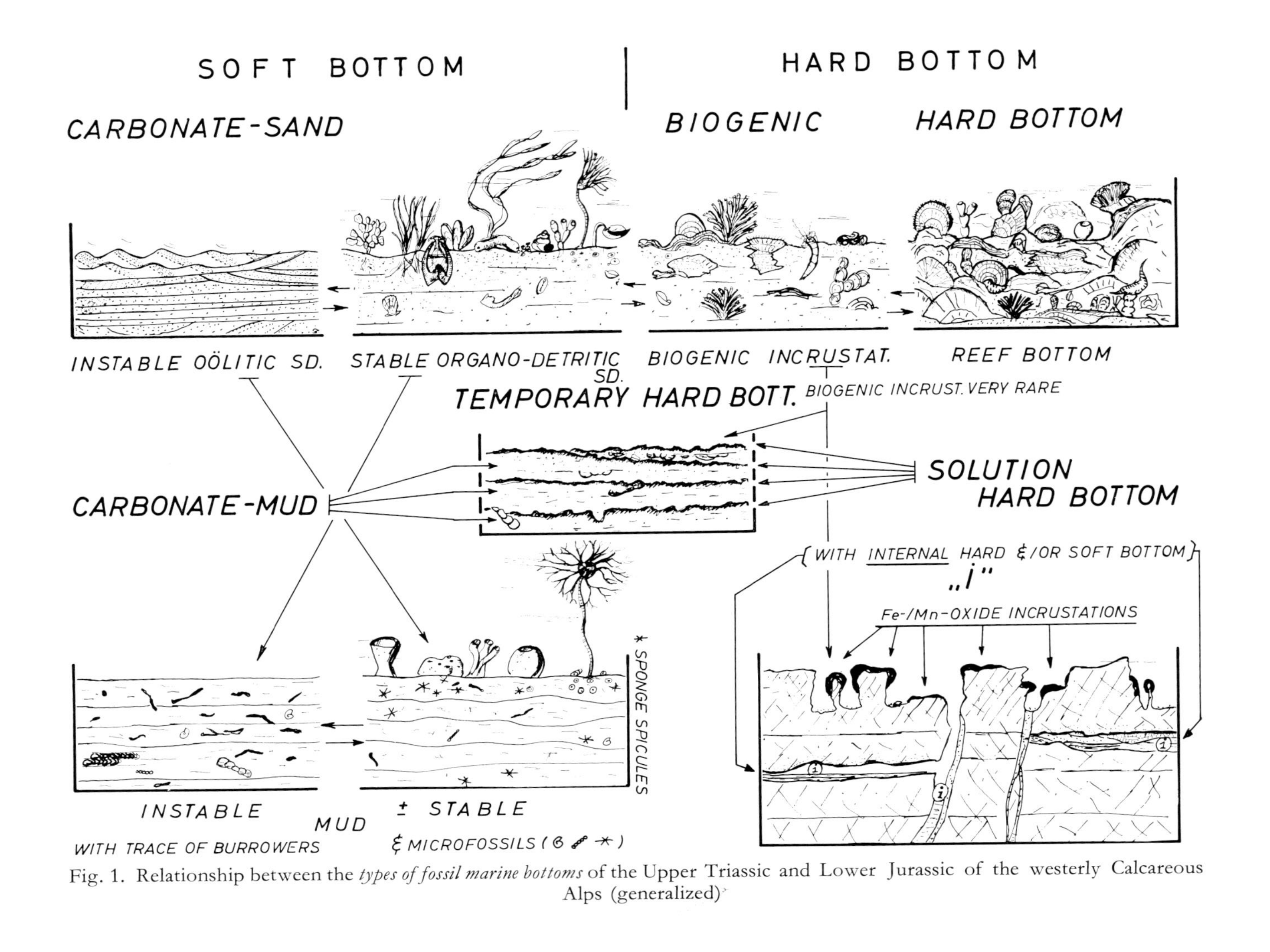

Fig. 1. Relationship between the *types of fossil marine bottoms* of the Upper Triassic and Lower Jurassic of the westerly Calcareous Alps (generalized)

II. Types of Sea Bottoms: Properties and Definitions

The main property of a sedimentary sea bottom is its mechanical consistence. This depends mainly on grain size and grain sorting, the mineralogical composition, and the degree of cementation. "Consistence" is here regarded as shear strength. But dealing with fossil sediments, we only can deduct to this former property from other evidences and parameters.

The sedimentary sea floors can be classified into "*solid or hard bottoms*"[1] and "*loose or soft bottoms*" with a large intermediate field of locally firm, or of semiconsolidated, or of stabilized sediments. The terms "loose" or "soft" are regarded appropriate to bottoms with unconsolidated sediments, where the grains can easily be moved out of their position; while "solid" or "hard" is used here in connection with bottoms of sedimentary origin, where the grains are ± fixed in their position to the adjacent ones.

In some cases the paleontological record also admits a specification of the bottom type; in others the traces of boring and burrowing organisms are the only indications recorded. But in any case, a classification based only on these data seems to be even more difficult. Although the influence of the substratum on the benthonic organisms must not be underestimated, it cannot be regarded as more important as turbidity, turbulance, O_2, CO_2, H_2S content, nutrient, and other bionomic factors (c.f. THORSON, 1957: 466).

In general we can link together: (1) A bottom with a loose sediment and a burrowing infauna; the loose consistance allows a fixation in the sediment only by means of anchoring or rooting. — (2) A solid bottom and boring features made by algae, sponges, echinids, or pelecipods etc.; in some cases organic incrustations and a superficial attachment to the firm surface can be observed.

In bottoms with unconsolidated sediments organisms may cause changes of grain size and sorting not only by adding their shells or skelets to the sediment, but also by their activities as mentioned later on. Organisms also can consolidate a loose bottom by their dense growth and rooting (SWINCHATT, 1965).

Loose sediments can also become stable by the shielding effect of a flat epifauna or a complete mat-like covering by algae. This is the first step towards a biogenically consolidated or "biogenic hard bottom".

On some hard bottoms marks of submarine solution appear as well on the sediment surface as on shell material. Such fabrics are typical for certain rocks, i.e. for some bottom facies, to which the term "solution bottom" may be given.

III. Specific Properties of Calcareous Sea Floors

Comparing non-carbonate bottom sediments with calcareous ones, we find several important differences, proving the specific nature of the calcareous sea floor. In general outlines these are:

(1) The low mechanical resistance of calcite and aragonite grains, especially when of skeletal origin;

[1] The term "hard ground", used on nautical charts, is no term of the English geological nomenclature. It is used in some German geological publications, e.g. VOIGT (1959).

(2) the importance of solubility of lime in sea water and in organic agencies as secreted externally by algae, boring sponges, etc., or internally in the intestinal tract by the majority of higher animals (c.f. KINDLE, 1919);

(3) free and intergranular precipitation and cementation as well by chemical as by biological means;

(4) early diagenetic alterations of the grains and/or of the sediment by inversion and/or recrystallization. This contributes to the hardening of the sediment. — Such a hard bottom can show little or no difference to a fossil sedimentary bedrock.

(5) Biogenic grain size reduction by biomechanic grinding and breaking, and by breakdown as a cause of organic boring and solution actions;

(6) biogenic grain size enlargement by pelleting of lime mud and ooze; the pellets may keep their shape by intergranular cementation (ILLING, 1954) and can therefore become fossilized.

B. Calcareous Bottoms in the Raetian and Lower Jurassic Sea

I. Loose Bottoms

1. Limy and Marly Mud Bottoms

During the Upper Raetian time the midwestern part of the Northern Calcareous Alps can be divided into two main facies: the reef complex and the basin. In the basin facies ("Kössen beds") the bottom was muddy and level. The amount of terrigenous components (mostly illitic minerals) is almost as important as the limy part. Typical bottom features cannot be expected. Traces of mud burrowers are generally less frequent than in the Liassic time. The occurence of "fromboidal" pyrite ("Rogen-pyrit") shows a tendency towards a gyttia formation [FABRICIUS, 1962 (2).]

In some sheltered and non-turbulent backside areas of the Raetoliassic reef complex muddy lime bottoms with big Foraminifera (FABRICIUS, 1966, plate 4/2) were deposited. Lime ooze with gypsum cristalls and a millimeter-stratification has formed the level and soft bottom of probably hypersaline ponds on the reef (l.c., plate 17/2, 18).

The limy and marly mud bottoms of the Liassic time covered vast areas. The existance of a probably primary differentiation of the colour of the bottom sediment was well preserved: red sediment in areas of slow subsidence and gray marly and limy muds in areas with rapid subsidence [FABRICIUS, 1962 (1)]. In the gray bottom facies the frequent burrowing tracks ("spotted limestone" of the Allgäu beds) prove the former rich infauna. Pelleting was not observed in this environment.

2. Stable Mud Bottoms

A certain stabilization of the bottom may have allowed the settling of sessil benthos, and vice versa such settlements can have increased the stability of the sediment. In the Raetian time this is indicated by the occurence of some sessil clams, by brachiopods, and by branched corals. During the Lower Jurassic stable mud bottoms are indicated by siliceous sponges, of which we find the remains in a great number in certain rocks.

3. Sand Bottoms

Raetian: Sandy bottoms were most frequent in the neighbourhood of reefs. They are formed of skeletal fragments of the flora and fauna of the reef complex. Sometimes they are mixed with oolitic sands. On the sea-side of the reef core we may find remains of coarse reef debris: former slope deposits which were covered with gravel and boulders (OHLEN, 1959). — The general lack of stratification of the bottoms of the back reef area is interpreted as a sign of an intense bioturbation.

In this area bottoms with pellet sands are also common. They interfinger with the above mentioned lime muds of the reef complex. It is likely that many of the more or less pure pellet sands are only a mud bottom facies, but pelleted by certain animals, e.g. polychaeta (CLOUD, 1962). In case of a predominance of one group of these animals the obvious increment in grain size is only a question of mud ingestion and expelling of equally calibrated fecal pellets. Probably they preferred the fine particles. Thus these animals transformed a mud to a sand, even a sand with a "very good" TRASK-sorting.

Lower Jurassic: Sandy bottoms are less frequent but paleogeographically important. The *crinoidal sands*, which partly have formed at the site of living, consists of desintegrated skelets of crinoids. In dense assemblages the "crinoidal meadows" may have stabilized the loose substratum by rooting.

Oolitic sands of Sinemurian age must have been an extremely mobile bottom in very shallow water. The striking lack of fossils and the preservation of stratification indicate an environment almost bare of life. These oolithic sand bottoms show many similarities to the Recent oolitic sand ridges of the Bahama Bank (FABRICIUS, 1967).

II. Solid Bottoms

1. Biogenic Hard Bottoms

a) *Bottoms with organic incrustations* often occur together with a sandy facies and also show transitions to the proper biogenic hard bottom. The stabilization and consolidation of the loose sediment is achieved by the settling of flat, or more or less voluminous organisms on the sea floor: sessil Foraminifera or algae as well as larger colonies of hydrozoans or corals etc., enable to cause this shealding and cementing effect. — On the former sea floor the frequency of solid substrates probably was higher than we may compute from the rock. Biological and mechanical desintegration of the dead skeletons works fast and efficiently, sparing only a small percentage of the organic structures for preservation.

b) *Reef grounds* are areas of the sea bottom with an ecologically important amount of framebuilding organisms. Loose sediment often settles in between the frames, or if sediment is lacking, voids are formed. The eroding strength of the waves is broken by the organic constructions. The microenvironment in a reef therefore often is that of a calm-water facies. On a reef ground many solid surfaces, dead and alive, form the substratum for further attachment and solid construction. The main organisms of the Raetoliassic reef core are calcareous algae, encrusting Foraminifera, hydrozoa, calcareous sponges, and worms with calcareous tubes. By the constant exposure to mechanical and biological attacks these organic structures were destroyed to a large part. It is no wonder therefore that even in a reef core facies generally more detritus than framework is to be found.

2. Solution Bottoms

When the sediment-forming potential ceases the always existing destructive forces become dominant. This means: An organically constructed reef ground will then be destroyed by biological breakdown, erosion, and subsequently by submarine solution if still exposed to the sea water. A loose sandy bottom will be eroded by the currents. A mud bottom, being formed in a calm environment, will mainly be exposed to a dissolution by sea water if sedimentation slacks. Such an omission therefore will be documented especially by dissolving features on solid material.

In some areas on top of the Raetoliassic reef limestone we find features forming small ponds with rather steep walls, or crests and pinnacles. The elevation of this relief reaches up to two or three decimeters [FABRICIUS, 1962 (1), Fig. 1.] The tops of the relief sometimes are covered by a crust of Fe/Mn-oxides. A selective dissolution, especially of branched coral colonies, was earlier described (FABRICIUS, 1966, plate 20). These dissolved corals do not only appear on the top of the reef sequence but also some meters below. This may indicate the increasing importance of dissolution towards the end of the reef formation.

The dissolution features on top of the reef limestone often were interpreted as a fossil karst phaenomenon, and the red Liassic sediments covering the reef limestone were thought to be fossil "terra rossa". Yet, no prove for emersion could be found, and the red Lower Jurassic sediment as well as the reef limestone are entirely of marine origin.

Submarine solution patterns in a less pure limestone, e.g. the red Liassic one, are less pronounced and show a more rounded, nodulous surface. Remnants of dissolution gather in the depressions of this centimeter-relief in form of a dark red (iron oxides) and less calcareous sediment. This can only accumulate when the bottom is not swept by currents. In the relict sediment we find the same and mostly uncorroded microfauna as in the primary lime mud, indicating that the ecological situation was not changed during the time of omission.

Encrusting or ball-like iron/manganese oxide concretions (plate 1e) are typical for the submarine solution bottoms of the Liassic time. Whether this can be compared with the Recent "manganese nodules" is not yet proved. — We also find corroded ammonite shells or belemnits (plate 1, Figs. a, c). Tiny boring tubes in organic fragments could be of boring algal origin (plate 1, Figs. b, d). One finding of serpulid tubes on and in oxide crusts is also of interest (plate 1, Fig. f).

3. Temporary Hard Bottom (Alternating Soft and Hard Bottoms)

The red nodulous limestones of the alpine Jurassic is interpreted as a changement between a bottom facies of soft, loose lime mud and a solid, hard bottom with subsolution marks (HOLLMANN, 1962, 1964; FABRICIUS, 1966). This cyclic changement happened in different parts of the Alps at different times: The "Red Nodular Limestone" ("Adneter Kalk") from the Lower to the beginning of the Middle Jurassic of the Northern and Southern Alps (FABRICIUS, 1966), and the "Ammonitico Rosso Superiore" of the Southern Alps during the Upper Jurassic (HOLLMANN, 1962).

The steps of this cycle are: (1) *Period of sedimentation:* One or several layers of soft red lime mud accumulated so slowly that sometimes the upper parts of ammonite shells were dissolved. The rate of deposition was larger than the rate of solution. — (2) *Period of omission and hardening:* Sedimentation and dissolution were more or less

Plate 1 a—f. *Features of the "Temporary Hard Bottom":* Red Liassic limestones of the Northern Calcareous Alps. — (All micrographs in normal transparent light and with enlargement as on Fig. a, if not otherwise indicated.) a Corroded surface of a belemnite with algal (?) borings and dark oxidic rim. — South of Hoch Iss., Sonnwend Mts., Tyrol (DS 2447). b As Fig. a: Boring of algae (?) in belemnite and Foraminifera. c Corroded echinid spine, oblique section. — Rotwand, Bavaria (DS 2217b). d Boring feature in pelecipod shells (recrystallized). — Maurach, Tyrol (DS 2465). e "Manganese nodule": Weathered cross section of a sphaeric concretion with concentric fabrics. — Hoch Iss., Sonnwend Mts., Tyrol. f Serpuloid tubes in cross section, surrounded by oxidic incrustations. — Bayrach Klause, border Bavaria/Tyrol (DS 2335b)

balanced. A hardening by compaction (and/or cementation?) started. The consistence of the bottom sediment changed from loose to solid. — (3) *Period of intense submarine solution:* The dissolution may become so important that layers of sediment were removed entirely. Solution remnants (muddy sediment or solid fragments) became the main constituents of the bottom surface, slowing down the dissolution activity on the hard bottom.

The exact causes of these cyclic repetions are not known; it is likely that several factors came together. To my opinion a repeated (oscillatory) change of the velocity of subsidence could be responsible as well for temporary deficiencies of sediment supply as for temporary increases of the subsolution. A small elevation of relief in comparison to the surrounding facies of gray mud bottoms could have existed during the periods of omission, hard bottom and submarine solution. — A comparable Recent facies to the "temporary hard bottom" (as to the following bottom-type also) is not known with certainty.

4. Internal Bottom (Sedimentary Fillings of Fissures)

Mainly in the Lower Liassic tectonic movements caused a tension fissuring of the reef limestone and the covering Liassic sediments. These vertical and horizontal fissures of an extension of sometimes several decameters were filled with sediment from the surface of the sea floor. Geopetal fabrics and fossils proof the sedimentary origin of the filling. These fissures often opened several time, giving space for new sedimentation, documented by the different lithology and age [Fabricius, 1962 (1), Fig. 3].

In the fissures an "internal sea bottom" with a distinct sediment-water interface was forming below the normal plane of the sea floor (c.f. Fabricius, 1966, plate 25 and 26). In general the internal bottom was muddy or sandy, but always loose—except for the very first time. Solution patterns are known from the ceiling of the horizontal, cave-like fissures; oxidic crusts are not known. — Similar features are described from the Middle Jurassic rocks of Sicily (Wendt, 1963).

C. Paleogeographic Distribution of Calcareous Sea Bottoms

The midwestern part of the Northern Calcareous Alps, i.e. a region between the meridians of Mittenwald (Bavaria) and Kufstein (Austria) was chosen as an example of the configuration of the described bottom types. — The structural North-South shortening which amounts in this part of the Alps to about 60% as a minimum, was compensated in the graphics (Fig. 2).

I. The Configuration of Bottom Types in the Raetian Sea (Fig. 2 left)

In the *Lower Raetian* the entire area was occupied by limy and marly mud bottoms. — An important paleogeographic change happened at the beginning of the *Upper Raetian* time, when the reef facies began to expand into this marginal part of the Tethys, coming from the realm of the Dachsteinkalk-facies. The marly-limy bottom kept only its position in two areas: In the narrow East-West belt next to the north boundary of the Alps and in the "Karwendel Basin". Between these two soft bottom realms the reef complex expanded.

On the reef complex the area with biogenic hard bottoms (i.e. the reef core facies) is very restricted to a discontinuous zone near to the Karwendel Basin. The broad back reef area shows sandy bottoms especially near the reef core facies. The sands are of skeletal, some also of oolitic origin. Further off the biogenic hard bottoms, pellet sands and limy mud bottoms interfinger. Lagoonal soft bottoms with lime ooze are very limited on the reef complex.

II. The Configuration in the Lower Jurassic Sea (Fig. 2 right)

The patterns of the bottom facies of the Early Hettangian time do not show much difference to those of the foregoing Raetian time. In general not earlier than

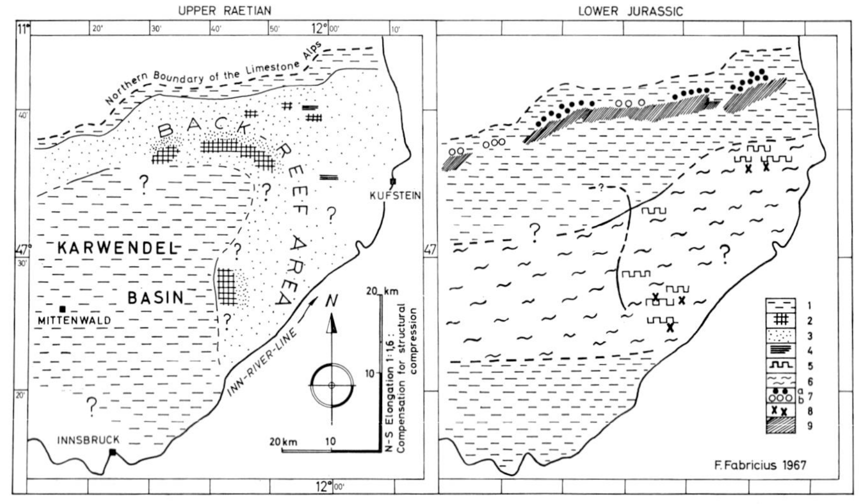

Fig. 2. *Paleogeographic distribution of the marine bottom types* in the middle part of the Calcareous Alps. 1. Limy and marly mud bottom. 2. Area of biogenic hard bottom: Reef core facies and patch reefs. 3. Loose sandy bottom, mainly of skeletal debris (back reef area). 4. Lime mud and ooze bottoms of the reef complex. 5. Solution bottoms on top of Raetoliassic Reef Limestone. 6. Facies of the "temporary solution bottoms" (red nodular limestone). 7. Loose sandy bottoms: (a) oolitic, (b) crinoidal sand. 8. "Internal bottoms": area of frequent fissure fillings. 9. Stabilized mud bottoms with siliceous sponges

during the Upper Hettangian to Lower Sinemurian the reef formation ceased and in many places a new configuration and new types of calcareous bottoms originated. The exemption is still the northern belt with marly-limy mud bottoms.

This quiet water environment is followed to the south by a turbulent water environment: A narrow belt of oolitic and also some crinoidal sands forming extremely unstable and loose bottoms which elongated from east to west. This ridge-like oolitic belt almost bare of any organic life, shows many similarities to the Bahamian oolite formation of today (Fabricius, 1967). Further to the south the oolitic sands are grading into a mud bottom facies which in some parts was probably stabilized by the abundance of siliceous sponges.

The bottom facies of gray limy mud was densely inhabited by an infauna of mud borrowers, producing the "spotted" appearance of the sediment. This facies also appears in the most southern part of the Northern Alps.

Between these two areas of gray mud bottoms with a fast sedimentation the zone of red bottoms is intercalated. In some areas this covers the top of the "dead" reef complex forming at that time a solution hard bottom. Here we also find the sediment fillings of the fissures with its "internal bottoms".

Since the Sinemurian the "temporary hard bottoms" dominate in this zone. They cover as well the sediments of the basin facies as a part of the former reef complex. The general tectonical characteristic of the zone of temporary hard bottoms is a very slow subsidence, perhaps with a superimposed oscillation of slow frequency and small amplitude. During the periods of hard bottoms this zone might have been a slight elevation relative to the adjacent areas of gray mud bottoms. During the soft bottom period the morphological differences have been less important.

References

Cloud, P. E., Jr.: Environment of calcium carbonate deposition west of Andros Island, Bahamas. U.S. Geol. Survey Profess. Papers **350**, 138 (1961).

Fabricius, F. H.: (1) Faziesentwicklung an der Trias/Jura-Wende in den mittleren nördlichen Kalkalpen. Z. deut. geol. Ges. **113**, 311—319 (1961).

— (2) Die Strukturen des „Rogenpyrits" (Kössener Schichten, Rät) als Beitrag zum Problem der „Vererzten Bakterien". Geol. Rundschau **51**, 647—657 (1961).

— Beckensedimentation und Riffbildung an der Wende Trias/Jura in den Bayerisch-Tiroler Kalkalpen. Intern. Sediment. Petrogr. Ser. **9**, 143 Leiden: Brill 1966.

— Die Rät- und Lias-Oolithe der nordwestlichen Kalkalpen. Geol. Rundschau **56**, 140—170 (1967).

Hollmann, R.: Über Subsolution und die „Knollenkalke" des Calcare Ammonitico Rosso Superiore im Monte Baldo. Neues Jahrb. Geol. u. Paläontol., Monatsh. **1962**, 163—179 (1962).

— Subsolutions-Fragmente. Neues Jahrb. Geol. u. Paläontol., Abhandl. **119**, 22—82 (1964).

Illing, L. V.: Bahaman calcareous sands. Bull. Am. Assoc. Petrol. Geologists **38**, 1—95 (1954).

Kindle, E. M.: A neglected factor in the rounding of sand grains. Am. J. Sci. **47**, 431—434 (1919).

Ladd, H. S.: Paleoecological evidence. In: H. S. Ladd (Ed.), Treatise on marine ecology and paleoecology. Geol. Soc. Am., Mem. **67**, vol. 2 (Paleoecology), 31—66 (1957).

Newell, N. D., J. Imbrie, E. G. Purdy, and D. L. Thurber: Organism communities and bottom facies, Great Bahama Bank. Bull. Am. Museum Nat. Hist. **117**, 177—228 (1959).

Ohlen, H. R.: The Steinplatte reef complex of the Alpine Triassic (Rhaetian) of Austria. Diss., Princeton Univ., 123 p. Princeton N.J. 1959.

Swinchatt, J. P.: Significance of constituent composition, texture, and skeletal breakdown in some recent carbonate sediments. J. Sediment. Petrol. **35**, 71—90 (1965).

Thorson, G.: Bottom communities (sublittoral or shallow shelf). In: J. W. Hedgpeth (Ed.), Treatise on marine ecology and paleoecology. Geol. Soc. Am. Mem. **67**, vol. 1 (Ecology), 461—534 (1957).

Voigt, E.: Die ökologische Bedeutung der Hartgründe („Hardgrounds") in der oberen Kreide. Paläont. Z. **33**, 129—147 (1959).

Wendt, J.: Stratigraphisch-paläontologische Untersuchungen im Dogger Westsiziliens. Boll. soc. paleont. ital. **2**, 57—145 (1963).

Ecology of Algal-Sponge-Reefs in the Upper Jurassic of the Schwäbische Alb, Germany

HERMANN ALDINGER*

With 1 Figure

Abstract

The algal-sponge-reefs of the Swabian Malm grew at lower energy conditions than the coral reefs of the Upper Malm. During the lower Malm Zeta the sea became shallow. The algal-sponge-reefs were now in a high-energy zone. Slumping occured and graded breccias were formed. Reef corals grew on the algal-sponge-reefs and calcareous muds containing characteristic faunal communities were deposited in lagoonal inter-reef basins. The graded breccias, containing broken and transported fossils, formed in two phases.

A. Introduction

The Upper Jurassic of the Schwäbische Alb consists of a repeated alternation of bedded limestone and calcareous marl-sequences with intercalated reef limestones. The stratigraphy of these rocks and the lithology of the reef limestones have been described in a number of papers (for literature see HÖLDER, 1964; HILLER, 1964). The present report discusses the ecology of the reefs and the origin of the graded breccias.

B. Regional Distribution, Lithology, and Paleontology of Algal-Sponge Reefs

The Malm reefs of the Schwäbische Alb are most widely distributed in the pseudomutabilis zone. At the end of this time, the sea bottom in the area of the Schwäbische Alb was almost completely covered by reefs. Bedded limestones were deposited in a few locally restricted areas (HILLER, 1964, Fig. 14; ZIEGLER, 1967, Fig. 10). This belt of reefs was locally more than 100 km wide because reef limestones of the same age are buried beneath Molasse rocks south of the river Danube as shown by drillings.

Extremely regular bedding characterizes the rocks deposited between the reefs during the lower and Middle Malm. Some beds can be traced over large distances in the Schwäbische Alb as in Franken (DIETRICH, 1940; v. FREYBERG, 1966; SEIBOLD, 1950; ZIEGLER, 1955, 1967). For example, a glauconite-bearing bed of the pseudomutabilis-zone can be recognized throughout the Schwäbische Alb from the Ries-area to Switzerland in the bedded as well as in the reef-facies. The alternation limestone-marl of the bedded series is therefore also reflected in the reefs, and the continuity of individual units of the bedded facies through the reefs (ROLL, 1934; GWINNER, 1962; v. FREYBERG, 1966) is only exceptionally interrupted by slumping or erosion.

Some reefs of the pseudomutabilis-zone are more than 100 m high. The reef-

* Geologisch-Paläontologisches Institut, Universität Stuttgart, Germany.

slopes possess an average dip of 10 to 15° but are locally > 50°. It is well known that siliceous sponges formed part of these reefs but they make out generally less than 10% of typical reef limestones. The most important reef-builders are variously-shaped calcareous crusts, knobs and pellets whose origin is partly due to symbiosis of lime-precipitating algae with sessile foraminifera of the genus Nubeculinella (HILLER, 1964). Additional components are stromatolites (in the middle and upper Malm) and oolitic and micritic limestone.

The association of algae with foraminifera and siliceous sponges as reef-builders is as unusual as the association of those organisms which generally occur within

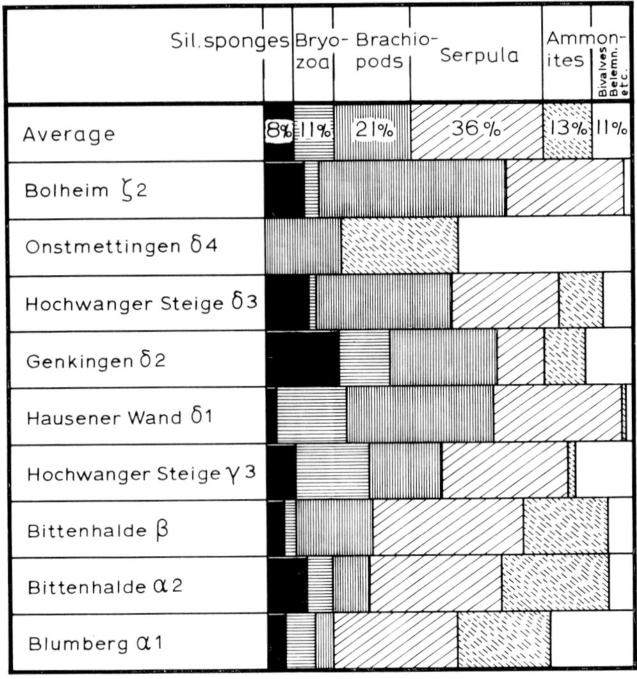

Fig. 1. Composition of the marcofauna of some algal-sponge-reefs, Schwäbische Alb (after WAGENPLAST, 1967)

the reefs but do not contribute to their structure. These organisms are predominantly serpula, brachiopoda and bryozoa (Fig. 1) most of which are fixed epibiontically to the underside of sponges and are represented by small forms (FRITZ, 1958; WAGENPLAST, 1967).

There is, therefore, a basic ecologic difference between the algal-sponge reefs of the middle Malm and the coral reefs of the Malm Zeta. In the algal-sponge reefs solitary and reef corals, and all the other characteristic members of a coral community of the Upper Jurassic, such as dasycladaceae, hydrozoa, nerinea and typical reef molluscs are lacking. Moreover, the almost complete absence of boring animals, so typical of coral reefs, such as boring molluscs, algae, sponges etc. is of some ecologic importance. The algal-sponge reefs must have formed in an environment with a low energy level under conditions differing appreciably from that of coral reefs. Depth of water, in which the algal-sponge reefs grew, can be inferred from several lines of evidence. There is no indication of reworking of the reefs by breakers, strong current or organisms; debris of reef rocks are absent over wide areas of the Schwäbische Alb in lower and middle Malm. Now some reefs exceeded 100 m in height; algae and stromatolites grew at times under optimum conditions; the top of the reefs was below the turbulent water zone and thus below the habitat of reef corals, the water depths between the reefs therefore probably exceeded 150 m during periods of most rapid reef growth [GWINNER, 1962 (2); HILLER, 1964]. Such a minimum depth is also indicated by the extreme regularity of the bedded facies.

C. Origin of Slumped Beds and Graded Breccias

There is a pronounced lithological change at the beginning of Malm Zeta in the inter-reef bedded sediments and in the reefs themselves. Bedded limestones are transported through submarine slumping and are now found as discordant intercalations within the bedded facies which on the whole is deposited under much more variable conditions. Graded breccias are generally associated with these slumped beds [GWINNER, 1962 (1)].

In small, well defined inter-reef basins, such as in the area between Reutlingen and Sigmaringen, part of the Malm Zeta is represented by laminated calcareous shales resembling the Solnhofen limestone. This sequence contains at Nusplingen a fossil community characteristic of calcareous and bituminuous shales: freely swimming crinoids, ammonites, belemmites, fish, reptiles, and terrestrial plants, but very little autochthonous benthos. The limestones must therefore have been deposited in a lagoonal basin with restricted water exchange. Growth of algal-sponge reefs decreases or ends during Malm Zeta. The dying reefs in the central and eastern parts of the Schwäbische Alb were colonized by reef corals and their characteristic accompanying fauna (PAULSEN, 1964). The malm sea had become so shallow that the top of the algal-sponge reefs emerged into the zone favorable for growth of reef corals. This decrease in depth of water also led to slumping, deposition of graded breccias, and sedimentation of calcareous shales in inter-reef basins in which many organisms are trapped.

Graded breccias are associated with undisturbed beds as well as with slumped strata. In the latter case, the irregular surface of a slumped complex is evened-out by the breccia [GWINNER, 1962 (1); TEMMLER, 1964, 1966]. The grain size of the breccias ranges from block- to silt-size and the clasts consist of various limestones. Generally, coarse breccias always overlay slumped beds. The Nusplingen outcrop shows this relation particularly well (TEMMLER, 1964, 1966). Coarse breccias are here folded or otherwise intimately mixed with the underlying slumped beds (TEMMLER, 1964, 1966). These breccias must, therefore, have been transported together with the underlying bedded rocks. The clasts are: fragments of micritic limestone, partly with ammonites; calcareous shale; broken stromatolites; fragments of carbonatized siliceous sponges; algal pellets; ooids and composite ooids; fragments of fossils (worms, molluscs, brachiopods, echinoderms); and skeletal elements of saccocoma.

The breccias are thus polymikt in contrast to the so-called "reef debris" of the older Malm. Fragments of algal-sponge-reef rocks are always most abundant, and some were indurated and silicified prior to formation of these breccias.

The breccias certainly did not form by brecciation of the slumped beds. Contact relationships and composition of the graded breccias indicate an origin in two phases.

At first a breccia of reef debris containing fragmental fossils was deposited on top of steeply dipping bedded lime muds on the reef slopes. Subsequently, these sediments slumped together with the breccia on top of them. Finally, the slumping led to folding, shearing and mixing of the entire moving complex. The finer grained detritus of the breccias was again suspended, transported beyond the slumped sediments and redeposited as graded bed.

References

DIETERICH, E.: Stratigraphie und Ammonitenfauna des Weißen Jura Beta in Württemberg, 6 Abb., 2 Taf. Jh. Ver. vaterl. Naturkunde Württemberg **96**, 1—40 (1940).

FREYBERG, B. v.: Der Faziesverband im Unteren Malm Frankens. Ergebnisse der Stromatometrie, 22 Abb., 8 Taf. Erl. geol. Abh. **62**, 1—92 (1966).

FRITZ, G. K.: Schwammstotzen, Tuberolithe und Schuttbreccien im Weißen Jura der Schwäbischen Alb. Eine vergleichende petrogenetische Untersuchung, 24 Abb., 5 Taf. Arb. geol. palaeont. Inst. TH Stuttgart N.F. **13**, 1—119 (1958).

GWINNER, M. P.: (1) Subaquatische Gleitungen und resedimentäre Breccien im Weißen Jura der Schwäbischen Alb (Württemberg), 6 Abb., 4 Taf. Z. deut. geol. Ges. **113**, 571—590 (1962).

— (2) Geologie des Weißen Jura der Albhochfläche (Württemberg), 22 Abb., 1 Tab., 4 Taf. N. Jb. Geol. Palaeont. Abh. **115**, 137—221 (1962).

HILLER, K.: Über die Bank- und Schwammfazies des Weißen Jura der Schwäbischen Alb, 38 Abb., 4 Tab., 26 Taf. Arb. Geol. palaeont. Inst. TH Stuttgart **40**, 1—190 + XIII S. (1964).

HÖLDER, H.: Jura. Handbuch der stratigraphischen Geologie, 158 Abb., 43 Tab. **4**, 1—603 (1964).

PAULSEN, S.: Aufbau und Petrographie des Riffkomplexes von Arnegg im höheren Weißen Jura der Schwäbischen Alb (Württemberg), 20 Abb., 22 Taf. Arb. geol. palaeont. Inst. TH Stuttgart, N.F. **42**, 1—98 (1964).

ROLL, A.: Form, Bau und Entstehung der Schwammstotzen im süddeutschen Malm, 18 Abb. Palaeont. Z. **16**, 197—246 (1934).

SEIBOLD, E.: Der Bau des Deckgebirges im oberen Rems-Kocher-Jagst-Gebiet, 17 Abb., 1 Taf., 12 Tab. N. Jb. Geol. Palaeont. Abh. **92**, 243—366 (1950).

TEMMLER, H.: Über die Schiefer- und Plattenkalke des Weißen Jura der Schwäbischen Alb (Württemberg), 18 Abb., 2 Tab., 24 Taf. Arb. Geol. palaeont. Inst. TH Stuttgart N.F. **43**, 1—106 (1964).

— Über die Nusplinger Fazies des Weißen Jura der Schwäbischen Alb (Württemberg), 4 Abb., 2 Tab., 5 Taf. Z. deut. geol. Ges. **116**, 891—907 (1966).

WAGENPLAST, P.: Die Begleitfauna der Schwammriffe des Weißen Jura der Schwäbischen Alb (Württemberg). Unveröffentlichte Diplomarbeit, TH Stuttgart 1967.

ZIEGLER, B.: Die Sedimentation im Malm Delta der Schwäbischen Alb, 7 Abb. Jber. Mitt. oberrh. geol. Ver. N.F. **37**, 29—55 (1955).

— Ammoniten-Ökologie am Beispiel des Oberjura, 20 Abb. Geol. Rundschau **56**, 439—464 (1967).

Continuous Gasometric Carbonate Determinations in Cuttings from Drill Holes

HERMANN SCHETTLER*

With 1 Figure

Continuous gasometric determination of carbonate in cuttings has been successfully used by the Mobil Oil A.G. in the Weser-Ems area since 1954. It is used to check changes in rock composition (facies development) during drilling and to compare the results with those of adjacent drillings (SCHETTLER, 1960, 1965).

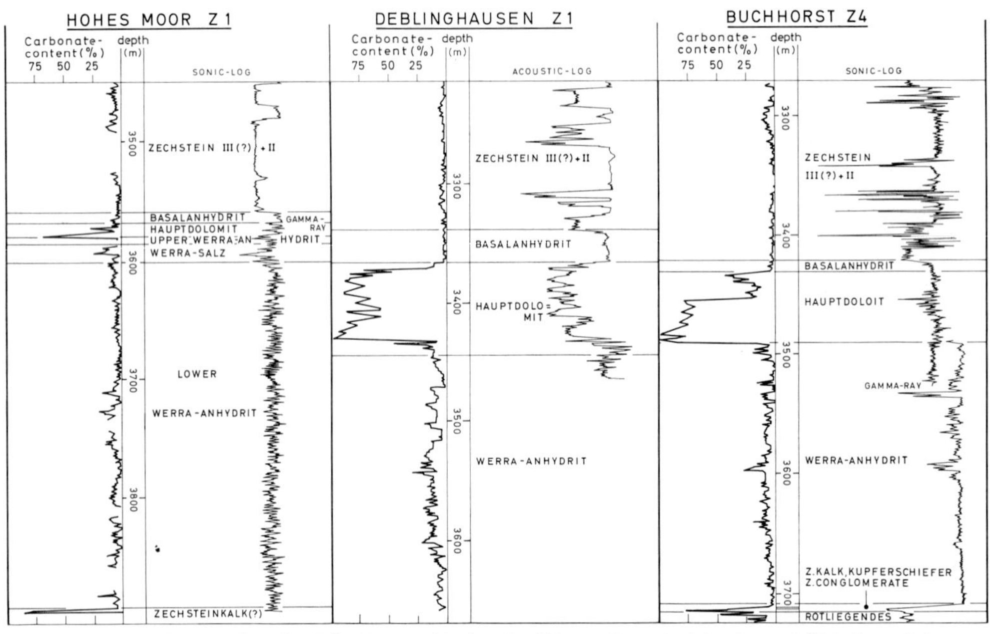

Fig. 1. "Carbonate-logs" of the Lower Zechstein (Upper Permian) in three drill holes of the Weser-Ems area (NW-Germany)

Cuttings are taken at intervals of 1 to 3 m at the site of drilling; they are washed, dried, and ground. The amount of CO_2 is determined with the "Scheibler-Gasometer" (described in MÜLLER, 1967) by treating $^1/_2$ gr of the sample with HCl.

Absolute carbonate content and the approximate calcite/dolomite ratio can be estimated if the first reading is taken after 15 seconds and the final reading after the gas development is completed (15 to 45 minutes). The resulting „carbonate-log" can

* Mobil Oil A. G., Lastrup i. O., Germany.

be compared with those of adjacent drill holes during drilling operation. Such logs are very useful for exploring carbonate reservoirs and for stratigraphic correlation of sections strongly influenced by tectonics and of rocks lacking fossils.

Fig. 1 shows a correlation using „carbonate-logs" of the "Hauptdolomit" and "Zechsteinkalk" units of the Zechstein (Upper Permian) in the Weser-Ems area of NW-Germany.

Thanks are expressed at this place to the Mobil Oil A. G. in Germany and to the Gewerkschaft Brigitta for permission to publish this paper.

References

MÜLLER, G.: Methods in Sedimentary Petrology, 283 p. New York: Hafner Publishing Company 1967.

SCHETTLER, H.: Einige methodische Fragen der Spülprobenbearbeitung. Z. deutsch. geol. Ges. **112,** 401—413 (1960).

— Untersuchungsmethoden und stratigraphische Ergebnisse von Trias- und Zechsteinbohrungen im Weser-Emsgebiet. Z. deutsch. geol. Ges. **115,** 214—227 (1965).

Druck: Carl Ritter & Co., Wiesbaden